K Walker

Applied Mathematical Sciences
Volume 21

Editors
F. John (deceased)   J.E. Marsden   L. Sirovich

Advisors
M. Ghil   J.K. Hale   J. Keller
K. Kirchgässner   B.J. Matkowsky
J.T. Stuart   A. Weinstein

Springer
*New York
Berlin
Heidelberg
Barcelona
Hong Kong
London
Milan
Paris
Singapore
Tokyo*

R. Courant
K. O. Friedrichs

Supersonic Flow and Shock Waves

Springer

R. Courant (1888–1972)
K. O. Friedrichs (1901–1982)

Editors

Fritz John (deceased)
Courant Institute of
Mathematical Sciences
New York University
New York, NY 10012

Lawrence Sirovich
Division of
Applied Mathematics
Brown University
Providence, RI 02912

J. E. Marsden
Control and Dynamical
Systems, 107-81
Caltech
Pasadena, CA 91125

Mathematics Subject Classification (1991): 76Jxx, 76L05

Library of Congress Cataloging in Publication Data
Courant, Richard, 1888–1972
   Supersonic flow and shock waves.
   (Applied mathematical sciences; 21)
   "Originates from a report issued in 1944 under the auspices of the Office of
Scientific Research and Development."
   Reprint of the ed. published by Interscience Publishers, New York, which was
issued as v. 1 of Pure and applied mathematics.
   Bibliography: p.
   Includes indexes.
   1. Aerodynamics, Supersonic. 2. Shock waves.
I. Friedrichs, Kurt Otto, joint author. II. Title. III. Series.
QA1.A647 no. 21 [QA930]   510'.8s [533'.275'0151]   76-57719

The present edition is an unchanged reprint of the original edition which was
published in 1948 by Interscience Publishers Inc., New York

All rights reserved.
No part of this book may be translated or reproduced in any form without written permission from Springer-Verlag.
© 1948, 1976 by R. Courant and K. O. Friedrichs respectively Nina Courant and
K. O. Friedrichs.

Printed and bound by Edwards Brothers, Inc., Ann Arbor, MI.
Printed in the United States of America.

9 8 7 6 5 (Corrected fifth printing, 1999)

ISBN 0-387-90232-5 Springer-Verlag New York Berlin Heidelberg

ISBN 3-540-90232-5 Springer-Verlag Berlin Heidelberg New York   SPIN 10696536

To

Warren Weaver

Preface

The present book originates from a report issued in 1944 under the auspices of the Office of Scientific Research and Development. Much material has been added and the original text has been almost entirely rewritten. The book treats basic aspects of the dynamics of compressible fluids in mathematical form; it attempts to present a systematic theory of nonlinear wave propagation, particularly in relation to gas dynamics. Written in the form of an advanced textbook, it accounts for classical as well as some recent developments, and, as the authors hope, it reflects some progress in the scientific penetration of the subject matter. On the other hand, no attempt has been made to cover the whole field of nonlinear wave propagation or to provide summaries of results which could be used as recipes for attacking specific engineering problems.

The book has been written by mathematicians seeking to understand in a rational way a fascinating field of physical reality, and willing to accept compromise with empirical approach. The authors hope that it will be helpful to engineers, physicists, and mathematicians alike, and that it will not be rejected by mathematicians as too heavily loaded with physical assumptions or by others as too strictly mathematical.

Dynamics of compressible fluids, like other subjects in which the nonlinear character of the basic equations plays a decisive role, is far from the perfection envisaged by Laplace as the goal of a mathematical theory. Classical mechanics and mathematical physics predict phenomena on the basis of general differential equations and specific boundary and initial conditions. In contrast, the subject of this book largely defies such claims. Important branches of gas dynamics still center around special types of problems, and general features of connected theory are not always clearly discernible. Nevertheless, the authors have attempted to develop and to emphasize as much as possible such general viewpoints, and they hope that this effort will stimulate further advances in this direction.

In a field which during recent years has attracted so many workers and in which such diverse practical and theoretical interests have asserted themselves, the authors found a balanced survey impossible; instead they have followed a path dictated largely by their personal interests and experience. The names of scientists with whom the authors happened to be in close contact appear frequently; names of others may have been omitted. No fair appraisal could be made of the merits of many recent contributions. This is true in particular of the large number of reports issued during the war by various agencies and still not freely accessible. In order to avoid further delay, the authors are publishing this book without a complete survey of the literature.

The book was prepared for publication with the cooperation of members of the staff of the Institute for Mathematics and Mechanics in New York University. The main burden of the editorial work, done for the original report by R. Shaw, has been carried by Cathleen Synge Morawetz, who has also contributed constructive criticism in many details, and whose understanding and competent assistance have been invaluable. L. J. Savage cooperated actively in rewriting the first chapter and other parts of the original report. D. A. Flanders has helped greatly by reading parts of the manuscript and suggesting important improvements. W. Y. Chen, W. M. Hirsch, E. Isaacson, A. Leitner, S. C. Lowell, and M. Sion have assisted in this publication by reading proofs and making useful suggestions. The drawings, many of which represent actual conditions, have been carried out by G. W. Evans and J. R. Knudsen. The preparation of the manuscript was in the competent hands of Edythe Rodermund and Harriet Schoverling.

Much more than a formal acknowledgement is due to the Office of Naval Research, not only for the generous support under Contract N6ori-201, Task Order No. 1, which made possible the preparation of the book, but also for the stimulating active interest of its staff members in the progress of the work.

Thanks also should be expressed to Interscience Publishers for the cooperative attitude of their staff, and for the genuine interest of their officers in the promotion of scientific publications.

The book is dedicated to Warren Weaver. As chief of the Applied Mathematics Panel during the war, he rendered very great services, not only for the problems of the day, but even more so for the lasting benefit of the mathematical sciences. For us personally his steady interest in the present work has been a source of encouragement. Thus the dedication of the book is as well a token of friendship as a tribute to a man whose energy and vision have contributed so much to the recent development of applied mathematics in this country.

R. COURANT and K. O. FRIEDRICHS

August, 1948

Contents

I. Compressible Fluids... 1

    1. Qualitative differences between linear and nonlinear waves   2

  A. General Equations of Flow. Thermodynamic Notions....... 3
    2. The medium... 3
    3. Ideal gases, polytropic gases, and media with separable energy.. 6
    4. Mathematical comments on ideal gases................. 8
    5. Solids which do not satisfy Hooke's law............... 10
    6. Discrete media... 12
    7. Differential equations of motion........................ 12
    8. Conservation of energy.................................... 15
    9. Enthalpy.. 17
    10. Isentropic flow. Steady flow. Subsonic and supersonic flow.. 18
    11. Acoustic approximation................................... 18
    12. Vector form of the flow equations...................... 19
    13. Conservation of circulation. Irrotational flow. Potential.. 19
    14. Bernoulli's law... 21
    15. Limit speed and critical speed........................... 23

  B. Differential Equations for Specific Types of Flow............ 25

    16. Steady flows.. 25
    17. Non-steady flows.. 28
    18. Lagrange's equations of motion for one-dimensional and spherical flow.. 30

  Appendix—Wave Motion in Shallow Water........................ 32

    19. Shallow water theory...................................... 32

II. Mathematical Theory of Hyperbolic Flow Equations for Functions of Two Variables... 37

    20. Flow equations involving two functions of two variables.. 37
    21. Differential equations of second order type............ 38
    22. Characteristic curves and characteristic equations... 40
    23. Characteristic equations for specific problems........ 45
    24. The initial value problem. Domain of dependence. Range of influence... 48
    25. Propagation of discontinuities along characteristic lines.. 53
    26. Characteristic lines as separation lines between regions of different types... 55

27. Characteristic initial values........................... 56
28. Supplementary remarks about boundary data.......... 57
29. Simple waves. Flow adjacent to a region of constant state.. 59
30. The hodograph transformation and its singularities. Limiting lines....................................... 62
31. Systems of more than two differential equations........ 70

Appendix... 75

32. General remarks about differential equations for functions of more than two independent variables. Characteristic surfaces................................. 75

III. One-Dimensional Flow... 79

33. Problems of one-dimensional flow..................... 79

A. Continuous Flow.. 80

34. Characteristics....................................... 80
35. Domain of dependence. Range of influence............ 82
36. More general initial data............................. 84
37. Riemann invariants................................... 87
38. Integration of the differential equations of isentropic flow... 88
39. Remarks on the Lagrangian representation............ 91

B. Rarefaction and Compression Waves....................... 92

40. Simple waves.. 92
41. Distortion of the wave form in a simple wave.......... 96
42. Particle paths and cross-characteristics in a simple wave 97
43. Rarefaction waves.................................... 99
44. Escape speed. Complete and incomplete rarefaction waves.. 101
45. Centered rarefaction waves........................... 103
46. Explicit formulas for centered rarefaction waves........ 104
47. Remark on simple waves in Lagrangian coordinates..... 106
48. Compression waves................................... 107

Appendix to Part B... 110

49. Position of the envelope and its cusp in a compression wave.. 110

C. Shocks... 116

50. The shock as an irreversible process................... 116
51. Historical remarks on non-linear flow.................. 118
52. Discontinuity surfaces................................ 119
53. Basic model of discontinuous motion. Shock wave in a tube... 120

54. Jump conditions... 121
55. Shocks... 124
56. Contact discontinuities... 126
57. Description of shocks... 126
58. Models of shock motion... 129
59. Discussion of the mechanical shock conditions... 130
60. Sound waves as limits of weak shocks... 131
61. Cases in which the mechanical shock conditions are sufficient to determine the shock... 131
62. Shock conditions in Lagrangian representation... 132
63. Shock relations derived from the differential equations for viscous and heat-conducting fluids... 134
64. Hugoniot relation. Determinacy of the shock transition 138
65. Basic properties of the shock transition... 141
66. Critical speed and Prandtl's relation for polytropic gases 146
67. Shock relations for polytropic gases... 148
68. The state on one side of the shock front in a polytropic gas determined by the state on the other side... 150
69. Shock resulting from a uniform compressive motion... 150
70. Reflection of a shock on a rigid wall... 152
71. Shock strength for polytropic gases... 154
72. Weak shocks. Comparison with transitions through simple waves... 156
73. Non-uniform shocks... 160
74. Approximate treatment of non-uniform shocks of moderate strength... 161
75. Decaying shock wave. N-wave... 164
76. Formation of a shock... 168
77. Remarks on strong non-uniform shocks... 171

D. Interactions... 172

78. Typical interactions... 172
79. Survey of results... 176
80. Riemann's problem. Shock tubes... 181
81. Method of analysis... 182
82. The process of penetration for rarefaction waves... 191
83. Interactions treated by the method of finite differences... 197

E. Detonation and Deflagration Waves... 204

84. Reaction processes... 204
85. Assumptions... 207
86. Various types of processes... 208
87. Chapman-Jouguet processes... 211
88. Jouguet's rule... 215
89. Determinacy in gas flow involving a reaction front... 218
90. Solution of flow problems involving a detonation process. 222
91. Solution of flow problems involving deflagrations... 224
92. Detonation as a deflagration initiated by a shock... 226

93. Deflagration zones of finite width.................... 227
94. Detonation zones of finite width. Chapman-Jouguet hypothesis... 231
95. The width of the reaction zone....................... 232
96. The internal mechanism of a reaction process. Burning velocity... 232

Appendix—Wave Propagation in Elastic-Plastic Material........ 235

97. The medium... 235
98. The equations of motion............................. 238
99. Impact loading...................................... 240
100. Stopping shocks..................................... 243
101. Interactions and reflections......................... 245

IV. Isentropic Irrotational Steady Plane Flow.................. 247

102. Analytical background............................... 247

A. Hodograph Method.. 248

103. Hodograph transformation............................ 248
104. Special flows obtained by the hodograph method....... 252
105. The role of limiting lines and transition lines...... 256

B. Characteristics and Simple Waves......................... 259

106. Characteristics. Mach lines and Mach angle.......... 259
107. Characteristics in the hodograph plane as epicycloids.... 262
108. Characteristics in the (u,v)-plane continued........... 264
109. Simple waves....................................... 266
110. Explicit formulas for streamlines and cross Mach lines in a simple wave.................................... 271
111. Flow around a bend or corner. Construction of simple waves... 273
112. Compression waves. Flow in a concave bend and along a bump.. 278
113. Supersonic flow in a two-dimensional duct............ 282
114. Interaction of simple waves. Reflection on a rigid wall.. 286
115. Jets.. 289
116. Transition formulas for simple waves in a polytropic gas. 290

C. Oblique Shock Fronts.................................... 294

117. Qualitative description............................. 294
118. Relations for oblique shock fronts. Contact discontinuities.. 297
119. Shock relations in polytropic gases. Prandtl's formula... 302
120. General properties of shock transitions.............. 304
121. Shock polars for polytropic gases.................... 306
122. Discussion of oblique shocks by means of shock polars.... 311
123. Flows in corners or past wedges..................... 317

D. Interactions—Shock Reflection 318

124. Interactions between shocks. Shock reflection 318
125. Regular reflection of a shock wave on a rigid wall 319
126. Regular shock reflection continued 327
127. Analytic treatment of regular reflection for polytropic gases 329
128. Configurations of several confluent shocks. Mach reflection 331
129. Configurations of three shocks through one point 332
130. Mach reflections 334
131. Stationary, direct and inverted Mach configuration 335
132. Results of a quantitative discussion 338
133. Pressure relations 342
134. Modifications and generalizations 343
135. Mathematical analysis of three-shock configuration 346
136. Analysis by graphical methods 347

E. Approximate Treatments of Interactions. Airfoil Flow 350

137. Problems involving weak shocks and simple waves 350
138. Comparison between weak shocks and simple waves ... 350
139. Decaying shock front 354
140. Flow around a bump or an airfoil 356
141. Flow around an airfoil treated by perturbation methods (linearization) 357
142. Alternative perturbation method for airfoils 364

F. Remarks about Boundary Value Problems for Steady Flow 367

143. Facts and conjectures concerning boundary conditions .. 367

V. Flow in Nozzles and Jets 377

144. Nozzle flow 377
145. Flow through cones 377
146. De Laval nozzle 380
147. Various types of nozzle flow 383
148. Shock patterns in nozzles and jets 387
149. Thrust 392
150. Perfect nozzles 394

VI. Flow in Three Dimensions 397

A. Steady Flow with Cylindrical Symmetry 397

151. Cylindrical symmetry. Stream function 397
152. Supersonic flow along a slender body of revolution 398
153. Resistance 404

B. Conical Flow 406

154. Qualitative description 406
155. The differential equations 408
156. Conical shocks 411
157. Other problems involving conical flow 414

C. Spherical Waves 416

  158. General remarks 416
  159. Analytical formulations 418
  160. Progressing waves 419
  161. Special types of progressing waves 421
  162. Spherical quasi-simple waves 424
  163. Spherical detonation and deflagration waves 429
  164. Other spherical quasi-simple waves 431
  165. Reflected spherical shock fronts 432
  166. Concluding remarks 433

Bibliography 435

Index of Symbols 453

Subject Index 455

CHAPTER I

Compressible Fluids

Violent disturbances—such as result from detonation of explosives, from the flow through rocket nozzles, from supersonic flight of projectiles, or from impact on solids—differ greatly from the "linear" phenomena of sound, light, or electromagnetic signals. In contrast to the latter, their propagation is governed by nonlinear differential equations, and as a consequence the familiar laws of superposition, reflection, and refraction cease to be valid; but even more novel features appear, among which the occurrence of *shock fronts* is the most conspicuous. Across shock fronts the medium undergoes sudden and often considerable changes in velocity, pressure, and temperature. Even when the start of the motion is perfectly continuous, shock discontinuities may later arise automatically. Yet, under other conditions, just the opposite may happen; initial discontinuities may be smoothed out immediately. Both these possibilities are essentially connected with the nonlinearity of the underlying equations.

Nature confronts the observer with a wealth of nonlinear wave phenomena, not only in the flow of compressible fluids, but also in many other cases of practical interest. One example, rather different from those mentioned above, is the catastrophic pressure in a crowd of panicky people who rush toward a narrow exit or other obstruction. If they move at a speed exceeding that at which warnings are passed backward, a pressure wave arises much like that behind a shock front receding from a wall. Related phenomena, such as congestion in traffic, seem to be essentially due to similar conditions. In this book, however, we shall concentrate primarily on the theory of compressible fluids.

Understanding and control of nonlinear wave motion is a matter of obvious importance. During a period beginning almost a hundred years ago, Stokes, Earnshaw, Riemann, Rankine, Hugoniot, Lord

Rayleigh, and later Hadamard and others wrote fundamental papers inaugurating this field of research. Then the development was left mainly to a small group of ingenious men in the fields of mechanics and engineering. During the last few years, however, when the barriers between applied and pure science were forced down, a widespread interest arose in nonlinear wave motion, particularly in shock waves and expansion waves.

It is the purpose of the present book to make the mathematical theory of nonlinear waves more accessible, giving particular attention to some recent developments.*

1. *Qualitative differences between linear and nonlinear waves*

Some characteristics of nonlinear wave motion can be described in general terms. In linear wave motion, as, for example, in the transmission of sound, disturbances are always propagated with a definite speed (relative to the medium) which may vary within the medium. This "sound speed" is a local property of the medium itself and remains the same for every conceivable linear wave motion in the medium. Such a sound speed also plays a role in nonlinear wave motion. Small disturbances or "wavelets," slightly modifying a given primary wave motion, are propagated with a certain speed, again called sound speed, though in this case the sound speed depends not only on the position within the medium but on the state of the medium induced by the primary motion.

The distinctive feature of nonlinear waves, however, concerns disturbances or discontinuities which are not necessarily small. In linear wave motion any initial discontinuity across a surface is preserved as a discontinuity and propagated with sound speed. Nonlinear wave motion behaves in a different manner: Suppose there is an initial discontinuity between two regions of different pressures, densities, and flow velocities. Then there are the following *alternative* possibilities: either the initial discontinuity is resolved immediately and the disturbance, while propagated, becomes continuous, or the initial discontinuity is propagated through one or two *shock fronts*, advancing not at sonic but at supersonic speed relative to the medium

---

* For the theory of compressible flow reference may be made to [3,4,5]; different approaches are given by Sauer [6] and Liepmann and Puckett [7].

ahead of them. As previously stated, shock fronts are the most conspicuous phenomena occurring in nonlinear wave propagation; even without being caused by initial discontinuities they may appear and be propagated. The underlying mathematical fact is that, unlike linear partial differential equations, nonlinear equations often do not admit solutions which can be continuously extended wherever the differential equations themselves remain regular.

Another striking difference between linear and nonlinear waves concerns the phenomenon of interaction: the principle of superposition holds for linear waves but not for nonlinear waves. As a consequence, for example, excess pressures of interfering sound waves are merely additive; in contrast to this fact, interaction and reflection of nonlinear waves may lead to enormous increases in pressure.

A. General Equations of Flow. Thermodynamic Notions

2. *The medium*

We shall be primarily concerned with a moving fluid, though many of the results apply to other moving media (e.g. to a solid slab in longitudinal wave motion). In this section we shall set forth the properties of the medium that will be assumed throughout the book and we shall describe certain idealized media of special interest. Moreover, since gas dynamics is thoroughly interwoven with thermodynamical concepts, it is appropriate to insert here a collection of basic notions of thermodynamics in a suitable mathematical form.*

Except where the motion is *discontinuous*, viscosity, heat conduction, and deviation of the medium from thermodynamic equilibrium (at any instant and any point) will be neglected. Some critical comments concerning the neglect of these phenomena will be made in later chapters. In particular it will be shown that viscosity and heat conduction play an important role in forming and maintaining shock discontinuities.

At each instant and each point of the fluid there is a definite state (of thermodynamic equilibrium) defined by:

p the pressure,
T the temperature,

*For textbooks on thermodynamics see Epstein [20] and Zemansky [21].

τ the specific volume (i.e. volume per unit mass),
ρ the density, with $\rho\tau = 1$,
S the specific entropy,
e the specific (internal) energy, and
i the specific enthalpy,* defined by $i = e + p\tau$.

It is known from thermodynamics that for any given medium only two of the parameters p, T, τ, e, and S are independent. In fact they may all be considered as functions of τ and S.

The internal energy gained by the medium during a change from one state to another is the heat contributed to the medium plus the work done on the medium by compressive action of the pressure forces. For a change from one state to an immediately neighboring one this fundamental fact is expressed by the relation

$$(2.01) \qquad de = TdS - pd\tau.$$

In a reversible process, TdS is the heat acquired by conduction; in an irreversible process, TdS is greater than the heat so acquired. If the irreversible process is one that can be described as determined by the action of viscosity, then the excess of TdS over the heat acquired by conduction may conveniently be interpreted as the heat produced by viscous forces.

Suppose that for some medium we know how the specific energy e depends on τ and S. Then the pressure p and temperature T may immediately be found on considering the meaning of relation (2.01). Thus

$$(2.02) \qquad p = -e_\tau, \; T = e_S,$$

the subscripts indicating partial differentiation.**

The functions giving p in terms of ρ, or τ, and S, occurring so frequently in the theory of fluid flow, will consistently be denoted by

$$(2.03) \qquad p = f(\rho, S); \; p = g(\tau, S).$$

Extending slightly the conventional nomenclature, we shall call either of these equations the *caloric equation of state* of the medium.

Neglecting viscosity and heat conduction is tantamount to assuming that as a particle of the medium moves about, the specific

* The notion of enthalpy will be discussed in Section 9.
** Nearly everywhere in the book, we indicate partial derivatives by subscripts.

entropy at the moving particle remains constant, i.e. the changes in state at the particle are *adiabatic*. We shall, therefore, often be interested in $f(\rho, S)$ and $g(\tau, S)$ considered simply as functions of ρ and τ respectively, with the specific entropy S fixed; indeed, in some cases these functions will be written in the abbreviated form $f(\rho)$ and $g(\tau)$. The equation $p = f(\rho) = g(\tau)$ is then called the *adiabatic equation*.

The word *isentropic* would perhaps be more accurate here than *adiabatic*. If, for example, heat conduction were absent but viscosity present, the changes would be adiabatic (heat not flowing to or from the particle), but not isentropic (the entropy at the particle generally increasing). But we are reserving the word *isentropic* for another concept, that of constant entropy throughout the medium.

It is a fundamental property of all actual media that, entropy remaining constant, the pressure increases with increasing density (or decreasing specific volume), that is,

(2.04) $$f_\rho(\rho, S) > 0; \; g_\tau(\tau, S) < 0,$$

except in the limiting case $\rho = 0$, in which $f_\rho = 0$. Because of (2.04) we can define a positive quantity c, with the dimension of speed, by

(2.05) $$c^2 = \frac{dp}{d\rho} = f_\rho(\rho, S), \qquad \rho^2 c^2 = -g_\tau(\tau, S).$$

This important quantity c is called the *sound speed*, a name which will be justified in Section 35, Chapter II; the quantity ρc is frequently called the *acoustic impedance*.

For any value of S, the function $g(\tau, S)$ is generally convex downward. We therefore assume throughout this book, except where the contrary is noted, that

(2.06) $$g_{\tau\tau}(\tau, S) > 0.$$

It is useful to recognize that together with (2.04), $f_{\rho\rho}(\rho, S) \geq 0$ implies (2.06).

We make the additional assumption that, for constant specific volume, the pressure increases with entropy,

(2.07) $$g_S(\tau, S) > 0.$$

From equations (2.02) we see that this assumption is equivalent to assuming that, at constant entropy, the temperature increases with increasing density.

For gases, for which the density may approach zero, we make the additional assumption

(2.08) $\quad e \to 0,\ \tau p \to 0,\ T \to 0,\ c \to 0,\ \text{as } \rho \to 0.$

The theory of nonlinear wave motion can be carried quite far without further assumptions about the medium. There are, however, various media of particular physical interest, which are described in Sections 3–6 (in somewhat more detail than is necessary for the subsequent mathematical treatment).

3. Ideal gases, polytropic gases, and media with separable energy

In practically all applications of the theory to gases the medium may, with sufficient accuracy, be assumed to be an *ideal gas*, that is a medium which satisfies the laws of Boyle and Gay-Lussac as expressed by the *equation of state*

(3.01) $\quad\quad\quad\quad\quad\quad p\tau = RT.$

Here the constant R may be taken to be the universal gas constant R_0 divided by the effective molecular weight of the particular gas.

In an ideal gas the internal energy is a function of the temperature alone, see Section 4. If, in particular, the internal energy is simply proportional to the temperature T, the gas is called *polytropic*. For such gases we may write

(3.02) $\quad\quad\quad\quad\quad\quad e = c_v T,$

where the constant c_v is the specific heat at constant volume. The assumption that a gas is polytropic is made in most applications of the theory; it leads, together with (3.01), to the entropic equation of state

(3.03) $\quad\quad\quad\quad\quad p = f(\rho, S) = A\rho^\gamma,$

in which the coefficient A depends on the entropy, S and the *adiabatic exponent* γ is a constant between 1 and $\frac{5}{3}$ for media most usually occurring. Air at moderate temperatures may be considered polytropic with $\gamma = 1.4$.

The equivalence of (3.02) and (3.03) will be proved in Section 4. Here we mention that c_v equals $R/(\gamma - 1)$, so that

(3.04) $$e = \frac{1}{\gamma - 1} p\tau;$$

and

(3.05) $$A = (\gamma - 1) \exp c_v^{-1}(S - S_0),$$

with an appropriate constant S_0.

The sound speed of a polytropic gas is, according to (2.05) and (3.03), characterized by the simple relations

(3.06) $$c^2 = \gamma A \rho^{\gamma-1} = \gamma p \tau = \gamma RT.$$

For later reference we note for polytropic gases:

(3.07) $$e = \frac{A}{\gamma - 1} \tau^{-(\gamma-1)} = \frac{A}{\gamma - 1} \rho^{(\gamma-1)},$$

(3.08) $$RT = A\tau^{-(\gamma-1)} = A\rho^{(\gamma-1)},$$

(3.09) $$p = A\rho^\gamma = A\tau^{-\gamma}.$$

It follows immediately from (3.09) that a polytropic gas satisfies the monotonicity and convexity conditions (2.04) and (2.06).

In terms of p and τ, the energy is given by (3.04), the temperature by (3.01), and the entropy by the relation

(3.10) $$S - S_0 = c_v \log \{p\tau^\gamma/(\gamma - 1)\},$$

which follows from (3.05) and (3.09).

In gases the pressure depends noticeably on the specific entropy. However, in some cases, especially if the medium is a liquid, the influence of changes in entropy is negligible, so that p may be considered a function of density (or specific volume) alone. In this case the entropic equation of state takes the form

(3.11) $$p = f(\rho) \quad \text{or} \quad p = g(\tau)$$

and as a consequence of (2.01) and (2.02)

(3.12) $$e = e^{(1)}(\tau) + e^{(2)}(S).$$

Conversely *separable energy* as expressed by (3.12) implies (3.11).

Similarly, the condition that T depends only on S is equivalent to that of separable energy.

The most important medium with approximately separable energy is water. For water the caloric equation of state has some resemblance to that of a perfect gas:

$$(3.13) \qquad p = A\left(\frac{\rho}{\rho_0}\right)^\gamma - B,$$

where ρ_0 is the density at 0° centigrade, and A, B, and γ are practically independent of the entropy. The particular values for water are $\gamma = 7$, $B = 3000$ atm., $A = 3001$ atm., approximately.

4. Mathematical comments on ideal gases

It seems worth while, though it is not essential for the subsequent sections, to prove quite generally the fact that *the internal energy of an ideal gas is a function of the temperature alone*, and to establish the relationship between this function and the entropic equation of state. For any ideal gas, we obtain, using the value $T = p\tau/R$ from (3.01) in (2.02), the linear partial differential equation for e,

$$(4.01) \qquad Re_s + \tau e_\tau = 0,$$

the general solution of which is

$$(4.02) \qquad e = h(\tau \mathrm{H}),$$

where h is an arbitrary function and

$$(4.03) \qquad \mathrm{H} = \exp(-S/R).$$

The entropic equation of state is therefore

$$(4.04) \qquad p = -e_\tau = -h'(\tau\mathrm{H})\mathrm{H} = -h'(\rho^{-1}\mathrm{H})\mathrm{H};$$

here h' is the derivative of h. For an ideal gas it follows from (4.04) that the monotonicity and convexity conditions (2.04) and (2.06) are respectively equivalent to the conditions that the second derivative of h be positive and the third derivative of h be negative,

$$(4.05) \qquad h''(\tau\mathrm{H}) > 0; \qquad h'''(\tau\mathrm{H}) < 0.$$

The temperature of an ideal gas is given by

(4.06) $$T = e_s = -\frac{1}{R}h'(\tau H)\tau H.$$

Equation (4.06) shows that T like e depends only on τH. For any actual medium, T is a strictly decreasing function of this variable. Therefore, it is possible to solve (4.06) for τH in terms of T, and then by (4.02) to express e as a function of T. In other words, *the specific energy of an ideal gas depends only on the temperature.*

According to (2.05) and (4.04) the sound speed for an ideal gas is given as a function of τ and S by

(4.07) $$c^2(\tau, S) = h''(\tau H)\tau^2 H^2;$$

thus the *sound speed*, like the specific energy, *depends only on the temperature*. The connection of sound speed with energy and temperature in the case of an ideal gas is brought out by the equation

(4.08) $$c^2(T) = \left(1 + R\frac{dT}{de}\right)RT$$
$$= \gamma(T)RT,$$

in which the dimensionless quantity $\gamma(T)$ is here introduced as a convenient abbreviation for $1 + R\frac{dT}{de}$. Equation (4.08) is easily derived from (4.07) by verifying $e_\tau + RT_\tau = -h''(\tau H)\tau H^2$ and $RT = -\tau e_\tau$ from (4.04) and (4.06).

We note that

(4.09) $$\frac{de}{dT} = \frac{R}{\gamma(T) - 1}$$

is the "specific heat at constant volume," see Section 9.

For a *polytropic gas*, that is an ideal gas for which e is simply proportional to T as expressed through (3.02), the function $\gamma(T)$ is constant. In fact from (4.08) and (3.02),

(4.10) $$\gamma = 1 + Rc_v^{-1},$$

or $c_v = R/(\gamma - 1)$, as mentioned earlier.

Since c_v and R are positive,

(4.11) $$\gamma > 1;{}^*$$

where the elementary kinetic theory of gases applies it is known that

(4.12) $$\gamma \leq \tfrac{5}{3}.$$

Inserting (4.02) and (4.06) in (3.02), we obtain by (4.10)

(4.13) $$h'(\tau H) = -(\gamma - 1)(\tau H)^{-1} h(\tau H).$$

Therefore

(4.14) $$h = \left(\tau \frac{H}{H_0}\right)^{-(\gamma - 1)},$$

where the presence of the constant H_0 reflects the fact that the specific entropy is defined only within an arbitrary constant.

From (4.04) we then have

(4.15) $$p = A(S)\tau^{-\gamma},$$

where $A(S)$ is given by

$$A = A(S) = (\gamma - 1)\left(\frac{H}{H_0}\right)^{-(\gamma-1)} = (\gamma - 1)\exp(c_v^{-1}(S - S_0)),$$

in accordance with (4.02) and (3.05). Relation (4.15) is equivalent to (3.03). This form of the caloric equation of state is thus derived from the basic assumption (3.02).

5. Solids which do not satisfy Hooke's law

In contrast to fluids, solids resist shear so that the thermodynamic description of a solid (involving, as it does, many components of stress and strain) is much more complicated than that of a fluid. But if attention is confined to plane longitudinal waves moving nor-

---

* Frequently, in particular in the theory of subsonic flow, see [12], the adiabatic equation $p = g(\tau)$ is approximated by the equation

$$p = -k^2(\tau - \tau_0) + g(\tau_0),$$

which does not satisfy condition (2.06). Aside from an additive constant, this relation between p and τ would correspond to a polytropic gas with

$$\gamma = -1.$$

mal to the faces of a slab, no shearing forces arise. Then there is quite a good analogy between elastic solids and fluids. Neglecting the effects of changes of entropy, the state at a cross section of the slab is characterized by two variables p and τ analogous to the pressure and specific volume of a fluid. In this context p denotes the negative of the "engineering stress," i.e., the negative of the force acting in the normal direction on the cross section of the slab divided by the original area of the cross section of the unstrained slab; and

$$\tau = \tau_0(1 + \epsilon),$$

where τ_0 is the specific volume of the unstrained slab and ϵ is the strain. The entropic equation of state almost always used for an elastic solid slab is expressed by Hooke's law, which in our notation is

(5.01) $$p = (\tau_0 - \tau)\frac{E}{\tau_0},$$

where E is Young's modulus.

When Hooke's law applies, the wave motion is linear; more accurately, but somewhat in anticipation of Chapter II, this motion is linear from the Lagrangian but not from the Eulerian point of view. Therefore, our special concern is with solids to which Hooke's law does not apply, that is, solids with a more general caloric equation of state than (5.01):

(5.02) $$p = g(\tau).$$

Just as in the case of a fluid, we shall assume p to decrease with increasing τ. We shall also assume that p, as a function of τ, is never convex away from the p-axis and that it vanishes at $\tau = \tau_0$. These assumptions are formally expressed by

(5.03)
$$g'(\tau) < 0$$
$$g(\tau_0) = 0$$
$$g''(\tau) \leq 0 \text{ for } \tau \leq \tau_0$$
$$\geq 0 \text{ for } \tau \geq \tau_0.$$

These conditions imply that, except for terms of higher than the second order, the solid satisfies Hooke's law in the neighborhood of τ_0. Note also that the last part of (5.03) is in contrast to (2.06).

Though not themselves plastic, these solids to which Hooke's law does not apply have played a role in some studies of plasticity. For more details see the appendix to Chapter III, in which there is a graph of a typical "non-Hookian" equation of state.

6. Discrete media

Wave motion can be studied in a medium consisting of a chain of masses connected with one another by springs which do not, in general, obey Hooke's law. There is a definite analogy between such *discrete media* and the continuous media which are our immediate object of interest. This analogy can be exploited in either direction; in some computations it proves advantageous to approximate a continuous medium by a discrete one and vice versa, see e.g. [58].

7. Differential equations of motion*

The questions to be studied in this book are intimately associated with the general framework of the differential equations of fluid dynamics, which govern the motion of the medium except at discontinuities. This system of equations expresses:
  a. The principle of conservation of mass,
  b. Newton's law of conservation of momentum,
  c. The condition that changes of state are adiabatic,
  d. The particular form of the equation of state.
The differential equations, together with appropriate initial and boundary conditions, determine an individual phenomenon.

In the remaining sections of this chapter a brief survey of classical results about the equations of hydrodynamics is given in a form suitable for our purposes.

The equations of fluid dynamics can be expressed in two different forms, *Lagrange's form* and *Euler's form*. The equations in Lagrangian form describe the motion in terms of the paths of the individual particles of the gas, i.e., the coordinates x, y, z of the particle as functions of the time t and three parameters a, b, c which

* For this section see Lamb [17] and Milne-Thomson [18] in the bibliography.

characterize the individual particle; a, b, c are often chosen as the coordinates of the particle at the time $t = 0$. In Lagrange's representation differentiation with respect to the time t will be denoted by a dot ($\cdot$).

In most cases, however, Euler's representation is preferable both from a mathematical and from a physical point of view. Attention is directed to definite points (x, y, z) and to what happens at these points in the course of time t. The motion is then described by giving as functions of x, y, z, and t the velocity components u, v, w of the particle that happens to be at the point (x, y, z) at the time t. In Euler's representation differentiation with respect to the independent variables x, y, z, t will be denoted by subscripts. The transition from Euler's representation to Lagrange's is effected by solving the system of ordinary differential equations

(7.01)
$$\dot{x} = u(x, y, z, t)$$
$$\dot{y} = v(x, y, z, t)$$
$$\dot{z} = w(x, y, z, t);$$

the constants of integration may be taken as parameters a, b, c. The converse transition is effected by eliminating a, b, c from the equations (7.01) and those obtained by differentiating (7.01).

In Lagrangian form the equations of fluid dynamics are:

(7.02)
$$(\rho \dot{\Delta}) = 0$$

(Conservation of mass)

(Here $\Delta = \dfrac{\partial(x, y, z)}{\partial(a, b, c)}$ denotes the Jacobian of the function $x(a, b, c, t)$, $y(a, b, c, t)$, $z(a, b, c, t)$.)

(7.03)
$$\rho\ddot{x} + p_x = 0$$
$$\rho\ddot{y} + p_y = 0$$
$$\rho\ddot{z} + p_z = 0$$

(Conservation of momentum)

(It is assumed here that no force other than the pressure gradient is acting. To be sure, the external force of gravity will generally be present, but in most applications it can either be neglected altogether or handled separately, see Milne-Thomson [18].)

(7.04) $$\dot{S} = 0$$

(Changes of state are adiabatic)

(7.05) $$p = f(\rho, S)$$

(Caloric equation of state).

The derivatives of the pressure p in equations (7.03) refer to x, y, z, t as independent variables. An explicit expression in the variables a, b, c, t by $p_x = p_a a_x + p_b b_x + p_c c_x$, etc. will lead to involved nonlinear terms, for a_x, b_x, $\cdots$ are to be expressed by the derivatives of the inverse functions $x(a, b, c, t)$, $\cdots$. Usually, therefore, the Lagrangian representation is too cumbersome. This objection does not hold for motions symmetrical enough to be characterized by merely one space coordinate; in these cases the Lagrangian representation is, as we shall see, often advantageous.

For motion involving more than one space coordinate it is generally preferable to write the equations in Eulerian form. Each of Euler's equations is derived directly from the corresponding equation of Lagrange by means of the identity

(7.06) $$\dot{F} = F_t + uF_x + vF_y + wF_z,$$

which applies to any function F defined for the particles of the medium. We further note the identity

(7.07) $$\dot{\Delta} = (u_x + v_y + w_z)\Delta,$$

which is easily verified.

In the absence of external forces Euler's equations are:

(7.08.1) $$\rho_t + u\rho_x + v\rho_y + w\rho_z + \rho(u_x + v_y + w_z) = 0$$

or

(7.08.2) $$\rho_t + (\rho u)_x + (\rho v)_y + (\rho w)_z = 0$$

(Conservation of mass),

CONSERVATION OF ENERGY

(7.09.1)
$$u_t + uu_x + vu_y + wu_z + \frac{1}{\rho}p_x = 0$$
$$v_t + uv_x + vv_y + wv_z + \frac{1}{\rho}p_y = 0$$
$$w_t + uw_x + vw_y + ww_z + \frac{1}{\rho}p_z = 0$$

or, by (7.08.2)

(7.09.2)
$$(\rho u)_t + (\rho u^2)_x + (\rho vu)_y + (\rho wu)_z + p_x = 0$$
$$(\rho v)_t + (\rho uv)_x + (\rho v^2)_y + (\rho wv)_z + p_y = 0$$
$$(\rho w)_t + (\rho uw)_x + (\rho vw)_y + (\rho w^2)_z + p_z = 0$$

(Conservation of momentum),

(7.10)
$$p = f(\rho, S)$$

(Caloric equation of state),

(7.11)
$$S_t + uS_x + vS_y + wS_z = 0$$

(Changes of state are adiabatic).

It is satisfying to note that the number of equations (six) both in Lagrange's and in Euler's system is equal to the number of unknowns. We may, therefore, expect that appropriate initial and boundary conditions lead to unique behavior (for the regions of continuous motion) without recourse to any further physical principles. It is generally more convenient to think of either system as consisting of only five equations in five unknowns, because p is easily eliminated from the equations of conservation of momentum by means of (2.05), since

(7.12)
$$c^2(\rho, S) = f_\rho(\rho, S);$$
$$p_x = c^2(\rho, S)\rho_x; \quad p_y = c^2(\rho, S)\rho_y; \quad p_z = c^2(\rho, S)\rho_z.$$

8. Conservation of energy

The condition (7.11) that changes of state are adiabatic can be derived from the assumption that the gain of total energy of a gas particle is due solely to the work done by the pressure force (through both a compressing and an accelerating action, see Section 2) with external forces absent. The total energy per unit mass is $e_{\text{tot}} = \frac{1}{2}(\dot{x}^2 + \dot{y}^2 + \dot{z}^2) + e$; the work done by the pressure force per unit volume per unit time is $-[(p\dot{x})_x + (p\dot{y})_y + (p\dot{z})_z]$. Thus our assumption leads to the relation

(8.01) $$\rho \dot{e}_{\text{tot}} + (p\dot{x})_x + (p\dot{y})_y + (p\dot{z})_z = 0$$

(Conservation of energy).

Carrying out the differentiations, using relation (2.01) in the form $de = \tau^2 p\, d\rho + T\, dS$, and relations (7.01) and (7.07), relation (8.01) becomes

$$\rho[\dot{x}\ddot{x} + \dot{y}\ddot{y} + \dot{z}\ddot{z} + \tau^2 p\dot{\rho} + T\dot{S}] + p_x\dot{x} + p_y\dot{y} + p_z\dot{z} + p\Delta^{-1}\dot{\Delta} = 0.$$

By (7.02) and (7.03) this relation reduces to relation (7.04), $\dot{S} = 0$, which expresses the fact that the changes of state are adiabatic.

In Euler's representation the law of conservation of energy can be given a concise form by employing the *flow speed* or flow velocity, $q = \sqrt{u^2 + v^2 + w^2}$, and the *specific enthalpy* $i = e + p\tau$, about which more will be said in Section 9. Relation (8.01) can then by (7.08) be reduced to

(8.02.1) $$\rho[\tfrac{1}{2}q^2]_t + \rho T S_t + \rho u[\tfrac{1}{2}q^2 + i]_x \\ + \rho v[\tfrac{1}{2}q^2 + i]_y + \rho w[\tfrac{1}{2}q^2 + i]_z = 0$$

or by (7.08.2)

(8.02.2) $$\{\rho[\tfrac{1}{2}q^2 + e]\}_t + \{\rho u[\tfrac{1}{2}q^2 + i]\}_x \\ + \{\rho v[\tfrac{1}{2}q^2 + i]\}_y + \{\rho w[\tfrac{1}{2}q^2 + i]\}_z = 0.$$

Incidentally, neither the total energy per unit mass, $e_{\text{tot}} = \frac{1}{2}(u^2 + v^2 + w^2) + e$, nor the work done per unit volume per unit time, $(pu)_x + (pv)_y + (pw)_z$, is invariant under translation; i.e. these expressions change when the motion is observed from a frame moving with the constant velocity (u_0, v_0, w_0). The left member of relation (8.02) when referred to this moving frame, however, differs from the original expression only by multiples of the left members of (7.09), and, therefore, the system of equations (7.08), (7.09), (8.02) does remain invariant.

9. Enthalpy

In terms of the *specific enthalpy i*, defined by

(9.01) $$i = e + p\tau,$$

the fundamental equation (2.01) takes the form

(9.02) $$di = \tau dp + TdS.$$

For adiabatic changes, $dS = 0$, we have by (2.05)

(9.03) $$di = c^2 \tau d\rho.$$

Our assumption for gases (2.08) leads to the relation

(9.04) $$i \to 0 \text{ as } \rho \to 0.$$

In an ideal gas, i is evidently a function of the temperature. From (4.09), (3.01), and (9.01) we have

(9.05) $$\frac{di}{dT} = \frac{\gamma R}{\gamma - 1}.$$

This quantity is called "specific heat at constant pressure." Comparison of formulas (9.05) and (4.09) shows that γ is the *ratio of the two specific heats*.

Specializing to a polytropic gas we obtain from (3.07), (3.08), (3.09), and (9.01)

(9.06) $$i = \frac{\gamma}{\gamma - 1} A(S)\rho^{\gamma-1} = \frac{\gamma}{\gamma - 1} RT$$
$$= \frac{\gamma}{\gamma - 1} p\tau = \frac{c^2}{\gamma - 1}.$$

Introducing the specific enthalpy proves particularly helpful in situations in which the entropy S is constant, for example, at a given particle, or along a streamline in steady flow, or throughout the medium in isentropic flow, see Section 14. In other fields the usefulness of the enthalpy concept lies in the fact that $di = TdS$ is the increment of heat acquired by the particle if the pressure remains constant. This relation explains why the term *heat content* is often used for enthalpy.

10. Isentropic flow. Steady flow. Subsonic and supersonic flow

Frequently a most important simplifying assumption can and will be made: *at the beginning of the process the specific entropy has the same value throughout the medium.* It follows directly from (7.04), unless discontinuities across which (7.04) breaks down are present, that in this case the specific entropy retains this same value throughout the medium for all time. So, thinking of S as a given constant, we may omit equation (7.04), or correspondingly (7.11), and leave five equations in five unknowns, or with pressure eliminated, four equations in four unknowns. Such a flow, in which the entropy is everywhere and always the same, will be called *isentropic*. Though it is true that isentropic flow frequently occurs, there are many important phenomena which involve non-isentropic flow.

Of great interest is another special type of flow: *steady flow*, defined as a motion in which flow velocity, pressure, density, and specific entropy remain unchanged in time at each point, i.e., depend only on x, y, z and not on t. For such a motion the terms in Euler's differential equations containing u_t, v_t, w_t, ρ_t, and S_t drop out. In a steady flow all the particles passing through a particular point have the same velocity, density, pressure, and entropy at this point and they will follow the same path, the *streamline* through the point. The medium is thus covered by streamlines which do not change in time.

A steady flow is called *subsonic, sonic,* or *supersonic at a point as the flow speed* $q = \sqrt{u^2 + v^2 + w^2}$ *at that point is less than, equal to, or greater than the sound speed c at that point*, see (2.05) and Section 3, or as the *Mach number*

$$(10.01) \qquad M = q/c$$

is less than, equal to, or greater than one.

11. Acoustic approximation

The system of linear partial differential equations which is generally used to describe ordinary *acoustical disturbances* follows from Euler's nonlinear equations of general fluid motion (7.08–11) as a limiting case. Consider a slight isentropic disturbance, i.e., an isentropic motion of the medium with $S = S_0$ and $\rho = \rho_0 + \delta\rho$, such that terms of order higher than first in $\delta\rho$, u, v, w, and their derivatives may be neglected. Neglecting these higher order terms and eliminating p, Euler's equations become

(11.01) $$\delta\rho_t + \rho_0(u_x + v_y + w_z) = 0$$

(11.02) $$u_t + \tau_0 c_0^2 \delta\rho_x = 0, \; v_t + \tau_0 c_0^2 \delta\rho_y = 0, \; w_t + \tau_0 c_0^2 \delta\rho_z = 0,$$

in which $c_0^2 = c^2(\rho_0, S_0)$. The system (11.01–.02) is easily seen to be equivalent to the single equation of second order

(11.03) $$\delta\rho_{tt} = c_0^2(\delta\rho_{xx} + \delta\rho_{yy} + \delta\rho_{zz}),$$

the familiar *wave equation* for the small disturbance $\delta\rho$.

12. Vector form of the flow equations

Occasionally it is convenient to rewrite the differential equations (7.09) in vector form denoting the velocity vector by $\vec{q}$. Thus

(12.01) $$\dot{\vec{q}} + \tau \,\text{grad}\, p = 0,$$

or using (9.02)

(12.02) $$\dot{\vec{q}} + \text{grad}\, i = T \,\text{grad}\, S.$$

Expanding, and rearranging terms, we have

(12.03) $$\vec{q}_t + \tfrac{1}{2} \text{grad}\,(q^2) - \vec{q} \times \text{curl}\,\vec{q} + \text{grad}\, i = T \,\text{grad}\, S;$$

here q is the magnitude of the flow velocity or the *flow speed*

(12.04) $$q = |\vec{q}| = \sqrt{u^2 + v^2 + w^2}.$$

The symbol $\times$ stands for vector multiplication. Of course, if the flow is isentropic, the term on the right member of equation (12.02), or of (12.03), drops out.

13. Conservation of circulation. Irrotational flow. Potential

Under various rather wide assumptions, the equations of gas dynamics admit important "integrals" or conservation laws which are easily deduced. We shall discuss first *conservation of circulation*.

Let $\mathcal{C}$ be an arbitrary closed curve moving with the fluid. We consider the *circulation* C around $\mathcal{C}$,

$$(13.01) \qquad C = \oint_\mathcal{C} u\,dx + v\,dy + w\,dz = \oint_\mathcal{C} \vec{q} \cdot d\vec{x}$$

as a function of t. For various important types of flow the circulation remains constant, i.e. $\dot{C} = 0$, as the closed curve moves. To obtain conditions for the conservation of circulation we compute $\dot{C}$, which can be done very easily if we represent $\mathcal{C}$ by a variable radius vector $\vec{x}(\sigma, t)$, σ being a parameter on $\mathcal{C}$ such that $\mathcal{C}$ is described for $0 \leq \sigma \leq 1$, and $\vec{x}(0, t) = \vec{x}(1, t)$. Then we have

$$(13.02) \qquad \dot{C} = \oint_\mathcal{C} (\dot{\vec{q}} \cdot \vec{x}_\sigma + \vec{q} \cdot \dot{\vec{x}}_\sigma)\,d\sigma.$$

By (12.02) and $\dot{\vec{x}}_\sigma = \vec{q}_\sigma$, we find

$$(13.03) \qquad \dot{C} = \oint_\mathcal{C} [TS_\sigma - i_\sigma + \tfrac{1}{2}(q^2)_\sigma]\,d\sigma.$$

And finally integrating the last two terms of (13.03),

$$(13.04) \qquad \dot{C} = \oint_\mathcal{C} TS_\sigma\,d\sigma = \oint_\mathcal{C} T\,\mathrm{grad}\,S\,\overrightarrow{dx}.$$

Therefore (by Stokes' theorem) $\dot{C}$ vanishes for all curves $\mathcal{C}$ whenever the vector

$$(13.05) \qquad \mathrm{curl}\,(T\,\mathrm{grad}\,S) = \mathrm{grad}\,T \times \mathrm{grad}\,S$$

vanishes identically. In particular this happens when:
a. The flow is isentropic. Here grad S vanishes.
b. The energy is separable. Here T is a function of S and grad T is therefore parallel to grad S.
c. The flow is so symmetrical that T and S depend on a single space coordinate. (The most important possibilities here are one-dimensional flow, two-dimensional flow with cylindrical symmetry, and spherical flow.) Here grad T and grad S are constrained to lie in the same direction at each point.

A flow for which the circulation around every curve is (and remains) zero, that is, a flow for which curl $\vec{q}$ vanishes identically in space and time, is called an *irrotational* flow. Irrotational flows occur frequently, because many flows start from rest and proceed under the conditions enumerated above.

The relative mathematical simplicity of irrotational flow is often exploited by realizing that the equation

(13.06) $\quad v_z - w_y = 0, \quad w_x - u_z = 0, \quad u_y - v_x = 0,$

or

$$\operatorname{curl} \vec{q} = 0,$$

implies the existence of a *velocity potential*, i.e., a function $\varphi(x, y, z, t)$ such that

(13.07) $\quad u = \varphi_x, \quad v = \varphi_y, \quad w = \varphi_z,$

or

$$\vec{q} = \operatorname{grad} \varphi.$$

Similarly, we see from (13.05) that there is a potential Ω for the vector field $T \operatorname{grad} S$,

(13.08) $\quad T \operatorname{grad} S = \operatorname{grad} \Omega,$

whenever circulation is conserved, in particular *for irrotational flow*. If the flow is isentropic, Ω may be taken to be zero. If the specific energy of the medium is separable, Ω may be identified with the specific entropy component of the specific energy, cf. (3.12), $\Omega(x, y, z) = e^{(2)}[S(x, y, z)]$.

14. Bernoulli's law

In this section we shall derive three closely related conservation laws each of which is sometimes called *Bernoulli's law*. The first form of this law refers to *steady flow*, see Section 10. The differential equations (12.02) and the vectorial form of the statement (7.04) that changes of state are adiabatic, imply

(14.01) $\quad S_t + \vec{q} \cdot \operatorname{grad} S = 0.$

We may immediately infer that on each streamline of a steady flow we have

$$\frac{d}{dt} (\tfrac{1}{2} q^2 + i) = \vec{q} \cdot \dot{\vec{q}} + \vec{q} \cdot \operatorname{grad} i = \vec{q} \cdot T \operatorname{grad} S = 0,$$

from which

(14.02) $$\tfrac{1}{2}q^2 + i = \tfrac{1}{2}(u^2 + v^2 + w^2) + i = \tfrac{1}{2}\hat{q}^2,$$

where $\hat{q}$ is constant along a streamline. Relation (14.02) is *Bernoulli's law for steady flow*.

The *Bernoulli constant* $\tfrac{1}{2}\hat{q}^2$ may, of course, have different values along different streamlines. The same is true of the *entropy S*, which is also constant along each streamline, by (14.01). The rates of change of the Bernoulli constant and the entropy across the streamlines are coupled with the *vorticity* of the flow. For, equation (12.03) for steady flow can by (14.02) be written in the form

(14.03) $$\operatorname{grad} \tfrac{1}{2}\hat{q}^2 - T \operatorname{grad} S = \vec{q} \times \operatorname{curl} \vec{q}.$$

An immediate conclusion to be drawn from this equation is that *in the case of steady flow which is irrotational, i.e., curl $\vec{q}$ = 0, and isentropic, i.e., grad S = 0, the Bernoulli constant $\tfrac{1}{2}\hat{q}^2$ is the same on every streamline in a connected region.* This is the *strong form of Bernoulli's law*.

It is remarkable that a different form of Bernoulli's law holds for irrotational flow, even if it is not steady. Using the concepts of velocity potential φ and the potential Ω of $T \operatorname{grad} S$, introduced at the end of Section 13, *Bernoulli's law for irrotational flow* is expressed by the relation

(14.04) $$\varphi_t + (\tfrac{1}{2}q^2 + i) - \Omega = \tfrac{1}{2}\hat{q}^2,$$

in which the quantity $\hat{q}$, which may depend on the time, is the same throughout the fluid. This relation follows immediately from the form (12.03) of the equations of motion by (13.07) and (13.08). In the important case of isentropic flow, of course, Ω drops out.

For steady, irrotational, and isentropic flow, relation (14.04) simplifies to the strong form of Bernoulli's law.

Incidentally, a still different variant of Bernoulli's law will play a fundamental role in the theory of shock discontinuities; see Chapter III, Section 55.

Bernoulli's law for steady flow of polytropic gases assumes by (9.06) and (14.02) the particularly simple form

(14.05) $$q^2 + \frac{2}{\gamma - 1} c^2 = \hat{q}^2.$$

Introducing instead of γ the constant

(14.06) $$\mu^2 = \frac{\gamma - 1}{\gamma + 1}$$

and a quantity c_* with the dimensions of velocity,

(14.07) $$c_* = \mu \hat{q},$$

we may write Bernoulli's law (14.05) in the form

(14.08) $$\mu^2 q^2 + (1 - \mu^2) c^2 = c_*^2 .$$

The quantity c_*, which plays an important role in the theory of steady flow of polytropic gases, is called the *critical speed;* its significance will be discussed in the next section.

15. Limit speed and critical speed

In this section we are considering steady flow in a gas. Relations (9.03) and (9.04) then hold along each streamline; they imply that the specific enthalpy i is always positive and decreases to zero if the density decreases along a streamline. It is then seen from Bernoulli's law (14.02) in the form

(15.01) $$q^2 + 2i(\tau, S) = \hat{q}^2$$

(which also holds along each streamline), that the speed q cannot exceed the value $\hat{q}$ and approaches this value if ρ approaches zero. Thus

(15.02) $$q \leq \hat{q};$$

and the equality applies only in the limiting case of vanishing density, $\rho = 0$. The quantity $\hat{q}$ may therefore appropriately be called the *limit speed.* Similarly, (15.01) implies

(15.03) $$i \leq \tfrac{1}{2}\hat{q}^2;$$

and the equality applies only where the gas is stagnant, $q = 0$.

The significance of the notion of *critical speed*, introduced for polytropic gases through relation (14.07), is recognized when one writes Bernoulli's equation (14.08) in the form

$$q^2 - c_*^2 = (1 - \mu^2)(q^2 - c^2).$$

It is then apparent that

(15.04)
$$q > c_* \text{ according as } q > c \text{ and vice versa,}$$
$$q < c_* \text{ according as } q < c \text{ and vice versa.}$$

In other words: Supersonic or subsonic character of the flow, see Section 10, can be ascertained by comparison of the flow speed q with the critical speed c_*, which remains constant along a streamline, while the sound speed c in general varies.

If at some point of the streamline the flow speed q and sound speed c agree then both agree evidently with the critical speed c_*, a fact which can also be used to characterize the critical speed.

The concept of critical speed is not restricted to polytropic gases. We state that for given values of S and $\hat{q}$ there is just one value τ_* of τ such that relation (15.01) implies $q = c$ for $\tau = \tau_*$. The value

(15.05) $$c_* = c(\tau_*, S) = \sqrt{2\hat{q}^2 - i(\tau_*, S)}$$

is then the *critical speed*. Moreover, the assertion (15.04) remains valid for this generalized notion of critical speed. Clearly, c_* depends on $\hat{q}$ and S.

To prove these statements it suffices to show that the difference $c^2 - q^2$ increases monotonically from the value $-\hat{q}^2$ to some positive value when τ decreases from infinity to a certain value for which $q = 0$. From (15.01) we have

(15.06) $$c^2 - q^2 = c^2 + 2i - \hat{q}^2;$$

from (2.05) and (2.02), we have $c^2 = p_\rho = -\tau^2 p_\tau = \tau^2 e_{\tau\tau}$; hence, by (9.01),

(15.07) $$i + \tfrac{1}{2}c^2 = e - \tau e_\tau + \tfrac{1}{2}\tau^2 e_{\tau\tau}.$$

By differentiating with respect to τ we obtain

(15.08) $$(i + \tfrac{1}{2}c^2)_\tau = \tfrac{1}{2}\tau^2 e_{\tau\tau\tau} = -\tfrac{1}{2}\tau^2 p_{\tau\tau}.$$

Since $p_{\tau\tau}$ is positive, according to the basic assumption (2.06), we conclude from (15.08) and (15.06) that $c^2 - q^2$ increases monotonically with decreasing τ. Since this quantity is negative when $\tau = \infty$ and hence $c = 0$, and positive when $2i(\tau, S) = \hat{q}^2$ and hence $q = 0$, it follows that there is just one value of τ for which $c = q$.

On a streamline to which Bernoulli's equation (15.01) applies,

STEADY FLOWS

the points (if any) at which q equals c_* divide the streamline into intervals of subsonic and supersonic flow. Of course it may happen that $q < c_*$ or $q > c_*$ along the whole streamline in which case the flow is respectively either subsonic or supersonic along the whole streamline.

It is remarkable that in the case of a polytropic gas the critical speed $c_* = \mu \hat{q}$, see (14.07), is independent of S. That is, in fact, true of any ideal gas. If a gas is ideal, c is a function of temperature, see (4.08), and $i = e + RT$ is, by (4.09), an increasing function of temperature. Therefore, c is a function of i, and c_* is the value of c which satisfies the equation

$$c^2 - q^2 = c^2(i) + 2i - \hat{q}^2 = 0,$$

so that c_* is determined by $\hat{q}$ alone.

B. Differential Equations for Specific Types of Flow

The general differential equations for flow in three-dimensional space, formulated in Section 7, present mathematical difficulties beyond the present power of analysis. However, in many instances of great interest simplifications arise, in particular, when the number of independent variables reduces to two. Such is the case for non-steady one-dimensional flow, for steady flow in two dimensions, for steady flow with cylindrical symmetry, and for non-steady flow with spherical symmetry.

16. *Steady flows*

Steady *plane* or *two-dimensional flow* is described in terms of two velocity components u and v as functions of two coordinates x and y by requiring that the component w vanishes and that all quantities characterizing the flow are independent of z and t. Density and pressure can then be considered functions of the flow speed

(16.01) $$q = \sqrt{u^2 + v^2}$$

along each streamline through *Bernoulli's law*

(16.02) $$q^2 + 2i = \hat{q}^2,$$

see (14.02), and the *adiabatic equation*

(16.03) $$p = f(\rho, S),$$

see (2.03), which hold along each streamline with constant limit speed $\hat{q}$ and entropy S, see Sections 14 and 10. The sound speed c enters through the relation

(16.04) $$u\,du + v\,dv = -c^2 d\rho/\rho,$$

which holds along each streamline, as follows from (16.02), (16.03), and

(16.05) $$di = c^2 d\rho/\rho,$$

see (9.03). If the values of $\hat{q}$ and S for each streamline are known, two of the four quantities u, v, ρ, S can be expressed in terms of the others and only two of the original differential equations (7.02–.05) remain.

If the flow is *irrotational*, equation

(16.06) $$v_x - u_y = 0$$

holds, see (13.06), and if it is in addition *isentropic*, $S = $ constant, the limit speed $\hat{q}$ is constant over the field of flow, see Bernoulli's strong law, Section 14. Relation (16.04) which holds then in the whole field of flow now serves to eliminate the density ρ from the continuity equation

(16.07) $$(\rho u)_x + (\rho v)_y = 0,$$

see (7.08). The result is the equation

(16.08) $$(c^2 - u^2)u_x - uv(u_y + v_x) + (c^2 - v^2)v_y = 0,$$

in which c^2 is to be considered a function of the flow speed q, and hence of u and v, by Bernoulli's law (16.02) and the adiabatic equation (16.03). In the case of a polytropic gas this function is, see (14.05) and (14.07),

(16.09) $$c^2 = \frac{\gamma - 1}{2}(\hat{q}^2 - q^2) = (1 - \mu^2)^{-1}(c_*^2 - \mu^2 q^2).$$

Equation (16.08) and equation (16.06) (which expresses the irrotational character of the flow) form a system of two equations in two functions u, v of two variables x, y.

Equation (16.06) may be satisfied by introducing a *velocity potential* $\varphi(x, y)$, see (13.07), so that

(16.10) $$\varphi_x = u, \quad \varphi_y = v.$$

Equation (16.08) then reduces to a single differential equation of second order

(16.11) $$(c^2 - u^2)\varphi_{xx} - 2uv\varphi_{xy} + (c^2 - v^2)\varphi_{yy} = 0.$$

Incidentally, three-dimensional steady irrotational flow can also be characterized in a similar manner by one differential equation of second order for a velocity potential.

Equation (16.06) may be satisfied by introducing a *stream function* $\psi(x, y)$ so that

(16.12) $$\psi_x = -\rho v, \quad \psi_y = \rho u.$$

Equation (16.06) then becomes an equation of second order for ψ, in which ρ is to be considered a function of $\psi_x^2 + \psi_y^2$.

The lines $\psi =$ constant are the *streamlines* and the difference of the values of ψ on two streamlines is the flux of mass through a cylinder of unit height in the z-direction over any curve connecting any two points A and B on these two streamlines; this can be seen from the relation

$$\psi_B - \psi_A = \int_A^B \rho(ux_n + vy_n)\, ds$$

in which s refers to the arc length and (x_n, y_n) is the unit normal vector at points of the curve.

Steady flow with *cylindrical symmetry* can also be described by two velocity components u and v as functions of two variables x and y. Here x is the abscissa along the axis and y the distance from it; u is the component in the axial, v that in the radial direction. Thus every velocity vector lies in a plane through the axis and can be obtained from a vector in one such plane by revolving it about the axis. It is also required that ρ and p depend only on x and y. Bernoulli's law (16.02), the adiabatic equation (16.03), and hence relation (16.04), hold along each streamline. Irrotational character is again expressed by equation (16.06), and if, in addition, the flow is isentropic, relation (16.04) holds throughout the field of flow.

The only difference from plane flow lies in the continuity equation which now takes the form

(16.13) $$(y\rho u)_x + (y\rho v)_y = 0,$$

or after eliminating ρ by (16.04)

(16.14) $$(c^2 - u^2)u_x - uv(u_y + v_x) + (c^2 - v^2)v_y + c^2 v/y = 0.$$

Of course, this equation can again be reduced to one equation of second order for a potential function φ defined through (16.10).

Later on we shall use the *stream function*, introduced by Stokes, which is now defined by the relations

(16.15) $$\psi_x = -y\rho v, \quad \psi_y = y\rho u,$$

so as to satisfy the continuity equation (16.13). The surfaces ψ = constant, generated by revolving a streamline, are the *stream surfaces*. The value $2\pi\psi(x, y)$ is the flux through the annular ring cut out from the circle with radius y and abscissa x by the stream surface $\psi = 0$.

For steady *rotational flow* of a polytropic gas in which the limit speed $\hat{q}$ (but in general not the entropy S) is constant in the whole field of flow, a modified stream function was introduced by Crocco [22]. It is defined by equations which result from equations (16.12) or (16.15) when the factor ρ is replaced in them by the quantity $c^{2/(\gamma-1)}$, which by Bernoulli's law depends only on q and $\hat{q}$ and is independent of the entropy.

17. Non-steady flows

One-dimensional flow occurs when all quantities characterizing the flow depend, in addition to the time t, on only one space coordinate x, and when the components v and w of the velocity in the other directions vanish. Equations (7.08–.11) then reduce to

(17.01) $$\rho_t + u\rho_x + \rho u_x = 0,$$

(17.02) $$\rho(u_t + uu_x) + p_x = 0,$$

(17.03) $$S_t + uS_x = 0;$$

the latter equation expresses the assumption that the changes in the state of each particle are adiabatic and reversible. The pressure p is here a function of ρ and S. Using

(17.04) $$\frac{\partial p}{\partial \rho} = c^2,$$

see (7.12), equation (17.01) may conveniently be replaced by

(17.05) $$p_t + up_x + \rho c^2 u_x = 0,$$

so that the three equations (17.02), (17.05), and (17.03) involve only the derivatives of u, p, and S. For *polytropic gases*, equation (17.03) may be replaced by

(17.06) $$(p\rho^{-\gamma})_t + u(p\rho^{-\gamma})_x = 0,$$

since $p\rho^{-\gamma}$ is then a function of the entropy. If the flow is *isentropic*, one may express p in terms of ρ or vice versa. Equations (17.02) and (17.01) or (17.05) then represent two equations for two functions of x and t.

Spherical flow occurs when all quantities depend only on the distance from one point, chosen as the origin 0, in addition to the time, and if the velocity is directed away from (or toward) this point. Denoting the distance from the origin conveniently by x and the radial velocity component by u, equations (7.08–.11) reduce to

(17.07) $$\rho_t + u\rho_x + \rho u_x + 2\rho u/x = 0,$$

(17.08) $$\rho(u_t + uu_x) + p_x = 0,$$

(17.09) $$S_t + uS_x = 0.$$

We note that the only difference from the equations for one-dimensional flow is the additional term occurring in the continuity equation (17.07). The modifications and simplifications just discussed in connection with one-dimensional flow apply just as well to spherical flow.

The same remarks apply to *cylindrical flow*, a two-dimensional flow in which all quantities depend only on the distance from the axis and the velocity is directed away from (or toward) the axis. The only difference is that in equation (17.07) the factor 1 occurs instead of the factor 2.

18. Lagrange's equations of motion for one-dimensional and spherical flow

In one-dimensional flow Lagrange's equations are not encumbered by functional determinants, and in this special case they are sometimes preferable to Euler's equations.

The Lagrangian point of view requires us to attach a number h to each plane section of particles normal to the x-axis, so that the changing position of each section is given by a function $x(h, t)$. The quantities ρ, p, S are then considered functions of h and t. This number h could be chosen in many ways; in fact an arbitrary function is at our disposal.

Customarily, one identifies h with the abscissa of the particle at some initial time, e.g. $t = 0$. But not in all problems is such an initial position at a common initial time given for all particles.

Another rather natural choice of h, based on the law of conservation of mass, suggests itself. Without any loss of generality we may think of the flow as taking place in a tube of unit cross section along the x-axis. Now attach the value $h = 0$ to any definite "zero" section (moving, of course, with the medium), and then for any other section let h be equal in magnitude to the mass of the medium in the tube of unit cross section area between that section and the zero section, the sign of h being taken positive or negative according as the zero section is to the left or right of the other section in question. It is clear that h as so defined is different for every section.

Analytically, the quantity h satisfies the relation

$$(18.01) \qquad h = \int_{x(0,t)}^{x(h,t)} \rho \, dx.$$

Here ρ is the density at the position x at the time t; in other words, the density is here regarded from the Eulerian point of view as a function of the independent variables x and t. Differentiating (18.01) with respect to h leads to the relation

$$(18.02.1) \qquad \rho(h, t) x_h(h, t) = 1$$

or

$$(18.02.2) \qquad x_h(h, t) = \tau(h, t),$$

in which $\rho(h, t)$ and $\tau(h, t)$ are respectively the density and specific volume and h is a function of time.

LAGRANGE'S EQUATIONS OF MOTION

Lagrange's equations of motion (7.02–.05) for one-dimensional flow then take the form:

(18.03) $$(\rho x_h)_t = 0$$

(Conservation of mass),

(18.04) $$\rho x_{tt} = -p_x$$
$$= -p_h/x_h$$

(Conservation of momentum),

(18.05) $$S_t = 0$$

(Changes of state are adiabatic),

(18.06) $$p = f(\rho, S) = g(\tau, S)$$

(Caloric equation of state)

(We have here abandoned the dot ($\cdot$) in favor of the subscript t, which will not lead to confusion between Eulerian and Lagrangian concepts.)

Lagrange's equations (18.03–.06) may be simplified considerably. In the first place it follows from (18.02) that (18.03) is superfluous and that (18.04) may be replaced by

(18.07) $$x_{tt} = -p_h .$$

According to (18.05), S depends only on h; we will henceforth always imply $S = S(h)$. The function $S(h)$ is considered as given among the initial conditions of the problem. By means of (18.02) and (18.04–.06) we can eliminate ρ, τ, and p and the whole system reduces to a single partial differential equation of the second order in x:

(18.08) $$x_{tt} = k^2 x_{hh} - g_S S_h ,$$

in which we have introduced the quantity

(18.09) $$k = \rho c = \sqrt{-g_\tau(\tau, S)},$$

the *acoustic impedance* of the medium. In interpreting equation (18.08) it must of course be remembered that k^2 and g_S are given functions of $\tau = x_h$ and $S(h)$.

If the velocity $u = x_t$ and the specific volume $\tau = x_h$ are taken as the dependent variables the single second order equation (18.08) is replaced by the first order system for u and τ

(18.10)
$$u_h = \tau_t$$
$$u_t = k^2 \tau_h - g_S S_h(h),$$

in which k^2 and g_S are functions of τ and $S(h)$.

If the flow is isentropic, i.e. if $S(h) = S_0$ is a constant, equations (18.08) and (18.10) simplify to

(18.11) $$x_{tt} = k^2 x_{hh}$$

(18.12)
$$u_h = \tau_t$$
$$u_t = k^2 \tau_h .$$

Note that if $k^2 = -p_\tau$ is constant, as it is for a solid which obeys Hooke's law, equations (18.11) and (18.12) are linear.

The formulations of this section can easily be extended to *spherical* and *cylindrical* flow, see Section 17. Let $4\pi h$ denote the mass of the medium inside a sphere of radius $y(h, t)$ about the center of a spherical flow. Then we have

$$y_{tt} = y^2[k^2(y^2 y_h)_h - g_S S_h],$$

analogous to (18.08). For a plane flow with cylindrical symmetry we have in corresponding notation

$$y_{tt} = y^2[k^2(y y_h)_h - g_S S_h].$$

APPENDIX

Wave Motion in Shallow Water

19. Shallow water theory

An analogue to the nonlinear wave motion of gases is encountered in the motion of water, or any other incompressible fluid, with a free top surface if the height of the top surface above the bottom surface is sufficiently small. One speaks then of "shallow water." More precisely the condition is that the height of the top surface above the bottom surface is small compared with some characteristic length of the motion such as the maximum radius of curvature occurring on the top surface. The differential equations governing the motion of such shallow water can in good approximation be replaced by equations which are completely equivalent to those for a polytropic gas with the exponent $\gamma = 2$. As a matter of fact all the phenomena of wave motion which we shall discuss in subsequent chapters have their strict analogue in the wave motion of shallow water.

We place an (x, y, z)-coordinate system in the space filled by the water in such a way that the bottom surface is the plane $z = 0$ and the top surface is given by a function $z = Z(x, y, t)$. We denote the components of velocity in the x-, y-, and z-directions by u, v, w respectively; u, v, w are functions of x, y, z. In the water the continuity condition

(19.01) $$u_x + v_y + w_z = 0$$

and Newton's law

(19.02) $$\rho \frac{du}{dt} = -p_x, \quad \rho \frac{dv}{dt} = -p_y, \quad \rho \frac{dw}{dt} = -p_z - g\rho$$

holds. Here g is the acceleration of gravity, ρ the density of the water, and p the excess pressure above atmospheric pressure so that

(19.03) $$p = 0 \text{ at the top surface, } z = Z.$$

The boundary conditions for the velocity are

(19.04) $$w = 0 \text{ at the bottom, } z = 0,$$

and

(19.05) $$Z_t + uZ_x + vZ_y = w \text{ at the top, } z = Z.$$

It is now possible to replace these equations, in good approximation, by equations involving only the top surface elevation Z and the velocities u and v at the top surface. In order to do so we first integrate the continuity equation (19.01) from $z = 0$ to $z = Z$, obtaining

$$W \Big|_{z=0}^{z=Z} + \int_0^Z (u_x + v_y)\, dz = 0$$

from which, by the boundary conditions (19.04) and (19.05),

$$(19.06) \qquad Z_t + \left(\int_0^Z u\, dz\right)_x + \left(\int_0^Z v\, dz\right)_y = 0.$$

Next we introduce the *basic assumption that the variation of the pressure along a vertical column is the same as in hydrostatics*

$$(19.07) \qquad p = g\rho(Z - z).$$

By (19.02) and (19.03) this assumption is equivalent to the condition that the vertical acceleration of the water, $\dfrac{dw}{dt}$, vanishes. Assumption (19.07) is not arbitrary; it can be shown to be correct in first order as a consequence of the basic assumption of shallowness by a systematic expansion of the equations in powers of the height z; see Stoker [27, Appendix].

We note that assumption (19.07) implies that the pressure gradient (p_x, p_y, p_z) is independent of z; by (19.02) the same is also true for the acceleration $\left(\dfrac{du}{dt}, \dfrac{dv}{dt}, \dfrac{dw}{dt}\right)$. As a consequence, the velocity (u, v, w) is also independent of z if this was ever the case at some time. We now introduce the assumption, that at some time the velocity was constant over every vertical column. This assumption does not seem to be a serious restriction; it would, for example, be satisfied if the water was at rest at some time. As a consequence u and v depend on x, y, z only; we shall ignore the vertical velocity w from now on.

The equations (19.06) and (19.02) now assume the simple form

$$(19.08) \qquad Z_t + (Zu)_x + (Zv)_y = 0$$

$$(19.09) \qquad \begin{aligned} \rho(u_t + uu_x + vu_y) &= -g\rho Z_x, \\ \rho(v_t + uv_x + vv_y) &= -g\rho Z_y. \end{aligned}$$

To show that these equations are equivalent to those for polytropic gases with $\gamma = 2$ we introduce the "density"

$$(19.10) \qquad \bar{\rho} = \rho Z,$$

which is evidently the mass per unit area, and the "pressure"

$$(19.11) \qquad \bar{p} = \tfrac{1}{2} g \rho Z^2 = \int_0^Z p\, dz.$$

SHALLOW WATER THEORY

Equations (19.08–.09) can then be written in the form of the equations for gases, see (7.08), (7.09),

(19.12) $$\bar{\rho}_t + (\bar{\rho} u)_x + (\bar{\rho} v)_y = 0,$$

(19.13) $$\bar{\rho}(u_t + u u_x + v u_y) = -\bar{p}_x,$$
$$\bar{\rho}(v_t + u v_x + v v_y) = -\bar{p}_y.$$

The relationship

(19.14) $$\bar{p} = \frac{g}{2\rho} \bar{\rho}^2$$

between "pressure" $\bar{p}$ and "density" $\bar{\rho}$, which follows from (19.10) and (19.11), evidently corresponds to the relationship between the real pressure and density for a polytropic gas with $\gamma = 2$.

CHAPTER II

Mathematical Theory of Hyperbolic Flow Equations for Functions of Two Variables

In the preceding part we have shown that for many specific cases the flow differential equations reduce to *systems of quasi-linear partial differential equations of the first order for functions of two independent variables*. For such systems a fairly complete mathematical theory can be developed provided they are of the *hyperbolic* type, in which case the notion of *characteristics* plays the dominant role. This theory becomes particularly simple when the number of functions and equations is two.

To prepare a deeper understanding of the treatment of specific flow problems in the following chapters we insert here a detailed theory of systems of two differential equations; supplementary remarks will be added about systems of more than two differential equations, as they occur for example in non-isentropic flow.

20. Flow equations involving two functions of two variables

For convenient reference we enumerate the specific types of flow which are governed by a system of two equations for two functions of two variables.

a) One-dimensional isentropic flow

(20.01)
$$\rho_t + u\rho_x + \rho u_x = 0,$$
$$\rho(u_t + uu_x) + c^2\rho_x = 0$$

see (17.01), (17.02), (17.04).

b) Spherical isentropic flow

(20.02)
$$\rho_t + u\rho_x + \rho u_x + 2\rho u/x = 0,$$
$$\rho(u_t + uu_x) + c^2\rho_x = 0$$

see (17.04), (17.07), (17.08).

c) One-dimensional isentropic flow in Lagrangian representation

(20.03)
$$\tau_t = u_h$$
$$u_t = k^2 \tau_h ,$$

see (18.12); here k is a given function of τ.

d) One-dimensional non-isentropic flow in Lagrangian representation

(20.04)
$$\tau_t = u_h$$
$$u_t = k^2 \tau_h - g_S S_h ,$$

see (18.10); here the distribution of the entropy $S = S(h)$ over the particles is assumed to be given, k^2 and g_S are given functions of τ and S.

e) Steady two-dimensional irrotational isentropic flow

(20.05)
$$v_x - u_y = 0$$
$$(c^2 - u^2)u_x - uv(u_y + v_x) + (c^2 - v^2)v_y = 0,$$

see (16.06) and (16.08); here c^2 is a given function of $u^2 + v^2$.

f) Steady irrotational isentropic flow in three dimensions with cylindrical symmetry

(20.06)
$$v_x - u_y = 0$$
$$(c^2 - u^2)u_x - uv(u_y + v_x) + (c^2 - v^2)v_y + c^2 v/y = 0,$$

see (16.06) and (16.14); again c^2 is a given function of $u^2 + v^2$.

21. Differential equations of second order type

In the general theory we denote by u, v the dependent and by x, y the independent variables, to be identified later with the variables in the specific differential equations of gas dynamics which we have enumerated in the preceding section. Then the general form of the system of differential equations is

(21.01)
$$L_1 = A_1 u_x + B_1 u_y + C_1 v_x + D_1 v_y + E_1 = 0$$
$$L_2 = A_2 u_x + B_2 u_y + C_2 v_x + D_2 v_y + E_2 = 0$$

in which A_1, A_2, $\cdots$, E_2 are known functions of x, y, u, v. Once and for all we make the assumption that all functions occurring in the theory of this chapter are continuous, and possess as many continuous derivatives as may be required. Without restriction we assume that nowhere $A_1 : A_2 = B_1 : B_2 = C_1 : C_2 = D_1 : D_2$. If $E_1 = E_2 = 0$ the system is *homogeneous*. If the coefficients A, B, C, D, E, are functions of x and y only, the equations are *linear* and consequently much easier to handle.

In another important case the system can be reduced to a linear one: *If the system is homogeneous, $E_1 = E_2 = 0$, and the coefficients $A_1, \cdots, D_2$ are functions of u, v alone, the equations are called reducible.* In this case for any region where the Jacobian

(21.02) $$j = u_x v_y - u_y v_x$$

is not zero, the system (21.01) can be transformed into an equivalent linear system by interchanging the roles of dependent and independent variables. If $j \neq 0$ for a solution $u(x, y)$, $v(x, y)$ of (21.01), we may consider x and y as functions of u and v. From

$$u_x = jy_v, \quad u_y = -jx_v,$$
$$v_x = -jy_u, \quad v_y = jx_u,$$

we then see that $x(u, v)$ and $y(u, v)$ satisfy the linear differential equations

(21.03)
$$A_1 y_v - B_1 x_v - C_1 y_u + D_1 x_u = 0$$
$$A_2 y_v - B_2 x_v - C_2 y_u + D_2 x_u = 0.$$

Vice versa every solution x, y of equations (21.03) leads to a solution of (21.01) if the Jacobian

(21.04) $$J = x_u y_v - x_v y_u$$

does not vanish.

The described transformation of the (x, y)-plane into the (u, v)-plane is frequently called a *hodograph transformation*.

Reducible equations occur in one-dimensional flow, see (20.01) and (20.03), and in two-dimensional steady flow, see (20.05).

Since the possibility of this reduction depends essentially on the assumption $j \neq 0$, solutions for which $j = 0$ cannot be obtained by the hodograph transformation. Yet, as we shall discuss in Section

29, these solutions, here called *simple waves*, are most important tools for the solutions of flow problems; simple waves and their generalizations apparently have not been sufficiently emphasized in mathematical studies of hyperbolic differential equations.

22. Characteristic curves and characteristic equations

The key to the general theory of systems of quasilinear partial differential equations of the form (21.01) is to distinguish between elliptic and hyperbolic equations, and for the latter, which will be our main object, to introduce the notion of characteristics. These concepts emerge naturally from the following considerations.

A linear combination $af_x + bf_y$ of the two derivatives of a function $f(x, y)$ means derivation of f in a direction given by $dx:dy = a:b$. If $x(\sigma)$, $y(\sigma)$ represent a curve with $x_\sigma:y_\sigma = a:b$, then $af_x + bf_y$ is a derivative of f along this curve. We consider specific functions $u(x, y)$, $v(x, y)$; then the coefficients of the differential equations (21.01) become functions of x, y only. In each of the two differential equations the functions u and v are, generally speaking, differentiated in two different directions. We now ask for a linear combination

$$L = \lambda_1 L_1 + \lambda_2 L_2$$

so that in the differential expression L the derivatives of u and those of v combine to derivatives in the same direction. Such a direction which depends on the point x, y as well as on the values of u, v at this point, is called *characteristic*. Suppose the direction is given by the ratio $x_\sigma : y_\sigma$ as above; then the condition, that, in L, u and v are differentiated in this direction, is simply

(22.01) $\lambda_1 A_1 + \lambda_2 A_2 : \lambda_1 B_1 + \lambda_2 B_2 = \lambda_1 C_1 + \lambda_2 C_2 : \lambda_1 D_1 + \lambda_2 D_2 = x_\sigma : y_\sigma;$

for, the coefficients of the derivatives u_x, u_y and v_x, v_y in L are given by the respective members of the proportion (22.01). The expression L can be written after multiplication with either x_σ or y_σ as

(22.02) $(\lambda_1 A_1 + \lambda_2 A_2)u_\sigma + (\lambda_1 C_1 + \lambda_2 C_2)v_\sigma + (\lambda_1 E_1 + \lambda_2 E_2)x_\sigma = x_\sigma L$

or as

(22.03) $(\lambda_1 B_1 + \lambda_2 B_2)u_\sigma + (\lambda_1 D_1 + \lambda_2 D_2)v_\sigma + (\lambda_1 E_1 + \lambda_2 E_2)y_\sigma = y_\sigma L.$

If at the point x, y the functions u and v satisfy the differential equations (21.01), we obtain four homogeneous linear equations for λ_1 and λ_2:

(22.04)
$$\lambda_1(A_1 y_\sigma - B_1 x_\sigma) + \lambda_2(A_2 y_\sigma - B_2 x_\sigma) = 0$$
$$\lambda_1(C_1 y_\sigma - D_1 x_\sigma) + \lambda_2(C_2 y_\sigma - D_2 x_\sigma) = 0$$
$$\lambda_1(A_1 u_\sigma + C_1 v_\sigma + E_1 x_\sigma) + \lambda_2(A_2 u_\sigma + C_2 v_\sigma + E_2 x_\sigma) = 0$$
$$\lambda_1(B_1 u_\sigma + D_1 v_\sigma + E_1 y_\sigma) + \lambda_2(B_2 u_\sigma + D_2 v_\sigma + E_2 y_\sigma) = 0.$$

If these equations are satisfied all determinants of two rows in the matrix of the coefficients of λ_1 and λ_2 vanish. Thus a number of *characteristic relations* follow.

From the first two equations, in particular,

(22.05)
$$\begin{vmatrix} A_1 y_\sigma - B_1 x_\sigma & A_2 y_\sigma - B_2 x_\sigma \\ C_1 y_\sigma - D_1 x_\sigma & C_2 y_\sigma - D_2 x_\sigma \end{vmatrix} = 0,$$

or

(22.06) $$a y_\sigma^2 - 2b x_\sigma y_\sigma + c x_\sigma^2 = 0.$$

Here

(22.07) $$a = [AC],\ 2b = [AD] + [BC],\ c = [BD]$$

with the abbreviation

$$[XY] = X_1 Y_2 - X_2 Y_1.$$

If $ac - b^2 > 0$, then (22.06) cannot be satisfied by a (real) direction. Characteristic directions do not exist in this case; the differential equations are then called *elliptic*. We disregard the case that $ac - b^2 = 0$ in which one characteristic direction exists through each point. If $ac - b^2 < 0$ we have two different characteristic directions $y_\sigma : x_\sigma$ through each point; the system is called *hyperbolic*. For the flow problems considered in this book the differential equations are mostly hyperbolic. We shall from now on assume the hyperbolic character* of equations (21.01) and accordingly suppose

(22.08) $$ac - b^2 < 0.$$

* For the *elliptic* equations characterized by $ac - b^2 > 0$, a treatment somewhat corresponding to the following can be given but is omitted here. See Courant-Hilbert [32, pp. 337–342].

This assumption excludes the exceptional case that all three coefficients vanish. Moreover, we assume for convenience

(22.09) $$a = [AC] \neq 0.$$

The latter condition can always be satisfied, if necessary by introducing new coordinates instead of x and y. Consequently, $x_\sigma \neq 0$ for a characteristic direction (x_σ, y_σ) as seen from (22.05); thus we are at liberty to introduce the slope

(22.10) $$\zeta = y_\sigma/x_\sigma.$$

Equation (22.06) becomes a quadratic equation for ζ

(22.11) $$a\zeta^2 - 2b\zeta + c = 0.$$

This equation has two different real solutions ζ_+ and ζ_-

(22.12) $$\zeta_+ \neq \zeta_-$$

by (22.08). Accordingly, the two different *characteristic directions* are given by $dy/dx = \zeta_+$ and $dy/dx = \zeta_-$ at the point (x, y). Since the roots ζ_+ and ζ_- of (21.11) are, generally speaking, functions of x, y, u, v, it ought to be noted that the hyperbolic character of the system (21.01) depends on the individual functions $u(x, y)$, $v(x, y)$ under consideration.

Once a fixed solution of equations (21.01) is inserted, the equations $dy/dx = \zeta_+(x, y, u, v)$ and $dy/dx = \zeta_-(x, y, u, v)$ are two separate ordinary differential equations of the first order which define two one-parametric families of *characteristic curves* or simply *characteristics*, C_+ and C_- in the (x, y)-plane, belonging to the solution $u(x, y), v(x, y)$.

These two families may be represented in the form $\beta(x, y) = $ constant and $\alpha(x, y) = $ constant, respectively, and form a curvilinear coordinate net. It is now natural to introduce new parameters α, β instead of x, y in such a way that β is constant along the curves C_+ and α is constant along C_-. To specify such *characteristic parameters* we may, for example, take any curve $\mathcal{I}$ given by $x = x(s)$, $y = y(s)$ which nowhere has a characteristic direction, that is,

(22.13) $$ay_s^2 - 2bx_s y_s + cx_s^2 \neq 0 \text{ on } \mathcal{I}.$$

Through any two points $s = \alpha$ and $s = \beta$ on $\mathcal{I}$ we pass the curves C_- and C_+ up to a point (x, y) where they intersect. (Such an intersection exists if $|\alpha - \beta|$ is sufficiently small since the directions of

C_+ and C_- are different according to (22.12).) The curvilinear coordinates α, β of the point (x, y) are then characteristic parameters.

Of course, instead of the parameters α and β so introduced, one may introduce any monotone functions $\alpha' = V(\alpha)$, $\beta' = W(\beta)$ as characteristic parameters. Such a transformation leaves the equations of the characteristics invariant, and the curves themselves unchanged.

In a domain in which characteristic parameters are introduced we have

(22.14)   I $\quad \begin{aligned} y_\alpha &= \zeta_+ x_\alpha \quad \text{along } C_+ \\ y_\beta &= \zeta_- x_\beta \quad \text{along } C_- \, . \end{aligned}$

According to our original objective we now have to determine the factors λ_1, λ_2 in order to form the combined relation $\lambda_1 L_1 + \lambda_2 L_2 = L = 0$; instead, we obtain more elegantly the relation $L = 0$ by eliminating λ_1, λ_2 from the first and third of equations (22.04). The result is

$$\begin{vmatrix} A_1 y_\sigma - B_1 x_\sigma & A_2 y_\sigma - B_2 x_\sigma \\ A_1 u_\sigma + C_1 v_\sigma + E_1 x_\sigma & A_2 u_\sigma + C_2 v_\sigma + E_2 x_\sigma \end{vmatrix} = 0.$$

Writing $y_\sigma = \zeta x_\sigma$, we obtain, after cancelling the factor x_σ, the equation

(22.15) $\quad T u_\sigma + (a\zeta - S) v_\sigma + (K\zeta - H) x_\sigma = 0$ along C_+,

in which

(22.16) $\quad T = [AB], S = [BC], K = [AE], H = [BE].$

This relation holds if we identify ζ with ζ_+ and σ with α, and likewise if we identify ζ with ζ_- and σ with β.

Thus we have arrived at the following four *characteristic equations:*

(22.17) $\quad \begin{aligned} \text{I}_+ & \quad y_\alpha - \zeta_+ x_\alpha = 0, \\ \text{I}_- & \quad y_\beta - \zeta_- x_\beta = 0, \\ \text{II}_+ & \quad T u_\alpha + (a\zeta_+ - S) v_\alpha + (K\zeta_+ - H) x_\alpha = 0, \\ \text{II}_- & \quad T u_\beta + (a\zeta_- - S) v_\beta + (K\zeta_- - H) x_\beta = 0, \end{aligned}$

which hold for every solution $u(x, y), v(x, y)$ and refer to its characteristics C_+ or C_-.

While so far we have envisaged a fixed solution u, v, the equations (22.17) no longer depend explicitly on this solution since all the coefficients are known functions of x, y, u, v.

We now introduce a slight but decisive change of interpretation: The system (22.17) can and will be considered as a system of four partial differential equations for the four quantities x, y, u, v as functions of α and β. Replacing the original system (21.01) by this *characteristic system* is the basis of the following theory.

The differential equations (22.17) are of a particularly simple form inasmuch as each equation contains derivatives with respect to only one of the independent parameters; moreover, the coefficients do not depend on the independent parameters. (Such a system is called *canonical* hyperbolic, see [32, p. 324].)

According to our derivation, every solution of the original system (21.01) satisfies this characteristic system. The converse is also true as is easily verified. Every solution of the characteristic system (22.17) satisfies the original system (21.01) provided the Jacobian $x_\alpha y_\beta - x_\beta y_\alpha = (\zeta_- - \zeta_+)x_\alpha x_\beta$ does not vanish.

The two particular cases mentioned before should be noted:

When the differential equations (21.01) are *linear*, then ζ_+ and ζ_- are known functions of x, y; the equations I in (22.17) are not coupled with the equations II and thus the equations I determine two *families* of *characteristic curves*, C_+ and C_-, independent of the solution.

When the differential equations are *reducible*, i.e. when $E_1 = E_2 = 0$, and $A_1, \cdots, D_2$ depend on u, v only, the situation is similar. Then ζ_+ and ζ_- are known functions of u and v and the differential equations II are independent of x and y. (The same, incidentally, remains true even if E_1 and E_2 do not vanish but depend on u and v only.)

For reducible equations, the *characteristic* curves Γ *in the (u, v)-plane*, the images of the characteristics C in the (x, y)-plane, are independent of the special solution $u(x, y), v(x, y)$ considered. They are characterized by the differential equations II which can be written:

(22.18)
$$\Gamma_+: \quad T\frac{du}{dv} = S - a\zeta_+,$$
$$\Gamma_-: \quad T\frac{du}{dv} = S - a\zeta_-.$$

While equations (22.17) form a complete system, it may be mentioned that another relation can be derived which is particularly useful when referred to the characteristics Γ_+ and Γ_- in the u, v-plane; it combines the equations II just as (22.06) combines the equations I. By eliminating λ_1 and λ_2 from the last two equations of (22.04) one obtains

$$(22.19) \quad \begin{vmatrix} A_1 u_\sigma + C_1 v_\sigma + E_1 x_\sigma & A_2 u_\sigma + C_2 v_\sigma + E_2 x_\sigma \\ B_1 u_\sigma + D_1 v_\sigma + E_1 y_\sigma & B_2 u_\sigma + D_2 v_\sigma + E_2 y_\sigma \end{vmatrix} = 0.$$

This equation is analogous to (22.05). We assume that of the coefficients

$$[AB] = T, [AD] + [CB], [CD]$$

of u_σ^2, $u_\sigma v_\sigma$, and v_σ^2 in (22.19) not all are zero, in particular, that

$$(22.20) \quad T \neq 0;$$

the latter condition can always be satisfied, if necessary by introducing new functions instead of u and v.

In the case of reducibility equation (22.19) becomes a simple quadratic differential equation for the ratio $u_\sigma : v_\sigma$, which can be used, in place of (22.18), to determine the two families of Γ characteristics.

23. Characteristic equations for specific problems

For the flow equations enumerated in Section 20 the characteristic equations are easily obtained either directly or by substitution in the general formulas.

We carry out the method directly in the case of one-dimensional isentropic flow by forming a linear combination of the two equations (20.01) and asking when this combination involves the derivatives u_σ, ρ_σ of u and ρ in only one direction given by (t_σ, x_σ). We write the combination in the form

$$(23.01) \quad u_t + (u + \lambda \rho) u_x + \lambda \rho_t + \left(\lambda u + \frac{c^2}{\rho} \right) \rho_x = 0.$$

The condition for (t_σ, x_σ) is then evidently

$$x_\sigma = (u + \lambda \rho) t_\sigma, \qquad \lambda x_\sigma = \left(\lambda u + \frac{c^2}{\rho} \right) t_\sigma,$$

from which

(23.02) $$\lambda^2 = c^2/\rho^2$$

Hence there are two characteristic directions given by

(23.03) $$x_\alpha = (u + c)t_\alpha, \quad x_\beta = (u - c)t_\beta.$$

From (23.01–.03) the characteristic equations for u and ρ become

(23.04) $$u_\alpha + \frac{c}{\rho}\rho_\alpha = 0, \quad u_\beta - \frac{c}{\rho}\rho_\beta = 0.$$

As will be explained in detail in Chapter III, equations (23.03) express the fact that the characteristic curves in the (x, t)-plane represent motions of possible disturbances (later called "sound waves") whose velocity,

(23.05) $$\frac{dx}{dt} = u + c \quad \text{or} \quad \frac{dx}{dt} = u - c,$$

differs from the particle velocity u by the sound velocity $\pm c$.

The differential equations of three-dimensional spherical flow (20.02) differ from those of one-dimensional flow by a term $2\rho u/x$, which does not involve the derivatives of u and ρ. It is therefore clear that the characteristic equations for x and t are the same as for one-dimensional flow, namely (23.03) (of course, this fact does not imply that the characteristic curves are the same, since these curves depend on the solution). The characteristic equations for u and ρ differ from (23.04); they are immediately found to be

(23.06) $$u_\alpha + \frac{c}{\rho}\rho_\alpha + 2\frac{cu}{x}t_\alpha = 0, \quad u_\beta - \frac{c}{\rho}\rho_\beta - 2\frac{cu}{x}t_\beta = 0.$$

For *steady two-dimensional irrotational isentropic flow* we follow somewhat more closely the procedure of the general theory. We form a linear combination of the equations (20.05)

$$\lambda_2(c^2 - u^2)u_x + (\lambda_1 - \lambda_2 uv)u_y - (\lambda_1 + \lambda_2 uv)v_x + \lambda_2(c^2 - v^2)v_y = 0.$$

The condition that this combination involve the derivatives u_σ, v_σ in only one direction, given by (x_σ, y_σ), entails

$$(\lambda_1 - \lambda_2 uv)x_\sigma - \lambda_2(c^2 - u^2)y_\sigma = 0,$$
$$\lambda_2(c^2 - v^2)x_\sigma + (\lambda_1 + \lambda_2 uv)y = 0,$$
$$\lambda_2(c^2 - u^2)u_\sigma - (\lambda_1 + \lambda_2 uv)v_\sigma = 0,$$
$$(\lambda_1 - \lambda_2 uv)u_\sigma + \lambda_2(c^2 - v^2)v_\sigma = 0,$$

CHARACTERISTIC EQUATIONS FOR SPECIFIC PROBLEMS 47

corresponding to (22.04). By eliminating λ_1 and λ_2 from the first two equations we obtain

(23.07) $\qquad (c^2 - v^2)x_\sigma^2 + 2uv x_\sigma y_\sigma + (c^2 - u^2)y_\sigma^2 = 0;$

from the last two equations we obtain similarly

(23.08) $\qquad (c^2 - u^2)u_\sigma^2 - 2uv u_\sigma v_\sigma + (c^2 - v^2)v_\sigma^2 = 0,$

while from the first and third equations we have

(23.09) $\qquad [(c^2 - u^2)u_\sigma - uv v_\sigma]x_\sigma - [uv x_\sigma + (c^2 - u^2)y_\sigma]v_\sigma = 0.$

Equation (23.07) written in the form

(23.10) $\qquad (c^2 - v^2) + 2uv\zeta + (c^2 - u^2)\zeta^2 = 0$

for $\zeta = y_\sigma/x_\sigma$ has two real roots ζ_+, ζ_- so that

(23.11) $\qquad y_\alpha = \zeta_+ x_\alpha, \qquad y_\beta = \zeta_- x_\alpha$

provided that $(c^2 - v^2)(c^2 - u^2) - u^2v^2$ is negative or

$$0 < c^2 < u^2 + v^2.$$

This condition, implying that the flow is supersonic, ensures that the system of differential equations (20.05) is hyperbolic. The characteristic curves determined from (23.11) will later on be called "Mach lines," see Section 31.

Equations (23.08) determine two characteristic directions in the (u, v)-plane independent of the solution $u(x, y)$, $v(x, y)$ considered. That such fixed characteristics in the (u, v)-plane exist is clear from the general theory since the system (20.05) is reducible. Comparing equations (23.08) with (23.07) we see that the two roots dv/du of (23.08) are $-\zeta_+$ and $-\zeta_-$. The proper coordination is derived from equation (23.09), which yields

$$(c^2 - u^2)u_\alpha - [2uv + (c^2 - u^2)\zeta_+]v_\alpha = 0$$

or, since by (23.07)

$$(c^2 - u^2)(\zeta_+ + \zeta_-) = -2uv,$$

(23.12.1) $\qquad u_\alpha = -\zeta_- v_\alpha.$

Similarly we find

(23.12.2) $\qquad u_\beta = -\zeta_+ v_\beta.$

Of course, equation (23.08) is a consequence of the two equations (23.12) and the definition of ζ_+ and ζ_- as roots of equation (23.10).

The significance of the characteristic equations so derived will be discussed and interpreted in detail in Chapter IV. Here we only mention that for steady three-dimensional flow with cylindrical symmetry, governed by equation (20.06), the characteristic equations for x_σ and y_σ are again equations (23.07), the same as for two-dimensional flow. The characteristic equations for u_σ, v_σ, are, however, different. Instead of equation (23.08) we have

(23.13)
$$u_\alpha + \zeta_- v_\alpha + \frac{c^2}{c^2 - u^2} \frac{v}{y} x_\alpha = 0,$$
$$u_\beta + \zeta_+ v_\beta + \frac{c^2}{c^2 - u^2} \frac{v}{y} x_\beta = 0,$$

as is easily verified.

24. The initial value problem. Domain of dependence. Range of influence

The initial value problem is the crucial problem in the theory of hyperbolic differential equations. Let a curve $\mathcal{I}$ be given in the (x, y)-plane in terms of a parameter s: $x = x(s)$, $y = y(s)$. (We assume that the derivatives $x_s(s)$, $y_s(s)$ are piecewise continuous along $\mathcal{I}$ and that $x_s^2 + y_s^2 \neq 0$.) Along $\mathcal{I}$ continuous values $u(s)$, $v(s)$ are arbitrarily prescribed. Then the initial value problem is: to determine, in the neighborhood of $\mathcal{I}$, a solution $u(x, y)$, $v(x, y)$ of equations (21.01) which takes on the prescribed initial values $u(s)$, $v(s)$ on $\mathcal{I}$. We assume that the curve $\mathcal{I}$ with its prescribed values of u and v has nowhere a characteristic direction, in other words that

$$a y_s^2 - 2b x_s y_s + b x_s^2 \neq 0$$

along $\mathcal{I}$.

By means of the characteristic form (22.17) of the differential equations this problem can be treated as completely as the corresponding one for ordinary differential equations.*

Referring to the plane of the characteristic parameters α and β we may consider $\mathcal{I}$ as the image of the special line Λ: $\alpha + \beta = 0$.

---
*The discovery of this important fact is due to Hans Lewy [29].

The characteristic parameters α and β were introduced in Section 22 with reference to a curve $\mathcal{I}$ on which $\alpha = \beta$; we need only replace β by $-\beta$. We restrict ourselves in the following discussion to a onesided neighborhood of $\mathcal{I}$.

The initial value problem can now be formulated for the differential equations (22.17) in the (α, β)-plane. On the line Λ the values of x, y, u, v are prescribed as continuously differentiable functions of $\alpha = -\beta = s$, and we seek in a onesided neighborhood of Λ a solution of the characteristic equations I and II in (22.17) which attains these given values on Λ. (We assume that the coefficients in (22.17) possess two continuous derivatives with respect to their arguments.)

To construct the solution we differentiate equations I_+ and II_+ with respect to β and equations I_- and II_- with respect to α, thus obtaining four equations linear in $x_{\alpha\beta}$, $y_{\alpha\beta}$, $u_{\alpha\beta}$, $v_{\alpha\beta}$. The determinant of these linear equations has the value $aT(\zeta_+ - \zeta_-)^2$, which is different from zero by (22.09), (22.12), (22.20). Hence we can solve for $x_{\alpha\beta}$, $y_{\alpha\beta}$, $u_{\alpha\beta}$, $v_{\alpha\beta}$ and obtain a system of equations of the form

(24.01) $\quad x_{\alpha\beta} = f_1, \quad y_{\alpha\beta} = f_2, \quad u_{\alpha\beta} = f_3, \quad v_{\alpha\beta} = f_4,$

in which the functions f have continuous first derivatives with respect to each of the quantities $x, y, u, v, x_\alpha, x_\beta, y_\alpha, y_\beta, u_\alpha, u_\beta, v_\alpha, v_\beta$. From the original initial values and equations (22.17) we can determine the values of these twelve quantities on the line Λ. The initial value problem for equations (24.01) can now be solved by the method of iterations* in a neighborhood of the initial line Λ. The iteration process, leading to the values of u, v, x, y at a point α, β, uses only integrations of the functions f_i over a triangle ABP as indicated in Figure 1 (page 50). Assuming, for example, that the initial values are zero, one defines a sequence of functions $x^{(n)}, y^{(n)}, u^{(n)}, v^{(n)}$ of α and β by the recursion formulas

$$x^{(0)} = 0, \qquad x^{(n+1)}(P) = \iint_{ABP} f_1(x^{(n)}, y^{(n)}, u^{(n)}, v^{(n)}) d\alpha d\beta$$

and so on for y, u and v. This sequence is then proved to converge to a solution of equations (24.01). The solution of equations (24.01)

---
* See [32, Chapter V, Section 5].

yields a solution of a characteristic system (22.17). Solving the characteristic system (22.17) is equivalent to solving the initial value problem for our original system (21.01) provided that the Jacobian $x_\alpha y_\beta - x_\beta y_\alpha$ does not vanish.

For the details of the proof, reference may be made to the literature quoted above. (For a slightly different and more general procedure see Section 32.)

From the iteration process it becomes apparent that the values u, v, x, y at the point $P = (\alpha, \beta)$ depend only on the initial values on

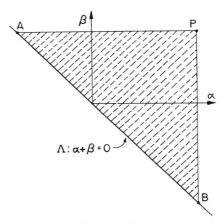

FIG. 1. Triangle in the (α, β)-plane to which the iteration process applies.

the segment between the points $A = (-\beta, \beta)$ and $B = (\alpha, -\alpha)$ indicated on the diagram. Suppose one has in the triangle ABP two solutions of the differential equations with the same initial values along the segment AB (but possibly differing outside). Then the two solutions coincide in the whole triangle ABP.

Our result has the following significant meaning concerning the (x, y)-plane. Suppose a solution $u(x, y)$, $v(x, y)$ with continuous second derivatives is given in a neighborhood of an initial curve $\mathcal{I}$. From the discussion in Section 22 we know that characteristic parameters can be introduced. Hence our solution leads to a solution of equations (24.01). Since the curves $\alpha = $ constant and $\beta = $ constant in the (x, y)-plane are characteristics, the values of the solutions u, v at a point P in the (x, y)-plane do not depend on the totality of initial values on $\mathscr{I}$, but only on the initial values on the section of $\mathscr{I}$ intercepted by the two characteristics through P. This interval on the line $\mathscr{I}$ intercepted by the two characteristics is called the *domain*

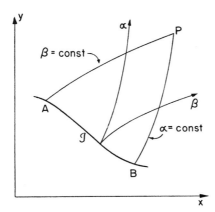

FIG. 2. Triangular domain in the (x,y)-plane in which the solution of the initial problem can be established.

of dependence of the point P. The significance of this notion is then embodied in the following:

Uniqueness theorem: Consider a solution (with continuous second derivatives) of equation (21.01) in the region ABP bounded by the two characteristics through the point P and the "domain of dependence" AB cut out by them from the initial curve $\mathscr{I}$. Suppose another solution (with continuous second derivatives) of equation (21.01) is given in ABP which assumes on AB the same values as the first one. Then the second solution is identical with the first one in the region ABP.

An important special case arises when E_1, E_2 in (21.01) vanish whenever $u = v = 0$. Then $u = v = 0$ in ABP is the only solution which vanishes in the domain of dependence AB of P.

The *range of influence* of a point Q on the initial line $\mathscr{I}$ is the totality of points in the (x, y)-plane which are influenced by the initial data at the point Q. This range of influence of the point Q consists of all points P whose domains of dependence contain Q; therefore it is just the angular region between the two characteristics drawn through Q. We shall realize later the significance of the preceding concepts both in the case in which one of the independent variables, say y, is interpreted as the time, see Section 35, and in the case of steady supersonic flows in which x and y are space variables and u and v components of velocity.

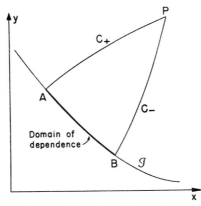

Fig. 3. Domain of dependence of a point.

It is the *existence of such domains of dependence and ranges of influence which characterizes phenomena of wave propagation* in contrast to states of equilibrium. In the latter the states of all points in the medium are interconnected. The differential equations are elliptic; their solutions as seen in the general theory of partial differential equations are analytic functions that are determined in the entire region by their values in any domain, however small. For problems of wave propagation, on the other hand, the solution of the differential equations is not necessarily analytic. Hence it may be composed of analytically different portions in different domains of the (x, y)-plane; therefore the construction of solutions by pieces is often possible, and this possibility makes the solution of hyperbolic problems often relatively easy, as we shall see in many instances.

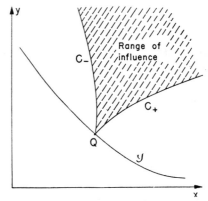

Fig. 4. Range of influence of a point on the initial curve.

The role of the concepts of domain of dependence and range of influence is implicitly referred to in such expressions as "the medium at a point P does not know of the state at a point Q," meaning that P does not belong to the range of influence of Q.

In our construction no side of the initial curve $\mathscr{I}$ is distinguished. The preceding construction therefore leads to the solution of the initial value problem for both sides of $\mathscr{I}$.

While for the existence and uniqueness stated above, the initial values of u and v along $\mathscr{I}$ have continuous second derivatives, solutions may exist which have *discontinuities in the first* (or higher) *derivatives* of these initial values. Suppose we have a continuous solution $u(x, y), v(x, y)$ interrupted by discontinuities of derivatives along certain curves; we further suppose that u and v have continuous second derivatives outside of these curves of discontinuity. At those points P whose domains of dependence do not contain the discontinuity points of the first (or higher) derivatives of the initial data, the solutions u, v have continuous first (and higher) derivatives. From our preceding construction it can be inferred: such *discontinuities of derivatives occur only along characteristics through the discontinuity points on the initial curve* $\mathscr{I}$. See [32, Chapter V, Section 7].

So far, the characteristics served for the theoretical discussion of the initial value problem. In addition, however, the characteristic form of the differential equation proves especially useful for actual numerical solutions. Such numerical solutions are often achieved with comparatively little labor, if the differential equations are replaced by equations for finite differences, as will be described in some detail in Chapter III, Section 83.

25. *Propagation of discontinuities along characteristic lines*

We insert supplementary remarks on the role of characteristics as the possible loci of discontinuities: if at a point A there is a discontinuity in some derivatives of the initial data on $\mathscr{I}$, then according to the statements made above this discontinuity is propagated along one or both of the two characteristics through A. Moreover, as we shall see presently, such discontinuities are propagated according to a definite law which implies that they can never disappear.

In case the variable y is identified with the time t, this can be interpreted as follows. Any discontinuity spreads through the one-dimensional x-region with velocities $\dfrac{dx}{dt}$ given by the slopes of the two characteristics through the corresponding point of discontinuity in the (x, t)-plane.

In two-dimensional steady flow small disturbances caused by a slight roughness of the boundary are indicated by characteristic curves, *Mach lines*, issuing from the boundary of flow. Such characteristic curves are often actually visible in flows along a slightly roughened wall.

The spreading of the discontinuities along characteristics can be described by the following mathematical considerations. Let us assume that characteristic parameters can be introduced and suppose that the discontinuity appears across the line $\alpha =$ constant, so that the (tangential) derivatives with respect to β remain continuous. Then consider the four jumps, or discontinuity intensities

$$[x_\alpha]_{\alpha-0}^{\alpha+0} = X(\beta), \qquad [y_\alpha]_{\alpha-0}^{\alpha+0} = Y(\beta),$$
$$[u_\alpha]_{\alpha-0}^{\alpha+0} = U(\beta), \qquad [v_\alpha]_{\alpha-0}^{\alpha+0} = V(\beta),$$

in which x, y, u, v are considered functions of α and β. We can now establish two homogeneous linear differential equations along $\alpha =$ constant for the discontinuity intensities.

Consider first equation I_+ and II_+ of (22.17) at points P_1 and P_2, one on each side of, and near to a point P on, the characteristic $\alpha =$ constant; subtract these equations from each other and then let P_1 and P_2 approach P. Since the coefficients and derivatives with respect to β are continuous, we conclude that

(25.01)
$$Y(\beta) - \zeta_+(\beta)X(\beta) = 0$$
$$U(\beta) + G_+(\beta)V(\beta) + R_+(\beta)X(\beta) = 0,$$

where ζ_+, $G_+(\beta) = (a\zeta_+ - S)/T$, and $R_+ = (K\zeta_+ - H)/T$ are known functions of β along $\alpha =$ constant, see (22.16). To obtain information from I_- and II_-, we first differentiate with respect to α and carry out the previous process again. We then find a differential equation of the form

(25.02)
$$Y_\beta - \zeta_- X_\beta + M(X, Y, U, V) = 0,$$
$$U_\beta + G_- V_\beta + R_- X_\beta + N(X, Y, U, V) = 0,$$

in which ζ_-, $G_- = (a\zeta_- - S)/T$, $R = (K\zeta_+ - H)/T$ and the coefficients of the linear forms M and N are known functions of β along $\alpha = $ constant.

The equations (25.01–.02) determine the discontinuity intensities as the solutions of linear homogeneous ordinary differential equations. Hence these discontinuities are uniquely determined and are not zero along the whole characteristic if they are known to be different from zero at any point of the characteristic.

It should be pointed out that the discussion of the propagation of discontinuities in the first derivatives*, as outlined in this section, does not apply to discontinuities in the functions u and v themselves. We shall see later, in chapter III, that discontinuities in the functions themselves are propagated as "shocks" in quite a different manner.

26. *Characteristic lines as separation lines between regions of different types*

A remark of basic importance should be repeated: whenever the flow in two adjacent regions is described by expressions which are analytically different (in particular when one is a region of rest or constant state while in the other region the state is not constant), then the two regions are necessarily separated by a characteristic. In general, the transition from one region into the other involves discontinuities of some derivatives; then the statement is an immediate consequence of the fact that only along characteristics can derivatives of u and v of any order change discontinuously. But, even if discontinuities of derivatives should not occur, the result follows easily by an alternative reasoning, based on the uniqueness of the solution of the initial value problem for a triangle formed by two characteristics and a section of the initial line, see [32, p. 297].

If the system (21.01) of differential equations is elliptic, no real characteristics exist, see [32, Chapter III, Section 2], and consequently continuous solutions do not possess discontinuities in their derivatives. If the coefficients of the differential equation are analytic the solutions are analytic functions of x and y and therefore cannot be constant in any region without being constant throughout.

* For a more detailed discussion of the propagation of discontinuities, see again [32, chapter V].

27. Characteristic initial values

Along a curve $\mathcal{I}$ which is not characteristic for initial values of u, v, our differential equations, as we have seen, permit the calculation of the derivatives of u and v (and similarly of all higher derivatives), and determine the solution uniquely on both sides.

What corresponding information do the differential equations yield for a line $\mathcal{I}$ with values u and v which make it characteristic? The answer is found immediately from the characteristic form (22.17) of the equations. Suppose $\mathcal{I}$ is a C_+-characteristic, with $\beta = $ constant. Equation II_+ shows that along $\mathcal{I}$ the values of

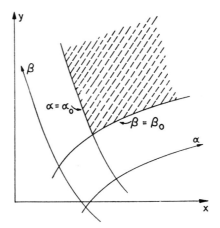

FIG. 5. Rectangular domain in the (x,y)-plane in which the solution of the characteristic initial problem can be established.

u and v cannot both be prescribed arbitrarily; or rather, II_+ establishes a relation between them since it is an ordinary differential equation in u and v along $\mathcal{I} = C_+$. We are consequently at liberty to prescribe only one function, e.g u and, at a single point, the value of the other function, v.

In many important applications the initial value problem is posed not for a non-characteristic initial curve $\mathcal{I}$, but for initial data along two intersecting characteristic arcs. This *characteristic initial value problem* is formulated for the characteristic differential equations as follows. Given compatible values of u and v along two characteristic segments $\alpha = \alpha_0$, $\beta = \beta_0$ as in Figure 5 (page 56), find the solution of equations I and II, (22.17), with these initial values for points α, β in one of the four angular domains, e.g., $\alpha > \alpha_0$, $\beta > \beta_0$. The solution is again uniquely determined and is obtained by the iteration method outlined above in Section 24. For this problem the method of finite differences is again an adequate tool for numerical computation.

28. Supplementary remarks about boundary data

Later on, in particular in connection with the theory of combustion and detonation, see Chapter III, part E, we shall encounter problems in which the data for the solutions of the differential equations (21.01) are prescribed on two non-characteristic arcs $\mathcal{I}$ and $\mathcal{J}$ meeting at a point O and enclosing an angular region $\mathcal{R}$ of the (x, y)-plane, see Figure 6. With a view to such applications we insert here a few remarks. Assigning a direction to each of the two families of characteristics, we assume that the two characteristics issuing from points on $\mathcal{I}$ enter region $\mathcal{R}$ while only one characteristic issuing from points on $\mathcal{J}$ enters $\mathcal{R}$. It is convenient to call $\mathcal{I}$ space-like, and $\mathcal{J}$ time-like, see [32, p. 355], for reasons given in III-A. Then we may state that *two data on $\mathcal{I}$ and one datum on $\mathcal{J}$ determine the solution in $\mathcal{R}$.* More precisely: draw from one

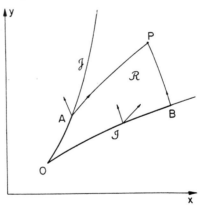

Fig. 6. Domain of dependence on a space-like and time-like arc.

point P in $\mathcal{R}$ the two characteristics in the negative direction until they intersect $\mathcal{I}$ or $\mathcal{J}$ in two points B and A. Suppose a second solution is given in $\mathcal{R}$ which assumes on the section OB on $\mathcal{I}$ the same values of u and v, and on the section OA on $\mathcal{J}$ the same values of one quantity,

u for example, as the original solution. Then the two solutions coincide in the region ABP and in particular at P. This is an important statement on uniqueness. It justifies the designation "domain of dependence" for the segment AB cut off from $\mathcal{J}$ and $\mathcal{K}$ by the two characteristics.

In formulating an existence theorem one encounters the difficulty that it depends on both u and v whether or not the arc $\mathcal{J}$ is "time-like," while only one of these quantities is prescribed there. However, the two data on $\mathcal{J}$ determine by continuity whether or not the direction of $\mathcal{J}$ at O is time-like (assuming continuity of the datum on $\mathcal{J}$ at O). If this is the case it can at least be stated that a solution exists in a neighborhood of $\mathcal{J}$ within which $\mathcal{J}$ is time-like.

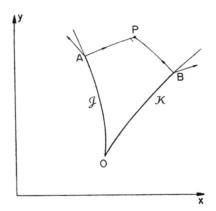

Fig. 7. Domain of dependence on two time-like arcs.

Another natural problem is one in which on two time-like arcs $\mathcal{J}$ and $\mathcal{K}$ one quantity is prescribed and two at the point of intersection O. Suppose a solution is given for which $\mathcal{J}$ and $\mathcal{K}$ are time-like, see Figure 7. Then the domain of dependence OA on $\mathcal{J}$, OB on $\mathcal{K}$ is again cut out by the two characteristics drawn through P in the negative direction. Consider any other solution for which one quantity, e.g. u, on OA and one quantity, u or v, on OB, and at O both u and v, are the same as for the original solution. Then the two solutions agree in the subregion ABP. This *uniqueness theorem* has apparently never been proved, although there is little doubt about its validity under appropriate conditions.

The same can be said about the *existence theorem* which states that, provided both the directions of $\mathcal{J}$ and $\mathcal{K}$ are time-like at O, a solution exists in a neighborhood of O if one quantity is prescribed along $\mathcal{J}$, one along $\mathcal{J}$, and two at O in such a way that the data are continuous at O.

The preceding statements form the basis of a complete treatment of the equations of gas dynamics in the cases characterized in Section 20. In all these cases the reduction of the differential equations to their characteristic form opens the way for theoretical or numerical procedure, as will be shown in Chapters III and IV. For the present we continue the general theory by discussing a point of major importance, the notion of *simple wave*.

29. Simple waves. Flow adjacent to a region of constant state

Very often the following situation is encountered. In a region (I) of the (x, y)-plane the solution (u, v) of the differential equation (21.01), or, as we shall simply say, the "flow" (u, v), is constant. Adjacent to this region of constant state is another region (II) in which u and v vary. Then, as we have seen in Section 26, the two regions are separated by a characteristic C. We shall show in this section that the flow in such a region (II), adjacent to a region (I) of constant state, is of a particularly simple character if the differential equation (21.01) is *reducible*, that is, if the coefficients A, B, C, D of the derivatives u_x, u_y, v_x, v_y depend on u and v only and the member E vanishes, see Section 21.

In the present section we shall be concerned only with reducible differential equations and we shall assume them to be hyperbolic. Such equations possess two fixed families of characteristics in the (u, v)-plane, $\Gamma_+ : \beta(u, v) = $ constant, and $\Gamma_- : \alpha(u, v) = $ constant. As was shown in Section 21, a reducible differential equation can be reduced to a linear one provided that the Jacobian $j = u_x v_y - u_y v_x$ does not vanish. Now, however, we are concerned with solutions $u(x, y)$, $v(x, y)$ for which the Jacobian j vanishes in a whole region and which therefore cannot be represented by solutions of the corresponding linear differential equation. Solutions or "flows" of this type will be called *simple waves*. Simple waves play a fundamental role in describing and building up solutions of flow problems. In particular, we shall show: *a flow in a region adjacent to a region of constant state is always a simple wave.*

In a region (I) of constant state both sets of characteristics C_+ and C_- are straight lines, since constant values of u and v imply constant values of β and α. We shall see that in a simple wave region at least one set of characteristics C consists of straight lines.

We start with a mathematical definition of a *simple wave* as a flow with the following properties: *The region of flow is covered by a set of arcs of characteristics C of one kind, say C_-, all of whose images in the (u, v)-plane fall on the same characteristic, e.g.* Γ_-^0. In other words, the u, v image of the whole simple wave zone lies on this one characteristic Γ_-^0.

It is clear that the image of any characteristic arc C_+ in the simple wave zone lies on a certain characteristic Γ_+. On the other hand, this image lies by assumption completely on Γ_-^0 since the image of the whole simple wave zone lies on Γ_-^0. Therefore the image of each characteristic C_+ of the second kind consists of just one point, the point of intersection of Γ_+ with Γ_-^0. This means that the quantities u and v are constant along each characteristic C_+ of the second kind. In particular, the slope $dy/dx = \zeta_+(u, v)$ of this characteristic is constant, and consequently this characteristic C_+ is a straight line. In other words: *a simple wave zone is covered by arcs of characteristics which carry constant values of u and v and hence are straight lines.*

An immediate consequence of the definition is the following:

Fundamental Lemma: If in a flow region a section of a characteristic carries constant values of u and v, then in regions adjacent to this section the flow is a simple wave or a constant state. Consider a region $\mathcal{R}$ in which u and v are continuous functions of x and y and which contains a section $\mathcal{S}$ of a C_+-characteristic on which u and v are constant. Through each point of the region $\mathcal{R}$ a characteristic C_- passes. Take the sub-region $\mathcal{R}'$ of $\mathcal{R}$ of all points of $\mathcal{R}$ for which the characteristic C_- intersects the section $\mathcal{S}$. The precise statement of the fundamental lemma is then that the flow in this sub-region $\mathcal{R}'$ is a simple wave. The proof is now immediate: the image of $\mathcal{S}$ in the (u, v)-plane is a point. The images of all characteristics C_- in $\mathcal{R}'$, therefore, lie on characteristics Γ_- passing through that point. There is only one characteristic of a particular kind through a point.

Hence these images all lie on the same characteristic Γ_-. By definition then the flow in $\mathcal{R}'$ is a simple wave.

An immediate consequence of the fundamental lemma is the *Fundamental Theorem:* The flow in a region adjacent to a region of constant state is a simple wave.

For, the line separating regions of constant and non-constant state is a section of a characteristic and, in a region of constant state, characteristics carry constant values of u and v. At the same time

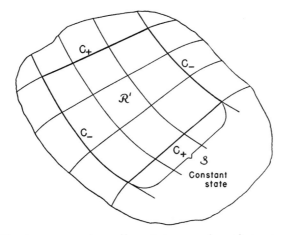

Fig. 8. Simple wave region adjacent to a region of constant state.

it is clear in precisely which adjacent region $\mathcal{R}'$ the character of the flow as a simple wave is established.

There exist a great variety of simple wave solutions of our reducible differential equation (21.01), and it is of interest to ascertain what data are suitable to specify an individual simple wave. A natural possibility, corresponding to typical physical problems, is this: one prescribes the values $u = u(s)$, $v = v(s)$ along a given curve $\mathcal{B}$: $x = x(s), y = y(s)$ such that the image in the (u, v)-plane lies on a specified Γ-characteristic, e.g. a Γ_--characteristic. Then the slopes of the straight C_+-characteristic issuing from any point $(x(s), y(s))$ on $\mathcal{B}$ are given by $dy/dx = \zeta_+(u(s), v(s))$ and the values of u and v on this C_+-characteristic are constant along it: $u = u(s)$, $v = v(s)$.

Thus the flow can be described by $x = x(s) + \sigma$, $y = y(s) + \sigma\zeta_+$, $u = u(s)$, $v = v(s)$ in terms of two parameters s and σ. We assume that the curve $\mathcal{B}$ is nowhere tangent to a C_+-direction:

$$y_s \neq \zeta_+ x_s .$$

Then the Jacobian $\partial(x, y)/\partial(\sigma, s) = y_s + \sigma(\zeta_+)_s - \zeta_+ x_s$ does not vanish on $\mathcal{B}$. Consequently, x and y can be introduced as independent variables in a neighborhood of $\mathcal{B}$. It is easily verified that the functions $u(x, y)$, $v(x, y)$ so obtained solution are simple wave solutions of the differential equations.

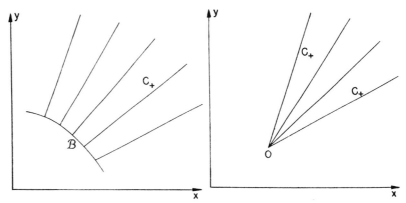

Fig. 9. Simple wave region adjacent to an arbitrary curve $\mathcal{B}$.

Fig. 10. Centered simple wave.

In the special case that the curve $\mathcal{B}$ degenerates into a point O all the straight characteristics C_+ issue from the point O; we then call the wave *centered* with the center O. Such a *centered simple wave* is evidently determined if the corresponding section of the Γ_--characteristic is prescribed.

30. The hodograph transformation and its singularities. Limiting lines

We add a few remarks about the "hodograph transformation," described in Section 21, as a means of reducing a reducible equation (21.01) for (u, v) as function of (x, y) to a linear one (21.03) for (x, y) as function of (u, v). In Section 29 we studied simple waves in a domain of the flow throughout which $j = u_x v_y - u_y v_x = 0$ and the hodograph transformation becomes impossible. We now consider the case in which $j = 0$ merely along a smooth curve. Similarly we are interested in the case in which the Jacobian $J = x_u y_v - x_v y_u$ vanishes along a curve in the (u, v)-plane in a region in which the transformed differential equations (21.03) possess a smooth solution. As we shall show in Chapter IV, see Section 105, both phenomena are of importance in the theory of steady two-dimensional flow. Here we concentrate on the underlying mathematical facts concerning singularities of mappings.

We consider the mapping of a (ξ, η)-plane into a (ξ', η')-plane, and suppose that at the point O, $(\xi = \eta = 0)$,

$$(30.01) \qquad j = \xi'_\xi \eta'_\eta - \xi'_\eta \eta'_\xi = 0,$$

while not all derivatives ξ'_ξ, ξ'_η, η'_ξ, η'_η vanish at O. By this requirement we exclude branch point singularities of the type occurring in potential theory or more generally with solutions (u, v) of any elliptic system of equations of the form (21.01). If, for instance, we assume $\eta'_\xi \neq 0$, then the condition

$$(30.02) \qquad \frac{\partial(\eta', j)}{\partial(\xi, \eta)} = \eta'_\xi j_\eta - \eta'_\eta j_\xi \neq 0 \quad \text{at} \quad O$$

assures that the locus of $j = 0$ is a smooth curve through O, the "critical" curve. How does the image of a neighborhood of O look in the neighborhood of the image point O', $(\xi' = \eta' = 0)$?

The answer is: *the image of a full neighborhood of O in the (ξ, η)-plane is not a full neighborhood of the image point O' in the (ξ', η')-plane, but only a partial neighborhood, doubly covered as if the original domain were distorted and folded over.*

The image domain consists of two sheets which meet along an edge. This edge is the image of the critical curve $j = 0$. There is one "exceptional" direction through the point O such that the images of any curves C passing through O in this exceptional direction have cusps at the image point O'; their direction at these cusps depends on the curvature of the critical curve at the point O. This exceptional direction $(\dot{\xi}, \dot{\eta})$ is characterized by the condition

$$(30.03) \qquad \dot{\xi}' = \xi'_\xi \dot{\xi} + \xi'_\eta \dot{\eta} = 0, \qquad \dot{\eta}' = \eta'_\xi \dot{\xi} + \eta'_\eta \dot{\eta} = 0.$$

All "non-exceptional" directions through O are mapped into one and the same direction, the "edge direction". That is: the image of any

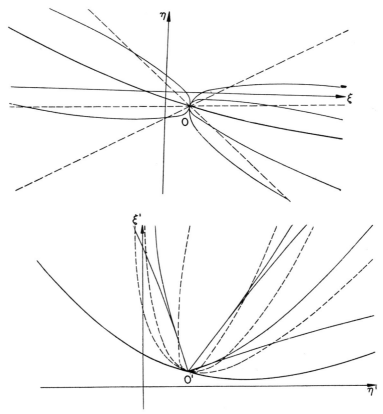

FIG. 11. Mapping of the neighborhood of a critical curve (heavy solid line). The figure shows three curves crossing the critical curve in the exceptional direction along with their images with cusps (light solid lines), and three curves crossing in the non-exceptional direction along with their images tangential to the edge (light broken lines).

curve passing through the point O in a non-exceptional direction is tangential to the edge at the point O', forms no cusp there and, in general, passes from one sheet of the fold into the other one.

These facts are most easily established if we subject both the (ξ, η)-plane and the (ξ', η')-plane to two suitable linear or affine transformations, so that the mapping can be written in the form

(30.04)
$$\sigma' = \sigma + F\sigma^2 + 2G\sigma\tau + H\tau^2 + \cdots$$
$$\tau' = L\sigma^2 + 2M\sigma\tau + N\tau^2 + \cdots,$$

in which σ, τ and σ', τ' denote the new coordinates and the dots indicate terms of higher than second order. The Jacobian, again denoted by j, then becomes

(30.05) $\quad j = 2M\sigma + 2N\tau + 4(FM - GL)\sigma^2 + \cdots,$

the dots indicating terms of higher than first order other than σ^2. The condition (30.02) at O is reduced to $N \neq 0$. The critical curve with the equation $j = 0$ is then mapped into the edge with the equation

(30.06) $\quad\quad\quad\quad \tau' = L\sigma'^2 + \cdots.$

The *edge direction* in the (σ', τ')-plane is therefore given by $d\tau' = 0$. As seen from (30.04), every curve in the (σ, τ)-plane through O with an equation

(30.07) $\quad\quad\quad\quad \tau = \beta\sigma + \cdots,$

is mapped into a curve with an equation

(30.08) $\quad\quad \tau' = (L + 2M\beta + N\beta^2)(\sigma')^2 + \cdots,$

which therefore is tangential to the edge. Consequently, every direction through O with $d\sigma \neq 0$ is non-exceptional.

Every curve through O with an equation

(30.09) $\quad\quad\quad\quad \sigma = \alpha\tau^2 + \cdots$

is by (30.04) mapped into a curve with the parametric representation

(30.10) $\quad \sigma' = (\alpha + H)\tau^2 + \cdots, \quad \tau' = N\tau^2 + \cdots,$

which, because of $N \neq 0$, has a cusp at O'. Obviously, therefore, *the exceptional direction through O is given by $d\sigma = 0$.*

We note that this image curve has in general infinite curvature at the cusp.

Incidentally if (30.02) does not hold and hence if $N = 0$, the critical curve itself has the exceptional direction in O; then its image, the edge, in general forms two branches meeting at a cusp at the point O', as is easily verified. The image of a full neighborhood of O in

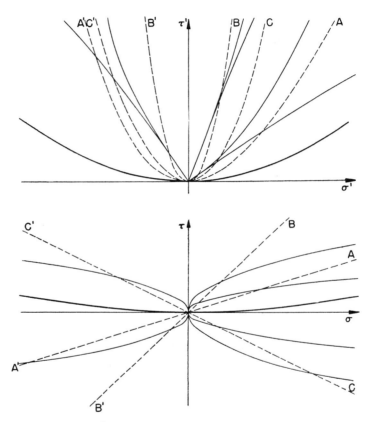

Fig. 12. Mapping of the neighborhood of a critical curve in special coordinates.

the (ξ, η)-plane then forms a "pleat" of three sheets. One of these sheets just covers the "cusp region" between the two edges and is connected at each of these edges with one of the other sheets which fold over and reach beyond the cusp region.

The facts which we have explained serve to illuminate important situations arising in *the mapping* of the (x, y)-plane into the (u, v)-plane and, vice versa, situations *furnished by a solution $u(x, y)$, $v(x, y)$ of the differential equations* (21.01) *of the reducible type.* We consider the quasi-linear differential equations (21.01) for (u, v) as functions of (x, y), the corresponding linear equations (21.03) for (x, y) as function of (u, v) and the characteristic equations (22.17)

(30.11)
$$\text{I} \quad y_\alpha = \zeta_+ x_\alpha, \qquad y_\beta = \zeta_- x_\beta,$$
$$\text{II} \quad Tu_\alpha = -(a\zeta_+ - S)v_\alpha, \qquad Tu_\beta = -(a\zeta_- - S)v_\beta,$$

in which ζ_+, ζ_-, a, T, S depend on (u, v) only. We assume that the coordinates are so chosen that ζ_+ and ζ_- are finite and $T \neq 0$, $a \neq 0$, see (22.09) and (22.20). Further we assume that in the domain considered $\zeta_- \neq \zeta_+$, see (22.12), so that two distinct characteristics pass through each point.

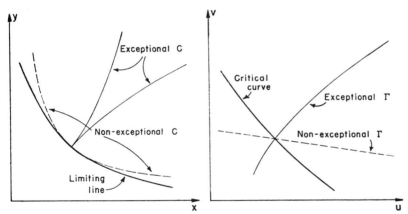

FIG. 13. Mapping induced by the solution of a reducible differential equation and involving a limiting line on whose image $J = 0$.

Firstly, we consider a solution (x, y) of the linear equation (21.03) and *the corresponding mapping of the (u, v)-plane into the (x, y)-plane,* assuming that the Jacobian

(30.12)
$$J = x_u y_v - x_v y_u$$

(but not each of the four derivatives individually) vanishes along a critical curve in the (u, v)-plane. The image in the (x, y)-plane of this critical curve, the edge of the fold, is called a *limiting line*. Since we have two fixed families of Γ-characteristics in the (u, v)-plane we may introduce the parameters (α, β) such that

$$u_\alpha v_\beta - u_\beta v_\alpha \neq 0.$$

Hence vanishing of J implies

(30.13) $\qquad x_\alpha y_\beta - x_\beta y_\alpha = 0$

or, by I in (30.11),

(30.14) $\qquad (\zeta_- - \zeta_+) x_\alpha x_\beta = 0.$

Since $\zeta_- \neq \zeta_+$ by assumption, either $x_\alpha = 0$ or $x_\beta = 0$ along the critical curve. Suppose $x_\alpha = 0$. Then by I also $y_\alpha = 0$. According to (30.03) this implies that *one characteristic direction*, in our case $d\beta = 0$, *is exceptional*. The corresponding "exceptional" C-char-

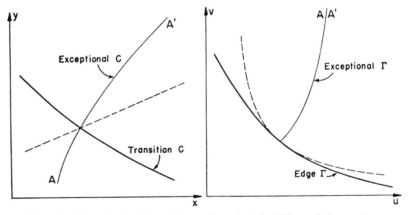

FIG. 14. Mapping by the solution of a reducible differential equation involving a transition characteristic at which $j = 0$.

acteristic has a cusp at the limiting line. The images of all curves in the (u, v)-plane which cross the critical curve in a direction different from the exceptional characteristic direction are tangential to the limiting line. This applies in particular to the other, non-exceptional, Γ-characteristics. Hence, *the limiting line is an envelope of a set of non-exceptional C-characteristics*.

Secondly, we consider *the mapping of the (x, y)-plane into the (u, v)-plane* furnished by a solution (u, v) of the original quasi-linear equation (21.01) assuming that the Jacobian

(30.15) $$j = u_x v_y - u_y v_x,$$

(but not each of the four derivatives) vanishes along a critical curve in the (x, y)-plane. We call such a curve a *transition curve*.

We maintain that *such a transition curve is a C-characteristic and its image, the edge of the fold in the (u, v)-plane, a Γ-characteristic. The C-characteristic of the other kind is exceptional; its image runs through a section of a Γ-characteristic up to the edge in one direction and then backwards through the same section in the opposite direction. The images of all curves in the (x, y)-plane which cross the transition curve in a direction different from the exceptional characteristic direction are tangential to the edge.*

To establish these facts we introduce characteristic parameters (α, β) in the neighborhood of the transition curve so that

$$x_\alpha y_\beta - x_\beta y_\alpha \neq 0$$

there. Then $j = 0$ implies $u_\alpha v_\beta - u_\beta v_\alpha = 0$ or, by II in (30.11),

$$a(\zeta_- - \zeta_+) v_\alpha v_\beta = 0;$$

since $a \neq 0$, $\zeta_- \neq \zeta_+$, either $v_\alpha = 0$ or $v_\beta = 0$ along the transition curve. Suppose $v_\alpha = 0$. Then by II also $u_\alpha = 0$. According to (30.03) this implies that the characteristic direction given by $d\beta = 0$ is exceptional. The image of the exceptional C_+-characteristic has a cusp at the edge of the fold. Since this image lies on a Γ-characteristic and the Γ_+-characteristics are fixed curves it is clear that the cusp of the image of C_+ is attained by running through the same section of Γ_+ forward and backward. That the images of all curves crossing the transition curve in the non-exceptional direction are tangent to the edge follows from the general theory. In particular, it follows that the edge is an envelope of Γ_--characteristics, having a characteristic direction at each of its points. Through any point there is only one curve with this property, viz. the characteristic curve itself. In other words, the edge itself is a Γ_--characteristic. Consequently, the transition curve is a C_--characteristic.

31. Systems of more than two differential equations

So far in this chapter, we have confined ourselves to systems of two differential equations, as they occur for isentropic flows with sufficient symmetry. However, the assumption of isentropy or irrotational character is sometimes not justified; then a system of three or of four differential equations governs the flow. For such quasilinear systems of more than two equations many relevant features of the theory of this chapter remain valid.

Characteristic directions can be defined in the same way as in Section 22. Again one seeks linear combinations of the differential equations which contain derivatives of all the unknown functions in only one direction. Such directions are called characteristic. A system of n equations for n functions of two variables leads to an algebraic equation of degree n for the characteristic directions at each point. If all are real and distinct, the system is called *totally hyperbolic*.

A few special cases may be mentioned:

For the three equations governing one-dimensional non-isentropic flow (17.01–.03) the three characteristic directions are characterized by

$$(31.01) \quad C_+ : dx/dt = u + c, \quad C_- : dx/dt = u - c$$
$$C_0 : dx/dt = u,$$

as will be derived in detail in Section 34. The three characteristics through a point are evidently the paths C_+, C_- of the two "sound waves" as described in Section 23, see (23.05), and in addition the particle path C_0, see Figure 15.

Initial value problems for such equations may, however, be reduced to problems for a system of merely two functions by introducing the Lagrangian representation, see (18.10), and assuming that the entropy is a given function of the Lagrangian parameter h.

Steady non-isentropic, rotational two-dimensional flow is governed by four differential equations for the four quantities u, v, τ, and S as functions of x and y, see (7.08–.11). The four characteristics through a point are the two "Mach lines" found in Section 23, see (23.11), and in addition the streamline, counted twice.

Again, initial value problems for these equations may be reduced to problems involving two functions only, by introducing the stream function ψ, see (16.12), and any other quantity, such as the speed q, as unknown functions, and considering the entropy S and the Bernoulli constant $\tfrac{1}{2}\hat{q}^2$ as given functions of ψ.

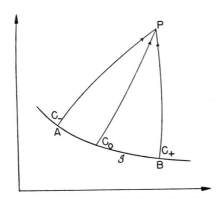

Fig. 15. Domain of dependence AB of a point P for the solution of a system with three characteristics.

There are, however, problems for which the reduction to the case of two unknown functions is not possible; therefore a study of general hyperbolic systems of quasilinear differential equations for n unknown functions $u^1, \cdots u^n, n > 2$, is not only of theoretical interest, but also relevant for applications. It is remarkable that the theory of integration developed in the preceding sections of this chapter can be modified and generalized, so that existence and uniqueness of the solution of the initial value problem as well as suitable numerical methods are ensured.

The methods of approach indicated in the present section appear somewhat simpler than earlier approaches, see [33, 34], and will be amplified in a forthcoming publication.

We consider n differential equations

$$(31.02) \qquad L_i \equiv a_{ij}\frac{\partial u^j}{\partial x} + b_{ij}\frac{\partial u^j}{\partial y} + c_i = 0 \qquad \begin{array}{l}(i = 1, \cdots, n)\\(j = 1, \cdots, n)\end{array}$$

with coefficients a, b, c depending on $x, y, u^1 \cdots u^n$.*

---

* As is usual, summation symbols are omitted; it is understood that in terms where an index is repeated summation should be carried out with respect to that index.

Again, as in Section 22, we start with a specific solution $u^1, \cdots, u^n$ and ask for curves $C: x(\sigma), y(\sigma)$, such that a linear combination $\lambda_i L_i$ of the differential equations can be formed in which differentiations occur only with respect to the curve parameter σ:

(31.03) $$\lambda_i L_i \equiv T_j u_\sigma^j + R = 0.$$

Exactly as in Section 22, we find that a necessary and sufficient condition for the existence of such a "characteristic direction" is the compatibility of the following homogeneous linear equations for λ_i:

(31.04) $$\begin{aligned} \lambda_i(a_{ij}y_\sigma - b_{ij}x_\sigma) &= 0, \\ \lambda_i(a_{ij}u_\sigma^j + c_i x_\sigma) &= 0, \quad (j = 1, \cdots, n), \\ \lambda_i(b_{ij}u_\sigma^j + c_i y_\sigma) &= 0. \end{aligned}$$

This condition is tantamount to the vanishing of all the nth order determinants of the matrix, yielding in the first place the determinant equation

(31.05) $$| a_{ij}y_\sigma - b_{ij}x_\sigma | = 0,$$

that is, an algebraic equation of n-th degree for the quotient

(31.06) $$\zeta = \frac{dy}{dx} = \frac{y_\sigma}{x_\sigma}.$$

We assume that there are n distinct real roots $\zeta_1, \zeta_2, \cdots, \zeta_n$, or that our system is *totally hyperbolic*. Then there exist n families of characteristics C_ν satisfying the ordinary differential equation

(31.07) $$\frac{dy}{dx} = \zeta_\nu,$$

each family covering the domain of the (x, y)-plane under consideration. As in Section 22 we set equal to zero the determinant of any n, except the first n, equations (31.04) and thus find that the original differential equations combine to linear combinations of the form

(31.08) $$M_{\nu j} du^j + N_\nu dy = 0 \text{ on } C_\nu, \quad \nu = 1, \cdots, n$$

in which d indicates differentiation along C_ν and where the coefficients $M_{\nu j}$ and N_ν are known functions of x, y, u^j with a non-vanishing determinant $| M_{\nu j} |$. Together with (31.05) these equations

(31.08) constitute the characteristic equations for the original differential equations.

For $n > 2$ these equations can no longer be interpreted as a system of canonical partial differential equations, since the n characteristics imply n parameters, while we have only two independent variables.

Nevertheless, the characteristic form (31.05) and (31.08) of the equations (31.02) lends itself to a theoretical treatment which at the same time suggests various numerical procedures by finite difference methods.

The result is existence and uniqueness of the solution of the initial value problem; likewise, the theory immediately yields a *domain of dependence*, for a point P in the plane, in form of the largest segment on the initial curve $\mathcal{I}$ intercepted by characteristics drawn backwards from P.

Linear equations: Let us first assume that the original system is linear, i.e. that the coefficients depend on x and y only. Then the n families of characteristics are fixed curves in the (x, y)-plane given by the ordinary differential equations (31.07). We envisage the *initial value problem:* given a curve $\mathcal{I}$, nowhere characteristic, with prescribed initial values of u^j, find a solution at all points P with coordinates x, y in a suitably small domain $\mathcal{R}$ adjacent to $\mathcal{I}$. The domain is assumed to be such that the n characteristics through a point P intersect $\mathcal{I}$ in distinct points P_ν and have different directions at P. Then the equations (31.08) can be written, after integration by parts, in the form

$$M_{ij} u^j |_P = -N_i y |_P + M_{ij} u^j |_{P_i} + \int_{P_i}^{P} u^j \, dM_{ij}$$

in which the integration is extended over the arc $\overline{P_i P}$ of C_i and in which the two first members on the right-hand side are known quantities. These relations immediately suggest an iteration scheme: We insert on the right-hand side, for u^j, an arbitrary first approximation u_1^j satisfying the initial conditions. Now identifying u^j on the left side with the next approximation u_2^j we obtain a system of n linear equations for these values at P. The determinant of the system is assumed to be bounded away from zero. The resulting values u_2^j as functions of x, y are now used on the right-hand side again to produce expressions for the next approximation, and so on. Convergence of the procedure to a solution u^j of the original system, as well as uniqueness, is not difficult to prove. At the same time the domain of dependence of P is seen to be the segment of $\mathcal{I}$ cut out by the two outer characteristics.

In the general case of a nonlinear system (31.02) various plausible procedures of iteration by linear processes present themselves. In the first place, one could start out with the first approximation u_1^j, substitute these values in the coefficients of the differential equations, and thus obtain a linear system. By the method given above this system will produce as solution a second approximation u_2^j; iterating this procedure one obtains, as can be proved, convergence to the desired solution in a suitably small domain see [34].

Another method would be to use (31.02) directly, taking into account that with each step the characteristics are to be modified.

Fig. 16. Strips and characteristic vectors occurring in the finite difference scheme.

As a basis for numerical computations, however, it seems possible and even preferable to proceed as follows: We consider a sequence of curves $\mathcal{I}$ depending on a parameter t, while each curve may be represented in terms of a parameter s. As the parameter t increases from 0, the curves $\mathcal{I}(t)$ are to cover the domain $\mathcal{R}$ smoothly; furthermore we assume that, for the given initial values and a sufficiently small neighborhood of them, the curves $\mathcal{I}$ are nowhere characteristic. We now select a small value τ and consider the curves $\mathcal{I} = \mathcal{I}(0)$, $\mathcal{I}_1 = \mathcal{I}(\tau)$, $\mathcal{I}_2 = \mathcal{I}(2\tau)$, etc., defining in this way narrow strips Σ_0 Σ_1, $\cdots$. Then we solve, by the method for linear equations, the problem for the small strip Σ_0. The first approximation u_1^j for this purpose is constructed by extending the given initial values as constants along a non-characteristic family of transversals of curves $\mathcal{I}(t)$, e.g. along the lines $s = $ constant. By substituting these functions u_1^j in the coefficients of the differential equations, we make them a system of linear equation whose solution produces values of u^j on $\mathcal{I}_1$. With these values we solve, in the same way, a linear initial value problem for Σ_1, and so on. Thus we obtain a function in the domain $\mathcal{R}$ which, for $\tau \to 0$, can be shown to converge to the desired solution of the nonlinear problem.

The method just described lends itself to a much simplified numerical procedure. By making the step τ small, one may omit altogether solving the initial value problem in the interior of the strips Σ, and instead simply construct values u^j on the curve $\mathcal{I}_{n+1}$ from those on $\mathcal{I}_n$ in the following manner: From a point P on $\mathcal{I}_{n+1}$ we draw n short straight segments backwards in the n characteristic directions, intersecting $\mathcal{I}_n$ in points P_ν and having their slopes determined by the values s_ν of s at P_ν. Then we simply replace equations (31.05) and (31.08) by difference equations, which immediately represent the values of u^j at P by n linear equations involving only the values of u^j obtained previously on $\mathcal{I}_n$.

APPENDIX

32. General remarks about differential equations for functions of more than two independent variables. Characteristic surfaces

In quite a natural way the concept of characteristics can be extended to differential equations for functions of n independent variables when $n > 2$. We shall briefly develop the notion of characteristic surfaces and of characteristic equations and then discuss why the usefulness of this notion is limited.

With a slight change of notation we consider quite generally a system of k equations for a system of k functions $u^{(\kappa)}$, $\kappa = 1, \cdots, k$ of n variables x_ν, $\nu = 1, \cdots, n$,

(32.01) $$L_\mu(u) = a^\nu_{\mu\kappa} u^{(\kappa)}_{x_\nu} + f_\mu = 0, \qquad \mu = 1, \cdots, k,$$

in which the $a^\nu_{\mu\kappa}$ and f_μ are functions of $u = \{u^{(\kappa)}\}$ and $x = \{x_\nu\}$. Through any point x we place elements φ of $(n-1)$-dimensional surfaces, characterized by a normal vector $\xi = \{\xi^\nu\}$. We call such a surface element φ *characteristic* if an appropriate linear combination

$L = \lambda_\mu L_\mu$ of the differential expressions L_μ involves derivatives of the "function" $u = \{u^{(\kappa)}\}$ only in the directions of this surface element. The property of a surface element φ to be characteristic at a point depends, of course, on the value of u at this point, but not on the values of the derivatives of u there.

The direction in which $u^{(\kappa)}$ is differentiated in the differential expression L is that of the vector with the components $\lambda_\mu a^\nu_{\mu\kappa}$; the condition that its direction lies in a plane perpendicular to the vector with the components ξ^ν is therefore

(32.02) $$\lambda_\mu a^\nu_{\mu\kappa} \xi^\nu = 0.$$

Since it is to be satisfied for $\mu = 1, 2, \cdots, k$ we have a linear homogeneous system of k equations for $\lambda_1, \cdots, \lambda_k$, and thus we obtain as the condition for the existence of multipliers λ_μ the determinant equation

(32.03) $$\| a^\nu_{\mu\kappa} \xi^\nu \| = 0.$$

This is the *characteristic equation*, homogeneous and of the order k in $\xi^1, \xi^2, \cdots \xi^n$ or, if the characteristic element is that of a surface $\varphi(x_1, x_2, \cdots, x_n) = 0$ with $\xi^\nu = \varphi_{x_\nu}$, in the partial derivatives φ_{x_ν}.

Suppose $\xi = \{\xi^\nu\}$ represents a characteristic surface element. Then multipliers λ can be found to satisfy equation (32.02). The differential equation $L = 0$ then implies differentiations of all functions $u^{(\kappa)}$ in some direction within this surface element.

To write $L = 0$ in a form such that this latter fact appears evident, it is useful to introduce the *normal derivative* g_ξ of a function g by

$$g_\xi = g_{x_\nu} \xi^\nu,$$

supposing $|\xi|^2 = 1$. Then $g_{x_i} = \xi^i g_\xi$ represents a differentiation of g within the surface $\varphi = 0$, and can be interpreted as differentiation along the projection of the x_i-direction onto the surface element. $L = 0$ then can be written in the form

(32.04) $$L = \lambda_\mu a^\nu_{\mu\kappa}(u^{(\kappa)}_{x_\nu} - \xi^\nu u^{(\kappa)}_\xi) + \lambda_\mu f_\mu = 0$$

since the terms with negative sign combine to zero by (32.02). Obviously (32.04) contains differentiations within $\varphi = 0$ only.

Suppose now that it is possible to determine k characteristic surface elements depending in a continuous manner on x and u such that the corresponding k systems of multipliers λ are linearly independent.

(The system of equations (32.01) is then totally hyperbolic, see [32].) Then the resulting k differential equations (32.04) are equivalent to the system (32.01). If each of these k "characteristic equations" involved derivatives in only one direction, a decisive simplification of the system (32.01) would have been achieved. In general, however, each equation (32.04) involves derivatives in several directions, their number being limited only by the condition to be $\leq k$ and $\leq n - 1$. If $k \geq 2$ and $n \geq 3$, reduction to characteristic form therefore in general does not yield an essential simplification of the differential equations.

It is natural to ask whether or not there is a useful generalization of the notion of simple waves for functions of several independent variables. Suppose k functions $u^{(1)}, \cdots, u^{(k)}$ of n variables $x_1, \cdots, x_n$ are defined in a domain D. We say that the "function" $u = \{u^{(k)}\}$ represents *a simple wave* if the domain D is swept by sections of a one-parametric set of $(n - 1)$-dimensional hyperplanes on which u is constant. We say that the function u represents a *double wave* if the domain D is swept by sections of a two-parametric set of $(n - 2)$-dimensional hyperplanes on which u is constant. It is then clear how to define an *n-tuple wave*.

Various questions arise in connection with such waves. How is the property of a function u to represent an n-tuple wave connected with the vanishing of the determinant of the matrix $\{u^{(k)}_{x_\nu}\}$ of the derivatives of u? Which differential equations admit n-tuple waves as solutions?

In particular, one should like to know whether or not it is true that a solution which is constant on one $(n - 1)$-dimensional hyperplane is a simple wave in an adjacent region, and that a solution which is constant on each $(n - 2)$-dimensional hyperplane of a one-dimensional set is a double wave in an adjacent region.

Some of these questions can be answered in some cases. For example, it is easily seen that every irrotational vector field $u = \{u^{(1)}, \cdots, u^{(n)}\}$ is a simple wave if the rank of the matrix $\{u^{(\kappa)}_{x_\nu}\}$ equals one. For, if this rank equals one, there is a function $s(x)$ and "function" $U(s)$ such that $u(x) = U(s(x))$. Let φ be the potential so that $u^{(\kappa)} = \varphi_{x_\kappa}$, and set

$$\Phi = x_\kappa u^{(\kappa)} - \varphi.$$

Then
$$\Phi_{x_\nu} = x_\kappa \frac{dU^{(\kappa)}}{ds} s_{x_\nu}.$$

Thus the gradient of Φ is proportional to the gradient of s; this fact implies that Φ is a function of s,

$$\Phi = F(s)$$

so that
$$\frac{dF}{ds} = x_\kappa \frac{dU^{(\kappa)}}{ds}.$$

This relation implies that the surfaces $s =$ constant are planes since dF/ds and $dU^{(\kappa)}/ds$ depend on s only.

In a similar fashion one shows that an irrotational vector field is an r-tuple wave if the rank of the matrix $\{u_{x_\nu}^{(\kappa)}\}$ is r.

Starting with the derivation of this fact Giese [35] has investigated in great detail the geometry of simple and double waves representing irrotational isentropic flow in three dimensions.

CHAPTER III

One-Dimensional Flow

33. Problems of one-dimensional flow

Isentropic flow of compressible fluids admits a fairly exhaustive mathematical treatment if the state of the medium depends only on the time t and on a single cartesian coordinate. The differential equations of motion then reduce to simple systems of the type studied in the preceding chapter. Without necessarily basing our discussions on these general theories, we now turn to an analysis of flow problems depending only on two variables. The first subject is one-dimensional flow.

As the model of one-dimensional flow we shall usually consider the flow of a gas in a long tube extending along the x-axis. The tube may be infinite, semi-infinite, or finite, i.e., open at both ends, closed by a piston or wall at one end, or by pistons or walls at both ends. Unless otherwise stated, we shall assume an initial state of uniform velocity u_0, uniform pressure p_0, and uniform density ρ_0. The motion of the gas is then caused by the action of the pistons at the ends.

It is convenient to represent the phenomena in an (x, t)-coordinate system, and to refer to the curve in the (x, t)-plane representing the motion of a particle as a "path." Let the x-coordinate of the piston at the left end of the gas-filled tube be $x = 0$ for $t = 0$. Then the motion of the piston is represented in the (x, t)-plane as a curve $\mathcal{P}$, the *piston path*, starting at the origin as indicated in Figure 1 for compressive and for expansive action of the piston.

In Part A we shall investigate general methods of solving the differential equations of motion; in Part B we shall study the simplest types of continuous motion of a gas, in particular, the *rarefaction waves* caused by receding pistons. Part C is devoted to a discussion of discontinuous motion involving *shock fronts*, which develop as a result of compressive action. With some sacrifice of conciseness, an attempt is made to illuminate the shock front theory from various points of view. In Part D it is shown how more general types of motion result from interaction of the elementary motions studied in

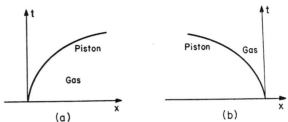

FIGURE 1.
(a) Piston path (compressive action). (b) Piston path (expansive action).

Parts B and C. In Part E we shall discuss the theory of discontinuous detonation and combustion processes, which is closely related to the theory of shocks.

A. Continuous Flow

34. Characteristics

In the present section we are concerned with the characteristic directions and curves of the differential equations

(34.01)
$$\rho_t + u\rho_x + \rho u_x = 0,$$
$$\rho u_t + \rho u u_x + p_x = 0,$$

of isentropic one-dimensional flow, see (17.01–.02), in which $p = f(\rho)$ is a given function. The characteristic equations derived in Chapter II, Section 23, see (23.03–.04), are

(34.02)
$$\text{I}_+ : \quad x_\alpha = (u + c)t_\alpha,$$
$$\text{I}_- : \quad x_\beta = (u - c)t_\beta,$$

(34.03)
$$\text{II}_+ : \quad u_\alpha = -\frac{c}{\rho}\rho_\alpha,$$
$$\text{II}_- : \quad u_\beta = \frac{c}{\rho}\rho_\beta$$

with $c^2 = f'(\rho)$, see (2.05).

With the pressure p instead of the density ρ as dependent variable, equations II take on the form

(34.04) $\quad$ II$_+$: $\quad u_\alpha = -\dfrac{p_\alpha}{\rho c},\quad$ II$_-$: $\quad u_\beta = \dfrac{p_\beta}{\rho c}$

in which the *impedance* ρc is to be considered a function of p.

Equations (34.03), (34.04) imply the relation

(34.05) $\qquad\qquad (du)^2 + dp\,d\tau = 0,$

between velocity, pressure, and specific volume in both characteristic directions; this equation does not involve coefficients depending on the solution.

Without reference to the formulas of Chapter II, the characteristic equations are easily derived directly. In doing so we shall not make the assumption that the flow is isentropic. Then the flow is characterized by three differential equations:

(34.06) $\qquad\begin{aligned}\rho_t + u\rho_x + \rho u_x &= 0,\\ \rho u_t + \rho u u_x + p_x &= 0,\\ S_t + uS_x &= 0,\end{aligned}$

see (17.01–.03), in which p is a given function of ρ and the specific entropy S. The characteristic equations to be derived assume a more concise form if, instead of ρ, p is considered as dependent variable; ρ is then regarded as a function of p and S, which is possible by (2.04). The relationship between these quantities, given by $p = f(\rho, S)$ or

$$dp = c^2 d\rho + f_S dS,$$

enables us to eliminate $d\rho$ from the continuity equation and to replace this equation by

$$p_t + up_x + \rho c^2 u_x = 0,$$

see (17.05). Adding and subtracting from this equation the second equation in (34.06) multiplied by c we find

$$p_t + (u + c)p_x + \rho c\{u_t + (u + c)u_x\} = 0,$$
$$p_t + (u - c)p_x - \rho c\{u_t + (u - c)u_x\} = 0.$$

These two equations together with the equation $S_t + uS_x = 0$ are equivalent to the original system of three equations. The new form of the three equations suggests introducing in the (x, t)-plane three directions $(+), (-), (0)$ by

(34.07)
$$I_+ : \quad dx = (u + c)dt, \quad I_- : \quad dx = (u - c)dt,$$
$$I_0 : \quad dx = u\,dt.$$

With reference to these three directions the three equations assume, respectively, the forms

(34.08)
$$II_+ : \quad dp = -\rho c\,du, \quad II_- : \quad dp = \rho c\,du,$$
$$II_0 : \quad dS = 0.$$

Since each of the equations (34.07–.08) contains differentiation only in its respective direction, the three directions I are characteristic in the sense of Chapter II, Section 22, and equations II are the characteristic equations. The third characteristic direction (0) corresponds to the particle velocity. For isentropic flow, when $S =$ constant is assumed beforehand, the characteristic equations reduce to the system of two pairs I and II previously formulated.

The characteristics C in the (x, t)-plane represent the paths of what we shall call *sound waves*, for reasons which will be discussed in the next section. The velocity of a forward and backward sound wave, corresponding to a C_+ or C_- characteristic, is by (34.02)

(34.09)
$$\frac{dx}{dt} = u + c, \quad \frac{dx}{dt} = u - c,$$

respectively.

35. Domain of dependence. Range of influence

The characteristic directions, aside from their value for the theoretical and numerical integration of the differential equations, are determining factors in discussing the dependence of the solutions on the given data. We restrict ourselves first to the case of isentropic flow. Suppose the values of u and p (or ρ) are prescribed at the time $t = 0$ as functions of x; and assume that, for $t > 0$, a solution exists which has these initial data. Consider any point P in the (x, t)-plane and draw the two characteristic curves C_+ and C_- through P until they intersect the x-axis in two points P_+ and P_-, see Figure 2. (We recall that the characteristic curves C_+, C_- represent the solutions of the differential equations I_+ and I_-, respectively, with reference to a solution $u(x, t)$, $p(x, t)$ of the flow problem.) Then the section P_+P_- on the x-axis is the *domain of dependence* of the point P, see Section 24. That means:

Suppose another solution of the flow problem exists (with continuous derivatives of u and p with respect to x and t) defined at least in the triangular region PP_+P_-, and having the same initial values on the section P_+P_- as the first solution. Then the second solution is identical with the first solution in the region PP_+P_-.

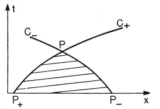 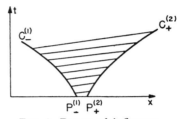

FIG. 2. Domain of dependence $P_+ P_-$ of a point P.

FIG. 3. Range of influence of a section $P_-^{(1)} P_+^{(2)}$.

In this sense, any disturbance of the initial data outside of the section P_+P_- does not influence the value at P. The disturbance of the initial data outside of P_+P_- is by no means confined to "infinitesimal" disturbances; the only restriction is that there should be a solution with continuous derivatives in PP_+P_- which has the disturbed initial data. (We shall see later, Section 48, that disturbances are possible which violate this condition of continuity.) This fact may also be interpreted as follows:

Suppose the initial data are modified on a section $P_-^{(1)}P_+^{(2)}$ of the x-axis. Then the two characteristics $C_-^{(1)}$ and $C_+^{(2)}$ issuing from the points $P_-^{(1)}$ and $P_+^{(2)}$, respectively, enclose a region outside of which the solution is not modified, see Figure 3. This region is what was called the *range of influence* of the section $P_-^{(1)} P_+^{(2)}$, see Section 24. That the solution is actually changed between the boundaries $C_-^{(1)}$ and $C_+^{(2)}$ of this range is a fact which does not follow from the general theory but which can be established for the equations considered here.

These two curves $C_-^{(1)}$ and $C_+^{(2)}$ represent the motion of the "head" of the "disturbance wave." The velocity of this motion is $u - c$ or $u + c$ respectively; this velocity, relative to the gas velocity u at each point in question, equals $\mp c$. Thus the "head" of a "disturbance wave" travels with sound velocity relative to the gas and the name *sound speed* for the quantity c is justified. It is for this reason that we have called a C-characteristic the path of a "sound wave."

36. More general initial data

Occasionally (see e.g. Part E of this chapter), problems occur in which data are prescribed on lines other than the x-axis, $t = 0$. We shall consider these problems in the present section although most of the discussion in this chapter does not depend on them.

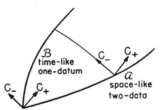

FIG. 4. Space-like and time-like arcs.

Before characterizing those problems which possess a unique solution, we must introduce the concepts of *space-like* and *time-like* directions. A direction (dx, dt) is called space-like if both characteristic directions with $dt > 0$ lie on the same side of it; a direction (dx, dt) with $dt > 0$ is called time-like if it separates the characteristic directions. A space-like direction corresponds to a supersonic velocity relative to the motion of the gas, $\left|\dfrac{dx}{dt} - u\right| > c$; a time-like direction to a subsonic velocity, $|\dfrac{dx}{dt} - u| < c$. Suppose that continuously differentiable data are prescribed on a curve in the (x, t)-plane such that this curve becomes space-like, i.e. that its direction everywhere becomes space-like (note that it depends on the data whether or not a curve is space-like). Then, according to the theory explained in Sections 24 and 35, a unique solution exists in a neighborhood of the curve and the domain of dependence of each point is cut out of the initial curve by the two characteristics through this point.

Now consider two arcs, $\mathfrak{A}$ and $\mathfrak{B}$, see Figure 4, which are given by two functions $x(s)$, $t(s)$ with continuous derivatives x_s, t_s, and $x_s^2 + t_s^2 \neq 0$, and which issue from a point O and enclose an angular region $\mathfrak{R}$. Suppose that data are prescribed in the following manner: *two* quantities (u and p) are prescribed in a continuously differentiable way on the arc $\mathfrak{A}$ such that $\mathfrak{A}$ becomes space-like and such that at all points

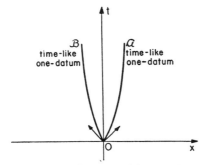

FIG. 5. Two time-like arcs.

of $\mathfrak{A}$ both characteristic directions with $dt > 0$ point into $\mathfrak{R}$. Furthermore, the direction of the arc $\mathfrak{B}$ at the point O is time-like and *one* quantity, such as u or p, is prescribed again in a continuously differentiable way, on $\mathfrak{B}$. This quantity should take the same value at O that it has already been given on $\mathfrak{A}$. Then, in a neighborhood of $\mathfrak{A}$ a unique solution again exists, see Section 24.

An example of data prescribed in this manner occurs in the determination of the flow resulting when a piston, beginning with velocity zero, is moved into a gas originally at rest. On the path of the piston in the (x, t)-plane the quantity u is prescribed as equal to the piston velocity; clearly the piston path is time-like everywhere.

Finally, let each of two arcs issuing from a point O carry one prescribed quantity; for example, let $\mathfrak{A}$ carry the value of u, $\mathfrak{B}$ the value of p. Let the values of u and p at O be such that both arcs $\mathfrak{A}$ and $\mathfrak{B}$ are time-like there; then a unique solution exists again in the enclosed angular region in a neighborhood of O, see Section 24.

Also in the case of non-isentropic flow, when three characteristics C_-, C_0, C_+ pass through each point P, the number of initial data pre-

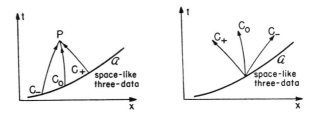

Fig. 6. Space-like arc with three data for non-isentropic flow.

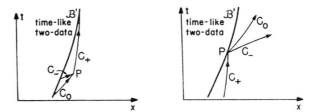

Fig. 7. Time-like arc with two data for non-isentropic flow.

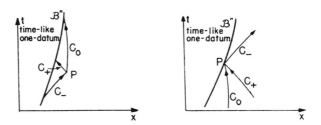

Fig. 8. Time-like arc with one datum for non-isentropic flow.

scribable on an arc $\mathcal{A}$ or $\mathcal{B}$ depends on the number of characteristics which, when drawn from a point P near to this arc in the direction of decreasing t, intersect this arc. Figures 6 to 8 illustrate the various possibilities.

The determinacy of a solution and the proper number of data to

37. Riemann invariants

be prescribed become, in general, readily apparent in a numerical procedure when the method of finite differences is used.

Assuming isentropic flow, equations II, see (34.02), for u and ρ can be integrated in the form

(37.01)
$$u + l(\rho) = 2r(\beta),$$
$$u - l(\rho) = -2s(\alpha),$$

in which $r(\beta)$ and $s(\alpha)$ may be considered arbitrary functions of α and β respectively, and in which the quantity $l(\rho)$ is given by

(37.02)
$$l(\rho) = \int_{\rho'}^{\rho} \frac{c\, d\rho}{\rho} = \int_{p'}^{p} \frac{dp}{\rho c},$$

ρ' or p' being arbitrary constants. For gases we may always assume $l = 0$ for $\rho = 0$, and hence $l > 0$ for $\rho > 0$. The quantities r and s, introduced by Earnshaw and Riemann, are frequently called *Riemann invariants*.

Equations (37.01) express the fact that the images Γ_+ and Γ_- of the characteristics C_+ and C_- in the (u, ρ) plane are two familes of curves which are independent of the solution under consideration. This is in agreement with the fact that these equations are reducible, see Section 21.

For polytropic gases we have $c = \sqrt{A\gamma}\rho^{(\gamma-1)/2}$, see (3.06). Hence

(37.03)
$$l(\rho) = \frac{2}{\gamma - 1} \sqrt{A\gamma}\, \rho^{(\gamma-1)/2},$$

(letting $\rho' = 0$), or simply

(37.04)
$$l = \frac{2}{\gamma - 1} c.$$

The Riemann invariants are therefore given by

(37.05)
$$r = \frac{u}{2} + \frac{c}{\gamma - 1}, \qquad -s = \frac{u}{2} - \frac{c}{\gamma - 1},$$

and we infer the following *basic statements*:

(37.06)
$$\frac{dx}{dt} = u + c, \quad \frac{u}{2} + \frac{c}{\gamma - 1} \text{ is constant along } C_+,$$
$$\frac{dx}{dt} = u - c, \quad \frac{u}{2} - \frac{c}{\gamma - 1} \text{ is constant along } C_-.$$

It is interesting that in the special case $\gamma = 3$ the characteristic velocities are $u + c = 2r$ for C_+ and $u - c = 2s$ for C_- ; hence these velocities are constant along the characteristics. In other words, the characteristics in the (x, t)-plane are straight lines when $\gamma = 3$.

The characteristics Γ_+ and Γ_- are fixed curves in the (u, ρ)-plane, namely,

(37.07)
$$u + \frac{2\sqrt{A\gamma}}{\gamma - 1} \rho^{(\gamma-1)/2} \text{ is constant for } \Gamma_+,$$
$$u - \frac{2\sqrt{A\gamma}}{\gamma - 1} \rho^{(\gamma-1)/2} \text{ is constant for } \Gamma_-.$$

If we consider, instead of u and ρ, u and the sound speed c, as dependent variables, the characteristics in the (u, c)-plane become *straight lines*, see Figure 9,

(37.08)
$$\frac{u}{2} + \frac{c}{\gamma - 1} = \text{constant along } \Gamma_+,$$
$$\frac{u}{2} - \frac{c}{\gamma - 1} = \text{constant along } \Gamma_-,$$

with $c \geq 0$.

38. Integration of the differential equations of isentropic flow

We must distinguish three types of solutions (with reference to a region $\mathcal{R}$ in the (x, t)-plane). Firstly, $\rho = $ constant, $u = $ constant in $\mathcal{R}$; then we speak of a *constant state* although the expression "steady" state would be correct. Secondly, either $r = $ constant, or $s = $ constant, in $\mathcal{R}$. The image in the (u, ρ)-plane of the region $\mathcal{R}$ then lies entirely on the curve $r = $ constant or on $s = $ constant, i.e. on a characteristic. According to Section 29 the flow in the region $\mathcal{R}$ is a *simple wave*. Such simple waves will be discussed in Section 40.

Finally, neither r nor s is constant in $\mathcal{R}$. More precisely, to every pair of values of r and s that occurs in $\mathcal{R}$ there corresponds only one point in $\mathcal{R}$. Then and only then may s and r be introduced as

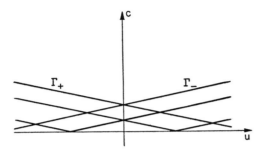

Fig. 9. Characteristics in the (u,c)-plane.

parameters instead of u and ρ. We observe that $dl/d\rho > 0$ implies that ρ, and hence c, can be considered a function of l. Since by (37.01)

(38.01) $\qquad l = r + s, \qquad u = r - s,$

we see that $u + c$ and $u - c$ are known functions of r and s. Thus the characteristic equations I in the form

(38.02) $\qquad x_s = (u + c)t_s, \qquad x_r = (u - c)t_r,$

see (34.02), may be considered a system of two *linear* differential equations for x and t as functions of r and s. Eliminating x, we obtain one linear partial differential equation of second order for $t(r, s)$

(38.03) $\qquad 2ct_{rs} + (u + c)_r t_s - (u - c)_s t_r = 0.$

Once the function $t(r, s)$ is found as a solution of this differential equation, the preceding equations immediately yield the function $x(r, s)$. In the case of a polytropic gas, with $c = \dfrac{\gamma - 1}{2}(r + s)$, $u = r - s$, see (37.04) and (38.01), equation (38.03) becomes

(38.04) $\qquad 2\mu^2 t_{rs} + \dfrac{1}{r + s}(t_r + t_s) = 0$

with, see (14.06),
$$\mu^2 = \frac{\gamma - 1}{\gamma + 1}.$$

An equivalent equation was first treated by Riemann [38]. As a matter of fact, it was the problem of one-dimensional gas flow which led Riemann to develop his famous theory of linear hyperbolic differential equations. In the special case of polytropic gases considered here, an explicit solution of the initial value problem is possible by means of the *hypergeometric function*, see Section 82.

For the special values

(38.05) $\quad \gamma = \dfrac{2N + 1}{2N - 1}, \quad \mu^2 = 1/2N, \quad N = 0, 1, 2, 3, \cdots,$

i.e. for
$$\gamma = -1, 3, 5/3, 7/5, \cdots,$$

equation (38.04) can even be integrated explicitly by means of elementary functions.

For $N = 0$, $\gamma = -1$, see Section 4, we have $2\mu^2 = \infty$, and equation (38.04) reduces to the linear wave equation $t_{rs} = 0$, with the general solution $t = f(r) + g(s)$ in terms of arbitrary functions f and g.

For $\gamma = 3$ we have $2\mu^2 = 1$, and equation (38.04) reduces to
$$(r + s)t_{rs} + t_s + t_r = 0$$
or
$$((r + s)t)_{rs} = 0,$$
which has the general solution

(38.06) $\quad t = \dfrac{1}{r + s}(f(r) + g(s)),$

with arbitrary functions f and g.

The general solution of equation (38.04) for the special values (38.05) of γ (with $N \geq 1$) is, as is easily verified,
$$t = k + \frac{\partial^{N-1}}{\partial r^{N-1}} \frac{f(r)}{(r + s)^N} + \frac{\partial^{N-1}}{\partial s^{N-1}} \frac{g(s)}{(r + s)^N},$$
with arbitrary functions f and g and any constant k. By the proper choice of $f(r)$, $g(s)$, and k the initial conditions of the problem can be satisfied.

It should be noted that the value $1.4 = 7/5$ of γ for air occurs among the special values (38.05); the value $\gamma = 11/9 \approx 1.2$ can frequently be used for the gases produced by combustion or other chemical processes.

As we shall see in Section 82 the preceding remarks have important applications in the theory of interaction of waves.

39. Remarks on the Lagrangian representation

No essentially new ideas occur in the *Lagrangian representation*, see Chapter I, Section 18. The independent variables $h = \int_{x_0}^{x} \rho(\xi)d\xi$ and t are connected with the dependent variables u and τ through the relation

$$(39.01) \qquad dx = \tau dh + u d\tau,$$

which follows from (18.02.2) and $x_t = u$. Inserting this relation in the characteristic equations (34.02–.03) we obtain the characteristic form for the differential equations (18.12) of isentropic flow in Lagrangian coordinates,

$$(39.02) \quad \text{I} \quad \begin{aligned} C_+ &: h_\alpha = k(\tau)t_\alpha, \\ C_- &: h_\beta = -k(\tau)t_\beta, \end{aligned} \quad \text{II} \quad \begin{aligned} \Gamma_+ &: u_\alpha = k(\tau)\tau_\alpha, \\ \Gamma_- &: u_\beta = -k(\tau)\tau_\beta, \end{aligned}$$

in which $k(\tau) = c(\tau)/\tau$ is the impedance, see (18.09). The characteristics Γ_+ and Γ_- in the (u, τ)-plane can again be explicitly described:

$$(39.03) \qquad u = \pm \int_0^\tau k(\tau)\,d\tau = \text{constant}.$$

For non-isentropic flow the impedance k depends on the specific entropy S in addition to the specific volume τ. The third characteristic equation is

$$\text{I}_0 \quad C_0 : \quad dh = 0,$$

expressing that the third characteristic curve is a particle path. The corresponding equation II is

$$\text{II}_0 \quad \Gamma_0 : \quad dS = 0$$

and can be integrated to
$$S = S(h).$$
Thus if the entropy S is known as a function of h beforehand, only two equations I remain to be solved, see Chapter I, Section 18.

B. Rarefaction and Compression Waves

40. Simple waves

In Section 38 three types of solutions were distinguished among isentropic flows:
 1. the constant state in which u and ρ are constant,
 2. the simple wave in which r or s is constant, and
 3. the general flow in which neither r nor s is constant.

Simple waves are frequently used for building up solutions of isentropic one-dimensional flow problems. In the present section we shall discuss such simple waves in general terms; in later sections we shall make use of them for the solution of specific problems.

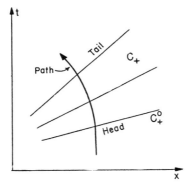
Fig. 10. Forward-facing expansion wave.

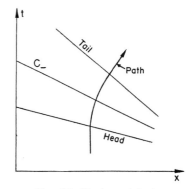
Fig. 11. Backward-facing expansion wave.

A basic property of simple waves derived in Section 29 was: the characteristics C of one kind are straight lines in an (x, t)-plane. In other words, these characteristics represent a propagation with constant velocity. Specifically, if the invariant $-2s = u - l(\rho)$ is constant in the wave region, the C_+-characteristic lines, $r =$ constant, are straight. The velocity of the corresponding sound wave, $u + c$, is greater than the particle velocity u; consequently the particle path enters each characteristic from the right, i.e., comes from the side with greater values of x. This fact is indicated by calling such waves *forward-facing*. If, on the other hand, $2r = u + l(\rho)$ is constant in a flow region, the C_--characteristics are straight, and the simple wave is called *backward-facing*.

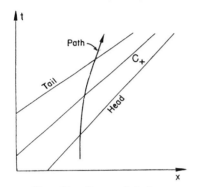

Fig. 12. Forward-facing compression wave.

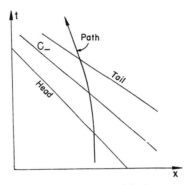

Fig. 13. Backward-facing compression wave.

According to the fundamental theorem of Section 29 the *flow adjacent to a constant state is a simple wave*. Clearly, the transition from the zone of constant flow to that of a simple wave takes place across a characteristic. Assume the simple wave to be *forward-facing*. The transition between wave region and region of constancy takes place across a straight characteristic, C_+^0, called the *head* of the wave if the gas enters the wave across it, or the *tail* of the wave if the gas leaves the wave across it. Let u_0 and ρ_0 be velocity and density in the region of constant flow, then throughout the simple wave region

(40.01) $$u - l = u_0 - l_0, \quad l_0 = l(\rho_0).$$

In particular, if the initial characteristic terminates a state of rest

(40.02) $$u - l = -l_0.$$

From $dl/d\rho > 0$ and $dp/d\rho > 0$ we see that *the density and the pressure change in the same sense as the gas velocity across a forward-*

facing simple wave (and in the opposite sense across a backward-facing simple wave).

A simple wave is called an *expansion* or *rarefaction wave* if pressure and density of a gas particle decrease on crossing it; if pressure and density increase, the wave is called a *compression* or *condensation wave*.

The propagation velocity dx/dt of the "sound waves" represented by the straight characteristics C_+ is, according to (34.09) and (40.01), given by

$$(40.03) \qquad \frac{dx}{dt} = c(\rho) + l(\rho) - l_0 + u_0.$$

The rate of change of this velocity with respect to the gas velocity $u = l(\rho) - l_0 + u_0$ is

$$(40.04) \qquad \frac{dc + dl}{dl} = \frac{\rho\, dc + c\, d\rho}{c\, d\rho} = \frac{d\,(\rho c)}{c\, d\rho} = -\frac{\tau g_{\tau\tau}}{2g_\tau} > 0,$$

by (2.04), (2.05), and the basic assumption (2.06). Hence, for a forward-facing simple wave,

$$(40.05.1) \qquad \frac{d(u + c)}{du} > 0.$$

Similarly for a backward-facing simple wave, in which $u + l = l_0$, we have, from (40.04),

$$(40.05.2) \qquad \frac{d(u - c)}{du} > 0.$$

In other words, if the gas velocity u increases across a simple wave zone, the propagation velocity of the sound waves, $u + c$ or $u - c$, also increases.

For *polytropic gases* we found, see (37.04), $l(\rho) = \dfrac{2c(\rho)}{\gamma - 1}$. Therefore the basic relation in a forward-facing simple wave, see (40.01), is

$$(40.06) \qquad u - \frac{2}{\gamma - 1} c = u_0 - \frac{2}{\gamma - 1} c_0;$$

or, in particular, if the initial state (0) is a state of rest,

$$(40.07) \qquad u - \frac{2}{\gamma - 1} c = -\frac{2}{\gamma - 1} c_0,$$

where c_0 is the sound speed in the quiet gas. With the abbreviation, see (14.06),

(40.08) $$\mu^2 = \frac{\gamma - 1}{\gamma + 1}, \quad 1 - \mu^2 = \frac{2}{\gamma + 1},$$

relation (40.06) can be written in the form

(40.09) $$\mu^2(u - u_0) = (1 - \mu^2)(c - c_0).$$

Incidentally, this last equation is just the equation of the single characteristic Γ_- in the (u, c)-plane which belongs to the simple wave in accordance with the general theory of Chapter II, Section 29. It happens to be a straight line also, see Figure 13 in Section 37.

The quantities p, ρ, and c in a forward-facing simple wave can easily be expressed in terms of the velocity u by using (40.06) and $p/p_0 = (\rho/\rho_0)^\gamma$, $c/c_0 = (\rho/\rho_0)^{(\gamma-1)/2}$, see (3.03), (3.06). Then

(40.10)
$$p = p_0 \left[1 + \frac{\gamma - 1}{2} \frac{u - u_0}{c_0} \right]^{2\gamma/(\gamma - 1)},$$

$$\rho = \rho_0 \left[1 + \frac{\gamma - 1}{2} \frac{u - u_0}{c_0} \right]^{2/(\gamma - 1)},$$

$$c = c_0 + \frac{\gamma - 1}{2} (u - u_0),$$

$$u + c = u_0 + c_0 + \frac{\gamma + 1}{2} (u - u_0).$$

For later purposes we note down the terms of first and second order in the expansions of $p - p_0$, $\tau - \tau_0$, and $\rho - \rho_0$ in powers of $u - u_0$,

(40.11)
$$p = p_0 + \rho_0 c_0 (u - u_0) + \frac{\gamma + 1}{4} \rho_0 (u - u_0)^2 + \cdots,$$

$$\tau = \tau_0 - \tau_0 c_0^{-1}(u - u_0) + \frac{\gamma + 1}{4} \tau_0 c_0^{-2}(u - u_0)^2 + \cdots,$$

$$\rho = \rho_0 + \rho_0 c_0^{-1}(u - u_0) + \frac{3 - \gamma}{4} \rho_0 c_0^{-2}(u - u_0)^2 + \cdots,$$

in which $\gamma p_0 = \rho_0 c_0^2$ has been used. They are useful in the treatment of *weak* or moderately strong simple waves, that is, those across which only relatively small changes take place, see Section 74.

41. Distortion of the wave form in a simple wave

To illustrate the gas flow in a simple wave we may describe how the wave form or distribution of the quantities u, c, p, ρ as functions of x changes with the time t. Suppose the simple wave is forward-facing and the distribution of u and c at the time $t = 0$ is given by two functions $u = F(x)$, $c = G(x)$ satisfying the relation $u - l =$ constant, see (40.01), in which l is a given function of c. The forward sound wave issuing from the point $x = \xi$ at the time $t = 0$ is propagated with constant velocity $u + c$ and carries constant values of u and c. Its path is therefore represented by

(41.01) $\qquad x = \xi + (u + c)t,\ u = F(\xi),\ c = G(\xi).$

The fact that u and c are constant in this path is expressed by the relations

(41.02) $\qquad u = F(x - (u + c)t),\ c = G(x - (u + c)t).$

These equations do not give representations of u and c as functions of x and t; to obtain such representations they would have to be solved for u and c. Nevertheless, these relations are quite illuminating when contrasted with the relation

(41.03) $\qquad u = F(x - c_0 t),\ c = c_0 = \text{constant},$

which expresses the propagation of an initial wave form $u = F(x)$ in linear wave motion. While *in a linear wave the wave form travels unchanged, in a nonlinear simple wave it becomes distorted.* For, the values of u and c are transmitted by the sound waves issuing from different points $x = \xi$ with velocities $u + c$ which are, in general, different.

To describe this distortion we investigate how the steepness of the velocity profile, measured by its derivative $u_x(x, t)$, varies in time. From (41.01) we have

(41.04) $\qquad u_x = \dfrac{\partial u}{\partial \xi} \Big/ \dfrac{\partial x}{\partial \xi} = F'(\xi) / \{1 + (F'(\xi) + G'(\xi))t\}.$

If the wave is a rarefaction wave, we have $F'(\xi) > 0$ and hence $F'(\xi) + G'(\xi) > 0$ by (40.05.1); consequently, the denominator in the right member of (41.04) increases with time. This means that in the course of time *the velocity profile in a simple rarefaction wave flattens out.* In a *compression wave*, on the other hand, *the velocity profile gradually steepens.* As a matter of fact the denominator in (41.04) may possibly approach zero in a compression wave. The significance of this possibility will be discussed in Sections 48 and 50.

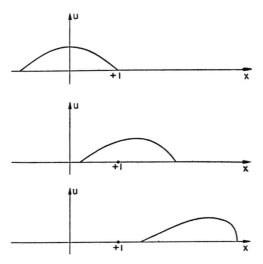

FIG. 14. Steepening of the compressive part and flattening of the expansive part in the velocity profile of a forward-facing simple wave entering gas at rest.

The flattening out and steepening of the velocity profile is illustrated in Figure 14.

42. *Particle paths and cross-characteristics in a simple wave*

The *straight characteristics* of a forward-facing simple wave may be described by the formula

(42.01) $$x = \xi + (u + c)t$$

in which $\xi = \xi(\beta)$, $u = u(\beta)$, $c = c(\beta)$, are given functions satisfying the relation $u + l =$ constant, l being a given function of c, see (40.01) and Section 41.

Any path may then be described parametrically by giving t as a function of β and x as a function of β through (42.01). For the *particle paths* the function $x(\beta)$ is to satisfy the condition $dx/dt = u$ or $x_\beta = ut_\beta$, from which, through (42.01), the condition

(42.02) $$ct_\beta + (u_\beta + c_\beta)t = -\xi_\beta$$

results, a *linear differential equation* for $t = t(\beta)$.

For the *cross-characteristics*, in the present case the C_--characteristics, we find, in a similar way, from $dx/dt = u - c$,

(42.03) $$2ct_\beta + (u_\beta + c_\beta)t = -\xi_\beta .$$

For polytropic gases, in particular, we have

(42.04) $$c = \mu^2(u + c) + (1 - \mu^2)c_0 ,$$

see (40.09), with $\mu^2 = (\gamma - 1)/(\gamma + 1)$ and assuming $u_0 = 0$. The linear differential equations therefore have an explicit solution,

(42.05) $$t = -c^{-\mu^{-2}}\left\{\int c^{\mu^{-2}-1}\xi_\beta \, d\beta + \text{constant}\right\}$$

for the particle paths, and

(42.06) $$t = \tfrac{1}{2}c^{-2\mu^{-2}}\left\{\int c^{2\mu^{-2}-1}\xi_\beta \, d\beta + \text{constant}\right\}$$

for the C_--cross-characteristics. Figures 19 and 20 illustrate a particular case.

These formulas are easily adapted to the case of a *simple wave produced by the action of a piston* in a tube of gas which, at the time $t = 0$, is at rest, and has constant pressure and sound speed. Suppose that originally the piston is at the point $x = 0$, and that the gas is to the right of it, $x > 0$, and has sound speed c_0. If the motion of the piston is given by

(42.07) $$x = X(t),$$

the resulting simple wave can be described by

(42.08) $$x = X(\beta) + (u + c)(t - \beta),$$

instead of by (42.01). Here $u = X(\beta)$ while c can be expressed in terms of u because of $u + l = l_0 = $ constant, in particular, for polytropic gases, by (42.04). Again the particle paths and the cross-characteristics may be described by giving t as a function of β. Adapting formula (42.05), or rederiving the analogous formula from (42.08), one obtains for polytropic gases the particularly simple representation

(42.09) $$t = \beta + t_0(c/c_0)^{-\mu^{-2}}$$

for the particle paths, in which t_0 is the time at which the particle considered crosses the head, $x_0 = c_0 t$, of the wave zone.

A similar but not quite so simple expression could be derived for the cross-characteristics.

43. Rarefaction waves

In this section and in subsequent sections the main subject will be the motion caused by a piston moving in a gas which is initially at rest.

No matter whether the piston recedes from or advances into the gas, not all parts of the gas are affected instantaneously. A "wave" proceeds from the piston into the gas and only the particles which have been reached by the wave front are disturbed from their initial state of rest. If this wave represents a continuous motion, as is always the case if the piston recedes from the gas, the wave front progresses with the sound speed c_0 of the quiet gas. If the piston moves into the gas the situation may become more complicated through the emergence of a supersonic discontinuous *shock wave* as we shall see in Part C. Here we shall be concerned only with continuous wave motion produced by a piston; such a wave motion, as we shall see, is always a *simple wave*, see Figure 15.

We distinguish between expansive and compressive motion and consider first the expansive action of a receding piston, assuming that the medium is a gas originally at rest with constant density ρ_0 and sound speed c_0. Furthermore, it is assumed that the piston, originally at rest, is withdrawn with increasing speed until ultimately the constant velocity $u_B < 0$ is attained. Then the "path" $\mathcal{P}$ in the (x, t)-plane, which represents the piston motion, bends backward from the origin O to a point B where the slope u_B with respect to the t-axis is reached and then continues as a straight line in the same direction, as shown in Figures 15, 16, 17.

The disturbance in the gas resulting from the motion of the

piston is propagated into the undisturbed gas with sound velocity c_0, corresponding to the state (ρ_0, p_0) of the undisturbed gas. This follows from the fundamental fact that the domain of dependence of the zone $x > c_0 t$ is the positive part $x > 0$ of the x-axis, so that there the initial state of rest implies a constant state of rest as the only solution of the differential equations, see Section 28. (The domain of dependence for a point of this region is obtained by drawing the characteristics C_+ and C_- through it to their intersection with the x-axis.) The flow resulting from the motion of the piston is thus confined to the region $x \leq c_0 t$. Since this region is adjacent to a region of constancy, the flow in it is a *simple wave*, see Section 40. Evidently, it is a *forward-facing* simple wave since the gas enters

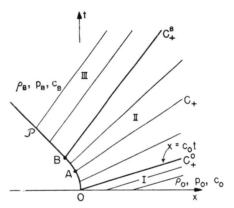

FIG. 15. Simple wave region (II) connecting two regions (I) and (III) of constant state $(-u_B < l_0)$.

this region from the right. In this wave region, therefore, $u - l = -l_c$ is constant, see (40.02). Since along the piston path $x = X(t)$ the gas velocity agrees with the piston velocity $\dot{x} = \dot{X}(t) = u_P(t)$, the density ρ, and hence pressure p and sound speed c are determined from $l = l_0 + u_P(t)$, since $dl/d\rho > 0$ and $dp/d\rho > 0$. The slope $u + c$ of the straight characteristics issuing from the piston path is likewise determined; hence the simple wave is determined as a whole. Since the piston was assumed to be withdrawn and its velocity u_P decreases, according to the statement made in Section 40 density and pressure also decrease across the wave. Thus the wave is a *rarefaction wave*. Also, as seen from Section 40, the velocity $dx/dt = u + c$ of the forward sound waves changes in the same sense as the gas velocity u and therefore, since the velocity u and hence $u + c$ decrease, the straight characteristics issuing from the piston curve fan out.

44. Escape speed. Complete and incomplete rarefaction waves

The above construction is to be modified if the final piston speed $-u_B$ exceeds a certain limit. The reason is that the law of rarefaction expressed by $u = l - l_0$ becomes meaningless as soon as $|u_B| < l_0$ since $l \geq 0$, see Section 37. The quantity l_0 is therefore called the *escape speed* of the gas originally at rest. For polytropic gases we have by (37.04)

$$(44.01) \qquad l_0 = \frac{2}{\gamma - 1} c_0 .$$

If $-u_P$ reaches the escape speed, the rarefaction thins the gas down to density zero; pressure and sound speed are likewise decreased to zero. If a rarefaction wave extends to this stage it is called a *complete rarefaction wave* as it then ends in a vacuum.

For the *tail* of the expansion wave there are two possible results, according to whether or not the terminal speed $-u_B$ of the piston is below the escape speed l_0.

If $-u_B < l_0$, the preceding construction of the simple wave yields the straight characteristics C_+ through every point A on the piston path from O to B. The rarefaction wave, covering the region (II), is incomplete and ends at the characteristics C_+^B through B with $u = u_B$. It is followed by a region (III) of constant state, u_B, ρ_B, p_B, c_B, between the tail of the incomplete rarefaction wave and the piston in which the characteristics C_+ are all parallel (as they are in the region (I) of constant state in front of the simple wave).

If $-u_B = l_0$ the characteristic C_+^B through the point $B = B_e$ is tangent to the piston curve, for at the point B the piston curve has the slope $f'(t) = u_B$, while that of the characteristic C_+^B is $\frac{dx}{dt} = u_B + c_B = u_B = -l_0$ since $c_B = 0$. In other words, the wave is just completed at the piston.

If $-u_B > l_0$ the completion of the wave is already achieved before the piston reaches the terminal speed. There is a point B_e on the piston curve L between O and B for which the characteristic $C_+^{B_e}$ is tangent to the piston curve and carries the value zero for density, pressure, and sound speed. In this case the rarefaction is completed with this line $C_+^{B_e}$, and beyond it we have a region (III') of *cavitation*, equivalent to a vacuum between the receding piston and the tail of the wave in the gas.

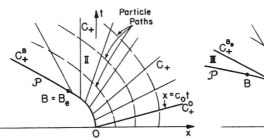
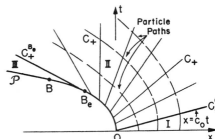

FIG. 16. Rarefaction wave just ending in a zone of cavitation $(-u_B = l_0)$.

FIG. 17. Rarefaction wave ending in a zone (III) of cavitation $(-u_B > l_0)$.

Physically speaking, the escape speed l_0 is the speed beyond which a piston cannot recede without separating from the thinned-out gas. If the speed of the piston exceeds l_0, then, as far as the motion of the gas is concerned, it does not matter what the actual value of u_B is. We might just as well consider $-u_B$ as infinite or imagine the piston as a wall suddenly removed, allowing the gas to escape into a vacuum, an interpretation to which the name "escape speed" alludes.

Summary. We can summarize our results qualitatively as follows. A piston receding from a gas at rest with speed which never decreases causes a rarefaction wave of particles moving toward the piston. At the head of the wave, which moves into the gas at sound speed, the velocity of the gas is zero. Through the wave the gas is accelerated. If the piston speed $-u_B$ is below the escape speed l_0, the gas expands until it has reached the speed $-u_B$ of the piston and then continues with constant velocity, density, and pressure. If, however, the piston speed exceeds the escape speed, the expansion is complete and the wave ends in a zone of cavitation between the tail of the wave and the piston. In any case the wave moves into the quiet gas, while the gas particles move at increasing speed from the wave head to the tail, i.e., from zones of higher pressure and density to zones of lower pressure and density.

45. Centered rarefaction waves

Of particular interest is the case in which the acceleration of the piston from rest to a constant terminal velocity u_B takes place in an infinitely small time interval, i.e., instantaneously. Then the family of characteristics C_+ forming the simple wave degenerate into a pencil of lines through the origin $O: x = 0, t = 0$, see Figure 18. In other

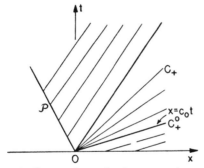

FIG. 18. Centered rarefaction wave $(-u_P < l_0)$.

words, the simple wave has degenerated into a *centered* simple wave. It is clear that such a centered simple wave is a rarefaction wave. For, u decreases on crossing the wave if it is forward-facing and increases if it is backward-facing: in both cases ρ and p decrease across the wave, as shown in Section 40, and, therefore, it is a rarefaction wave.

At the center O the quantities u, ρ, p as functions of x and t are discontinuous, but this *discontinuity is immediately smoothed out* in the subsequent motion. Here we have the first and typical example of an initial discontinuity which is immediately resolved into continuous flow.

46. Explicit formulas for centered rarefaction waves

A centered simple wave may be described by the equation

(46.01) $$x = (u + c)t$$

in which

(46.02) $$u = l - l_0$$

may be considered a given function of c. Inversely we may express u and c in terms of $u + c$ and thus in terms of x/t. For a polytropic gas we have by (42.04) the relations

(46.03)
$$c = \mu^2 \frac{x}{t} + (1 - \mu^2)c_0,$$

$$u = (1 - \mu^2)\left(\frac{x}{t} - c_0\right),$$

giving the distributions of u and c in a centered simple wave explicitly.

Comparing (46.01) with (42.01) we see that the linear differential equations (42.02) and (42.03) for *particle paths and cross-characteristics* become homogeneous. The solutions (42.05) and (42.06) for a polytropic gas simply reduce to, see (42.09),

(46.04) $$t = t_0 \left(\frac{c}{c_0}\right)^{-\mu^{-2}},$$

and

(46.05) $$t = t_0 \left(\frac{c}{c_0}\right)^{-2\mu^{-2}},$$

respectively where t_0 is the time at which the particle path or the cross-characteristic begins at the line $x = c_0 t$. Using relation (42.04) in the form

$$u + c = (1 - \mu^{-2})c_0 + \mu^{-2}c$$

and expressing c in terms of t by (46.04) or (46.05) we obtain from (46.01)

(46.06) $$x = -(\mu^{-2} - 1)c_0 t + \mu^{-2} c_0 t_0 \left(\frac{t}{t_0}\right)^{1-\mu^2},$$

for the particle paths, and

(46.07) $\quad x = -(\mu^{-2} - 1)c_0 t + \mu^{-2} c_0 t_0 \left(\dfrac{t}{t_0}\right)^{1-\frac{1}{2}\mu^2}$

for the cross-characteristics. These formulas are valid in the rarefaction zone.

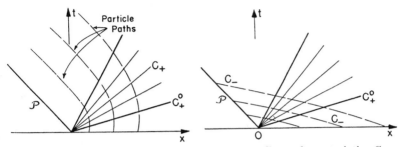

Fig. 19. Particle paths in a centered rarefaction wave.

Fig. 20. Cross-characteristics C_- in a centered rarefaction wave.

In the case of a *complete rarefaction* ending with zero density, $-u_B \geq l_0$, formula (46.07) holds for arbitrarily large values of t, and we have, for large t, the asymptotic representation

$$x \sim -(\mu^{-2} - 1)c_0 t$$

for the particle paths. As remarked previously, the gas remains in the zone (II) of rarefaction, and in the (x, t)-diagram the particle paths acquire asymptotically the direction of the characteristic C_+^c on which the escape speed l_0 is attained, see Figure 21.

For $-u_B < l_0$ the rarefaction wave terminates at the characteristic C_+ on which the velocity has the value $u = u_B$, and all the particle paths emerge from region (II) parallel to the terminal direction of the piston path $\mathscr{P}$ and remain parallel in region (III).

The cross-characteristics C_- after emerging from an incomplete rarefaction region continue as straight lines which meet the piston line $\mathscr{P}$: $x = u_B t$. For $-u_B \geq l_0 = (\mu^{-2} - 1)c_0$ the characteristics C_- remain within the "complete rarefaction region," and, since $x \sim -(\mu^{-2} - 1)c_0 t$, approach the particle paths asymptotically, see Figures 21 and 22.

Obviously these considerations can be generalized to non-polytropic equations of state.

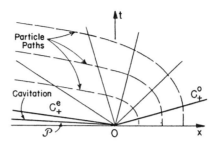

FIG. 21. Particle paths in a centered rarefaction wave ending in cavitation.

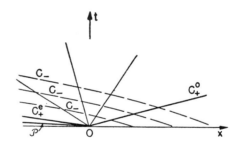

FIG. 22. Cross-characteristics, C_- in a centered rarefaction wave ending in cavitation.

47. Remark on simple waves in Lagrangian coordinates

We could just as well have developed the theory of simple waves in Lagrange's coordinates, using the equations (39.02) developed in Part III-A. The characteristics C_+ are given by

(47.01) $$\frac{dh}{dt} = \rho c = k,$$

and the characteristics C_- by

(47.02) $$\frac{dh}{dt} = -\rho c = -k.$$

For a forward-facing simple wave the lines C_+ in the (h, t)-plane are straight since the slope k is constant on each of them. For a centered simple wave in particular, $h/t = k$ is therefore constant. In other words, the impedance is always simply $k = h/t$, no matter

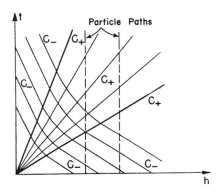

FIG. 23. Characteristics in Lagrangian coordinates for a centered simple wave.

what the adiabatic equation of state is. Consequently, the curves C_- satisfy the equation $dh/dt = -h/t$, which can be immediately integrated to

(47.03) $\qquad ht = \text{constant}.$

Thus, for isentropic flow of any fluid, the *cross-characteristics in centered rarefaction waves are equilateral hyperbolas in Lagrangian coordinates*.

48. Compression waves

If a piston is not withdrawn, but is moved *into* the gas-filled tube, or if a receding piston is slowed down or stopped, then a simple *compression or condensation wave* originates at the piston. The qualitative statements and formulas pertaining to rarefaction waves also apply to compression waves, except that density, pressure, and sound speed at the piston increase and that the forward characteristics C_+ no longer diverge from the piston curve. Because of this, the simple wave does not exist indefinitely, for all times t. For if the straight characteristics C_+ each carrying a particular value of u, could be continued indefinitely within the flow, they would converge and form an *envelope* on which the values of u would conflict.

At the earliest time, $t = t_c$, that such an envelope appears, it forms a cusp at some point $x = c_c$, see Section 49. The two branches of the envelope meeting at the cusp enclose an angular region covered three times by the C_+-characteristics. Thus, a unique continuation of the flow through a simple wave beyond the time t_c at the point x_c is mathematically impossible. We may also interpret this phenomenon, apparently first noticed by Stokes [40] in 1848, in the following manner. The influence of the piston motion is propagated into the gas through sound waves traveling with sound speed c relative to the gas. To greater piston velocity there corresponds greater sound speed c; therefore later influences of the piston motion travel faster and tend to overtake those sent out earlier. The wave form, described by the velocity as a function of x, see Section 41, becomes steeper and tends to become vertical at some point. The character of the resulting discontinuity is similar to the breaking of water waves which become steeper and steeper as more slowly progressing parts are overtaken by faster ones, see [27].

A particularly direct illustration of the inevitable development of discontinuities is afforded by a piston moving into a gas at rest with a speed ultimately exceeding the sound speed c_0. If the gas flow remained continuous the gas would be at rest in the zone of points, $x \geq c_0 t$, that cannot be reached from the original position of the piston with a speed less than that of sound. Since the piston moves eventually with a speed greater than that of sound, it must eventually enter this zone. Consequently the motion cannot remain continuous.

We have now seen that no simple wave flow exists that offers a unique continuation beyond the time $t = t_c$. From our theorem that any continuous flow adjacent to a constant state is a simple wave, see Chapter II, Section 29, it follows that there exists no alternative, everywhere continuous solution of the flow problem. This fact should be emphasized because of some argument about it in the literature, see [39].

It is thus impossible that under all circumstances the flow could remain continuous, isentropic, and governed solely by pressure forces. It is remarkable that this breakdown of a plausible hypothesis about the mechanism governing the flow results as a purely mathematical consequence. What alternative hypothesis is adopted will be seen in Part C of this chapter, in which "shock discontinuities" are discussed.

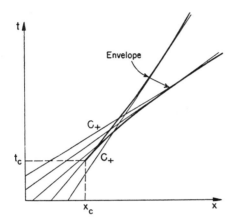

FIG. 24. The formation of the envelope of the straight characteristics of a simple compression wave.

The possible occurrence of an envelope in a solution of the flow problem affords a striking illumination of the theorem stated in Section 28 on the *unique existence* of the solution of the differential equations (21.01) that assumes given values on a line $t = t_c =$ constant. Suppose these values are taken as just the values of a solution which forms an envelope beginning with a cusp at a point $x = x_c$, $t = t_c$. The fact that no unique solution exists which assumes these initial data seems at first sight to be a contradiction to the theorem just mentioned. The resolution of this paradox lies in the fact that the theorem required initial values with continuous derivatives with respect to x. The values on $t = t_c$ taken from a solution with a cusp at $x = x_c$, $t = t_c$, though being continuous at $x = x_c$, have infinite derivatives there with respect to x as seen from the discussion in Section 41.

Appendix to Part B

49. Position of the envelope and its cusp in a compression wave

Here a few remarks are added regarding the geometry of the envelopes formed by the straight characteristic C_+ of a compression wave, see for example Hadamard [28].

We write the analytical representation of a forward-facing simple wave (42.01) in the form

(49.01) $\qquad x = \xi(\beta) + \omega(\beta)t, \qquad \omega(\beta) = u(\beta) + c(\beta).$

At the envelope the derivative of x with respect to β vanishes so that

(49.02) $\qquad t = -\dfrac{d\xi}{d\omega}, \qquad x = \xi - \omega \dfrac{d\xi}{d\omega}$

is the parametric representation of the envelope. Here we have used ω instead of β as parameter, which is permissible if $\omega_\beta \neq 0$ in the region considered. If $\dfrac{d\xi}{d\omega}$ has an extremum for some value β_e of β, then, assuming that $\xi(\beta)$ has a continuous second derivative, $\dfrac{d^2\xi}{d\omega^2} = -\dfrac{dt}{d\omega}$ changes sign at $\beta = \beta_e$ and hence $\dfrac{dx}{d\omega} = -\omega \dfrac{d^2\xi}{d\omega^2}$ also changes sign if $\omega_0 \neq 0$, which we assume for simplicity. Therefore such an extremum of t takes place at a *cusp* of the envelope and we have at the cusp

(49.03) $\qquad\qquad \dfrac{d^2\xi}{d\omega^2} = 0.$

It is useful to describe the formation of the envelope in greater detail. Assume that the cusp occurs at $t = 0$, $x = 0$ with a value $\omega = \omega_0$. Since $d\xi/d\omega = 0$ and $d^2\xi/d\omega^2 = 0$ at $x = t = 0$ we have $\xi(\omega) = k(\omega_0 - \omega)^3 + \cdots$. The compressive character of the wave or the condition that the envelope is formed for $t \geq 0$ is then expressed by the condition $k \geq 0$. It is convenient to describe the simple wave and the envelope first in an $(\omega_0 - \omega, t)$-plane. In that plane the C_+-characteristics are the straight lines $\omega_0 - \omega = $ constant and according to (49.02), the representation of the envelope is

(49.04) $\quad t = t^E(\omega) = -\dfrac{d\xi}{d\omega} = 3k(\omega_0 - \omega)^2 + \cdots .$

The two branches of the envelope in the (x, t)-plane, corresponding to $\omega_0 - \omega > 0$ and $\omega_0 - \omega < 0$ are at the same time edges of the fold of the mapping of the $(\omega_0 - \omega, t)$-plane into the (x, t)-plane, see Section 30. This is seen when one considers the image of a line $t = t_1 = $ constant. From $dx/d(\omega_0 - \omega) = -t_1 - d\xi/d\omega$ we see that x increases as $(\omega_0 - \omega)$ increases until $\omega_0 - \omega$ reaches the negative value $\omega_0 - \omega = \omega_0 - \omega_{-1}$ at which $-d\xi/d\omega = t_1$; then x decreases until $\omega_0 - \omega$ reaches the positive value $\omega_0 - \omega = \omega_0 - \omega_{+1}$ at which $-d\xi/d\omega = t_1$; from then on x increases again. It is then clear that *the cusp region* between the two branches of the envelope, that is the image of the region $t > t^E(\omega)$, *is covered three times* in the mapping.

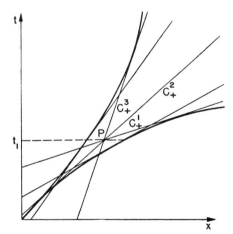

Fig. 25. The cusp region in a simple compression wave is covered three times by the straight characteristics.

We now shall investigate more specifically the formation of the envelope in the compression wave produced by the motion of a piston described by

(49.05) $\qquad\qquad x = X(t).$

We want to show that *an envelope is always formed if a simple wave occurs when a piston at the left-hand side is pushed into a gas with*

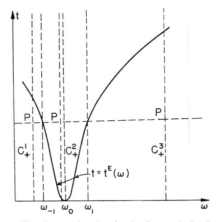

Fig. 26. The image in the (ω, t)-plane of the intermediate sheet of the cusp region of a simple compression wave.

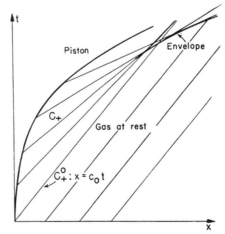

Fig. 27. Envelope with cusp formed by the straight characteristics of a compression wave produced by an accelerated piston.

positive acceleration and that this envelope always forms a cusp inside the wave region if the accelerated phase of the piston motion begins with the acceleration zero. We assume without restriction $X(0) = \dot{X}(0)$

$= 0$, $\ddot{X}(\beta) > 0$ for $\beta > 0$, and $u = 0$, $c = c_0$ on the characteristic $x = c_0 t$. For the resulting simple wave we may take as our parameter β the time at which the C_+-characteristic through the point (x, t) started at the piston. Thus the simple wave is to be described by relation (42.08),

(49.06) $$x = X(\beta) + \omega(\beta)(t - \beta).$$

Here $\xi = X(\beta) - \beta\omega(\beta)$, see (49.01). The representation (49.02) of the envelope then becomes

(49.07) $$t = \beta + \frac{\omega(\beta) - \dot{X}(\beta)}{\dot{\omega}(\beta)}, \qquad x = X(\beta) + \omega(\beta)\frac{\omega(\beta) - \dot{X}(\beta)}{\dot{\omega}(\beta)}.$$

Clearly the velocity of the piston is $\dot{X}(\beta) = u(\beta)$. Since $\omega = u + c = c + l - l_0$ is a given function of $l = u + l_0$, $\omega(\beta)$ is also determined, see (40.05.1). Using

(49.08) $$\frac{d\omega}{du} = \frac{d(c + l)}{dl} = \lambda > 0,$$

see (40.05.1), we have

$$\dot{\omega} = \lambda(\beta)\ddot{X}(\beta).$$

For the envelope we then obtain from (49.07)

(49.09) $$t = t^E(\beta) = \beta + \frac{\omega(\beta) - \dot{X}(\beta)}{\lambda(\beta)\ddot{X}(\beta)} = \beta + \frac{c(\beta)}{\lambda(\beta)\ddot{X}(\beta)}.$$

Now $c(\beta)$ and $\lambda(\beta)$ are positive and $\ddot{X}(\beta) > 0$ for $\beta > 0$ by assumption. Hence $t^E(\beta) > \beta$ for $\beta > 0$. Further,

$$x^E(\beta) = \omega(\beta)(t^E(\beta) - \beta) + X(\beta) > X(\beta).$$

Therefore an envelope is formed in the flow region.

Formula (49.09) shows that, if $\ddot{X}(0) = 0$, then $t^E(0) = \infty$. If $\ddot{X}(t) > 0$ for all $t > 0$ then $t^E(\beta)$ increases indefinitely as $\beta \to \infty$. If $\ddot{X}(t) = 0$ for a time $t = t_1 > 0$, then $t^E(\beta) \to \infty$ as $\beta \to t_1$. In any case it is clear that $t^E(\beta)$ first decreases and then increases again. Consequently, $t^E(\beta)$ has a minimum t_c for some value of β and therefore the envelope forms a cusp at the time t_c.

For a decelerated piston, $\ddot{X}(\beta) < 0$, the situation is different; there is no point of the envelope in the domain $x > X(\beta)$ corresponding to the interior of the (x, t)-domain of the flow.

Doubts may arise about our statement that the characteristics C_+ always form an envelope if the piston is accelerated in forward motion or decelerated in backward motion during an interval of time. For, on looking at Figure 13, one might think it possible for the envelope to form outside of the flow region or, in other words, for the piston path itself to cut off the envelope. A simple analysis shows that this not the case.

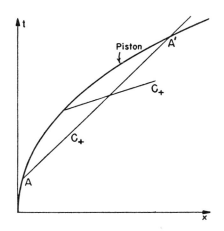

FIG. 28. The impossible situation in which a straight characteristic issuing from one point of the piston path intersects it again at another point.

Suppose a characteristic issuing from a point A on the piston path is intersected by the piston path in another point A'; then the characteristic issuing from a point near A' on the arc AA' of the piston path intersects the characteristic AA'. Therefore the envelope of the characteristics enters the segment between the arc AA' on the piston path and the arc AA' on the characteristic.

Whenever the initial acceleration of the piston is positive the envelope begins at a point $t = t_c$, $x = c_0 t_c$ on the straight characteristic C_+ starting at the origin. From (49.09) with $\beta = 0$ we have

(49.10) $$t_c = \frac{c_0}{\lambda_0 \ddot{X}(0)}.$$

In this case, the part of the characteristic $x = c_0 t$ beyond $t = t_c$ may be considered the second branch of the envelope, so that again a cusp is formed. This situation is illustrated in Figure 29, which is given below.

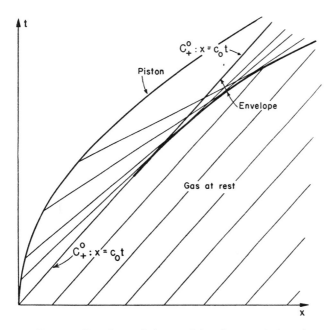

Fig. 29. Envelope of the straight characteristics of the compression wave produced by a uniformly accelerated piston.

The envelopes can take a wide variety of shapes corresponding to various motions of the piston. One can, for example, move the piston in such a way that the characteristics converge in a point. However, since the fine features of the geometry of the envelope depend on the local behavior of the second and higher derivatives of the function $X(t)$, we must expect that the actual behavior of the flow is not strongly affected by geometrical complexities of the envelope.

C. Shocks

50. The shock as an irreversible process

As we have seen, initial discontinuities are sometimes smoothed out, as in the case of centered rarefaction waves, while other motions starting as perfectly continuous waves cannot be maintained without a discontinuity. The fact is that any forward acceleration or backward deceleration of the piston, however slow, leads ultimately to discontinuities of velocity, pressure, density, specific entropy, and temperature.

Hence, for a mathematical description of motions caused by advancing pistons and of many other motions as well, we must abandon, or rather supplement, the physical hypotheses employed so far (as we already indicated in Section 48).

One possibility suggests itself immediately. We might try to obtain the necessary generalization from the differential equations of motion directly. In Chapter II, Section 24, we saw that these differential equations allow discontinuities of the first and higher derivatives of u and ρ across characteristics in the (x, t)-plane. Such "sonic discontinuities" arise from a natural extension of the differential equations; for example, they arise in initial value problems by passage to the limit from initial values with continuous derivatives to initial values with local jump discontinuities of the derivatives. In the case of linear differential equations the same type of limiting process leads to a "sonic propagation" even of discontinuities of the dependent functions themselves, see [32, pp. 360–361]. For our nonlinear differential equations, however, no such sonic transmission of discontinuities of ρ and u is deducible by a passage to a limit from continuous solutions.

Hence to arrive at an adequate theory we must give up as oversimplified our original description of the physical situation and seek a closer approximation to the actual situation by accounting for physical facts neglected in the original differential equations.

We have assumed so far that the forces in the gas are due to variations in the pressure $p = p(\rho, S)$ and not at all to friction, and that the entropy of a particle remains unchanged. These assumptions are justified only if the gradients of velocity and temperature are small. Otherwise a mathematical description of the physical situation must take into account the effect of irreversible thermodynamical processes caused by friction and heat conduction, always present whenever velocity and temperature are not constant. Such a theory would involve almost insuperable mathematical complications if it were not for a fortunate fact: Actual phenomena show that irreversible processes occur in gases only in narrow zones where the gradients of velocity and temperature become very large while outside of these transition zones the flow obeys the laws established for adiabatic reversible processes, i.e. the differential equations as discussed before. Thus the empirical facts suggest a further mathematical idealization, which will be the basis of our analysis.

Irreversible processes are to be described by sudden jump discontinuities, occurring across certain sharply defined surfaces in the fluid. Such discontinuities, with infinite gradients in some of the quantities, replace in the mathematical idealization the narrow zones of noticeable irreversibility.

In reality very considerable changes of velocity and temperature occur across such surfaces; thus the assumption of sharp discontinuities is indeed an idealization which agrees with the facts rather better than we might hope.

Naturally we require that the three laws of conservation of mass, momentum, and energy also hold for this irreversible process. Outside of the discontinuity surface the only force acting is, according to our assumptions, due to the pressure, and the only gain or loss of energy present is due to the work done by these pressure forces. Hence in these regions our basic differential equations are valid.

For continuous processes, as we recall from Section 8, the constancy of the specific entropy for each gas particle, i.e. reversibility, follows from the law of conservation of energy. For discontinuous processes subject to the same conservation law this is no longer the case. The thermodynamic condition expressing the irreversible character of the process is that the entropy does not decrease in the discontinuous process, and this entropy condition must be added to the conservation laws.

It is by no means obvious, but may be safely assumed, that a flow involving such a discontinuous process is completely determined by the three conservation laws and the entropy condition. The original differential equations, valid in the region of continuous flow, together with the conditions expressing the conservation laws and the entropy condition across a discontinuity surface, suffice to determine the flow without describing in detail the irreversible process across a discontinuity surface.

To clarify the situation we shall discuss in Section 63 a rational description of the internal mechanism of this irreversible process in terms of heat conduction and viscosity, taking into account the finite extension of the zones of such processes. In the following sections, however, we shall proceed strictly from our mathematical assumptions, for which the preceding arguments serve as a motivation.

51. *Historical remarks on non-linear flow*

A few historical facts may be inserted here.

Poisson (1808) [36] determined what was, in effect, a simple wave solution of the differential equation of flow in an isothermal gas:

$$u = F[x - (u + c)t],$$

F being an arbitrary function, see Section 41.

Challis (1848) [41] observed that such an equation can not always be solved uniquely for the velocity u. To obtain a unique solution, Stokes (1848) [40] proposed to assume that a discontinuity in the velocity begins at the time when $\dfrac{\partial u}{\partial x}$ becomes infinite (cf. Section 41). Using the laws of conservation of mass and momentum he then deduced two discontinuity conditions for an isothermal gas. Stokes argued that discontinuities would never occur physically because any tendency to a discontinuity would be smoothed out by viscous forces. Furthermore he indicated that flows involving a discontinuity must also involve some phenomena of reflection.

Earnshaw (1858) [37] developed the simple wave solution for the flow of gases satisfying any relation $p = p(\rho)$. He reasoned that since the local velocity of propagation increases across a compression wave, such a wave would be perpetually "gaining" on its front, and eventually a "bore" or discontinuity would form, see Section 41.

Independently, Riemann (1860) [38] developed the theory of the simple wave and the general solution of the flow problem by using "Riemann invariants," see Section 37. He rediscovered and elaborated the theory of shocks but made tacitly the incorrect assumption that the transition across a shock is adiabatic and reversible.

Rankine (1869) [42] showed that no steady adiabatic process in which the only forces are pressure forces can represent a continuous change over a small finite region from one constant state to another. He proposed instead that across this region a non-adiabatic process occurs subject to the condition that heat may be communicated from one particle to another but that no heat is received from outside.

Rankine's condition agrees with the principle of conservation of energy. But Rayleigh [39] and Hugoniot (1887) [43] were the first to point out clearly that an adiabatic reversible transition in a shock would violate the principle of conservation of energy. In fact, Hugoniot showed that in the absence of viscosity and heat conduction (outside the shock) conservation of energy implies conservation of entropy in continuous flow and also implies a change of entropy across a shock. From the conservation of energy he also deduced the third shock condition in its customary form, see (54.10), which is preferable to Rankine's form, although in the case of a perfect gas, Rankine's three shock conditions are equivalent to those of Hugoniot.

Rayleigh (1910) [39] pointed out that the entropy must increase in crossing a shock front and that for this reason a rarefaction shock can not occur in a perfect gas.

52. *Discontinuity surfaces*

We distinguish two types of discontinuity surfaces, *contact surfaces* and *shock fronts*. *Contact surfaces* are surfaces separating two parts of the medium without any flow of gas through the surface; *shock fronts* are discontinuity surfaces which are *crossed by the gas*. The side of the shock front through which the gas enters will be called the *front side* of the shock or the side *ahead* of the shock, the other, the *back side*. As we shall see in Section 65, the shock front, always moves with *supersonic speed* as observed from the *front side*, and with *subsonic speed* as observed from the *back side*. The zone of flow behind a shock front is frequently called *shock wave*. In this chapter we are concerned with one-dimensional motion. Hence the shock fronts and contact surfaces are assumed to be planes perpendicular to the x-axis, represented on the x-axis by points or in the (x, t)-plane by lines, henceforth called *shock lines*, or *contact lines*, respectively.

53. Basic model of discontinuous motion. Shock wave in a tube

Let us first describe the simplest case of a motion involving a shock front. The centered expansion wave caused by a piston receding at constant speed was studied as a basic type of motion. Just as basic and typical is the *motion caused by a piston* starting from rest and suddenly *moving with constant velocity* u_P *into the quiet gas.* No

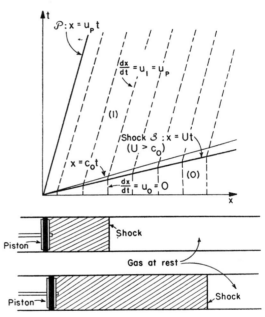

FIG. 30. Shock wave produced by moving a piston with constant velocity into a gas at rest.

matter how small u_P is, the resulting motion cannot be continuous because a continuous motion would imply a forward-facing simple wave, specifically, a centered simple wave, in order to achieve a discontinuous change of velocity at the origin. However, the gas velocity through a centered simple wave becomes negative if it vanishes ahead of the simple wave. Therefore, no adjustment to the positive piston velocity is possible by continuous motion.

What happens? The answer is: Immediately a shock front

JUMP CONDITIONS 121

appears, moving away from the piston with a constant and, as we shall prove, supersonic speed U, uniquely determined by the density and sound speed in the quiet gas and by the piston speed u_P. Ahead of the shock front the gas is at rest, while behind the shock it moves with the constant velocity u_P. In the (x, t)-plane this very simple motion is represented by Figure 30. For a sequence of decreasing values of u_P the shock line approaches the characteristic $x = c_0 t$ and the jump of velocity, pressure, and density across the shock approaches zero. The shock becomes *weak* and approaches a "sonic disturbance."

Before we can substantiate this qualitative description, we must discuss the discontinuity or jump conditions across the shock.

54. Jump conditions

We shall derive the jump conditions from the "caloric equation of state," see Section 2, and the following basic laws of physics:
(1) Conservation of mass,
(2) Conservation of momentum,
(3) Conservation of energy,
(4) Increase or conservation of entropy.

If the further assumption of continuous velocity, density, and pressure are made, the first two laws would lead to Euler's (or Lagrange's) equations for isentropic flow, see Section 7. Application of these principles to discontinuous motions leads to the corresponding first two jump conditions for shocks. The energy law (3) takes care of a more delicate point. Our original system of differential equations, see (7.08–.11), was supplemented by the caloric equation of state in which we assumed constant entropy in keeping with the supposed adiabatic reversible character of our processes and in agreement with the law of conservation of energy. Across a shock front, however, the third law of conservation, as we shall show presently, implies entropy changes and leads to the third "thermodynamical" jump condition, formulated by Rankine and Hugoniot, which replaces the assumption of adiabatic changes made for continuous motions.

We shall now derive the discontinuity condition by applying the three general principles to a column of gas in a tube. The column covers at the time t the interval $a_0(t) < x < a_1(t)$, where $a_0(t)$ and $a_1(t)$ denote the positions of the moving particles that form the ends of the column, and the flow is supposed to be continuous at the ends of the column. By e we denote the internal energy of the gas per unit mass, so that the total energy per unit mass is $e + \frac{1}{2}u^2$. Then, for the column, the four basic principles are expressed by the relations

$$(54.01) \qquad \frac{d}{dt} \int_{a_0(t)}^{a_1(t)} \rho \, dx = 0$$

(Conservation of mass),

$$(54.02) \qquad \frac{d}{dt} \int_{a_0(t)}^{a_1(t)} \rho u \, dx = p(a_0,t) - p(a_1,t)$$

(Conservation of momentum),

$$(54.03) \qquad \frac{d}{dt} \int_{a_0(t)}^{a_1(t)} \rho\{\tfrac{1}{2}u^2 + e\} \, dx = p(a_0,t)u(a_0,t) - p(a_1,t)u(a_1,t)$$

(Conservation of energy),

$$(54.04) \qquad \frac{d}{dt} \int_{a_0(t)}^{a_1(t)} \rho S \, dx \geq 0$$

(Increase or conservation of entropy).

Relation (54.01) needs no comment. Relation (54.02) expresses the assumption that the only forces acting are pressure forces, and that consequently the rate of change of momentum of the column equals the total resultant force exerted on the column by the pressure on the two ends. Relation (54.03) expresses the assumption that the gain of energy is due only to the action of the pressure forces, in other words that the rate of increase of energy contained in the column is equal to the "power-input," i.e., the work done in unit time by the pressure against the end surfaces of the column (whose velocities are $\dot{a}_0 = u(a_0, t)$ and $\dot{a}_1 = u(a_1, t)$. Relation (54.04) states that the column gains or maintains its entropy.

As long as we assume u, ρ, p, and S continuous and differentiable in the whole column, we can easily deduce from the first three of these equations the differential equations of motion, see Sections 7 and 8, conservation of entropy being a consequence. In the present analysis, however, we assume that in the moving column there is a point of discontinuity whose coordinate $x = \xi(t)$ moves with the velocity $\dot{\xi}(t)$, $= U(t)$ and we shall derive from equations (54.01–.04) relations between the quantities at both sides of this point.

All of our integrals have the form

$$J = \int_{a_0(t)}^{a_1(t)} \Psi(x, t)\, dx,$$

the integrand Ψ being discontinuous at $x = \xi$. Differentiation leads to

(54.05)
$$\begin{aligned}\frac{d}{dt} J &= \frac{d}{dt}\int_{a_0(t)}^{\xi(t)} \Psi(x, t)\, dx + \frac{d}{dt}\int_{\xi(t)}^{a_1(t)} \Psi(x, t)\, dx \\ &= \int_{a_0(t)}^{a_1(t)} \Psi_t(x, t)\, dx + \Psi_0 \dot{\xi}(t) - \Psi(a_0, t) u(a_0, t) \\ &\quad + \Psi(a_1, t) u(a_1, t) - \Psi_1 \dot{\xi}(t).\end{aligned}$$

The quantities Ψ_0 and Ψ_1 are the limits of $\Psi(x, t)$ as x approaches ξ from the sides $x < \xi$ and $x > \xi$ respectively. This formula holds no matter how short the column is, so long as it contains $x = \xi$ as an interior point. We now perform the limiting process, letting the length of the column approach zero. Since the first integral on the righthand side of (54.05) then tends to zero, $\Psi(a_1, t) \to \Psi_1$, and $\Psi(a_0, t) \to \Psi_0$, we obtain

(54.06)
$$\lim_{a_1 - a_0 \to 0} \frac{d}{dt} J = \Psi_1 v_1 - \Psi_0 v_0,$$

where

(54.07)
$$v_i = u_i - U, \quad i = 0, 1,$$

is the flow velocity relative to the discontinuity surface. Thus we derive from the four basic equations the following jump conditions:

(Conservation of Mass)

(54.08.1)
$$\rho_1 v_1 - \rho_0 v_0 = 0$$

or

(54.08.2)
$$\rho_0 v_0 = \rho_1 v_1 = m;$$

m is the *mass flux* through the surface.

(Conservation of Momentum)

(54.09.1) $\qquad (\rho_1 u_1)v_1 - (\rho_0 u_0)v_0 = p_0 - p_1;$

by (54.07) and (54.08) this relation is equivalent to

(54.09.2) $\qquad m u_0 + p_0 = m u_1 + p_1$

or

(54.09.3) $\qquad \rho_0 v_0^2 + p_0 = \rho_1 v_1^2 + p_1 = P,$

which involves only the relative velocities v. The quantity P, defined by (54.09.3), is occasionally called the *total momentum flux*.

(Conservation of Energy)

(54.10.1) $\qquad \rho_1(\tfrac{1}{2} u_1^2 + e_1)v_1 - \rho_0(\tfrac{1}{2} u_0^2 + e_0)v_0 = p_0 u_0 - p_1 u_1$

or

(54.10.2) $\qquad m(\tfrac{1}{2} u_0^2 + e_0) + u_0 p_0 = m(\tfrac{1}{2} u_1^2 + e_1) + u_1 p_1.$

By (54.07–.09) this relation is equivalent to

(54.10.3) $\qquad m(\tfrac{1}{2} v_0^2 + e_0 + p_0 \tau_0) = m(\tfrac{1}{2} v_1^2 + e_1 + p_1 \tau_1).$

(Increase of Entropy)

(54.11.1) $\qquad \rho_1 S_1 v_1 - \rho_0 S_0 v_0 \geq 0$

or by (54.08),

(54.11.2) $\qquad m S_0 \leq m S_1.$

All these relations hold across both shock fronts and contact surfaces. The two types of discontinuity surfaces are distinguished by the property that *gas flows across a shock front*, $m \neq 0$, and that *no gas flows across a contact surface*, $m = 0$.

We shall consider shock discontinuities in Section 55 and postpone the discussion of contact surfaces to Section 56.

55. Shocks

For shocks ($m \neq 0$) relation (54.10.3) reduces to

(55.01) $\qquad \tfrac{1}{2} v_0^2 + e_0 + p_0 \tau_0 = \tfrac{1}{2} v_1^2 + e_1 + p_1 \tau_1 = \tfrac{1}{2} \hat{q}^2$

SHOCKS

where $\hat{q}$ is the limit speed introduced in Sections 14 and 15. Remembering the definition of the enthalpy $i = e + p\tau$ from Section 9, we can write (55.01) in the form

(55.02) $$\tfrac{1}{2}v_0^2 + i_0 = \tfrac{1}{2}v_1^2 + i_1 = \tfrac{1}{2}\hat{q}^2.$$

Thus we see, *the third shock condition has exactly the form of Bernoulli's law*. It differs, however, from the three forms of it considered previously in Section 14, inasmuch as the function which represents the enthalpy i in its dependence on ρ is now discontinuous across the shock, since the values i_1 and i_0 correspond to different values S_1 and S_0 of the entropy, as will be seen. In other words, the change in enthalpy $i_1 - i_0$ across a shock is not equal to $\int_{(0)}^{(1)} \dfrac{dp}{\rho}$ but equals $\int_{(0)}^{(1)} \left(\dfrac{dp}{\rho} + T\,dS\right)$.

Each of the three shock conditions (54.08.–.10) has a form in which only the relative velocities $v = u - U$ are involved and not the velocities u and U separately. It is thus clear that the shock conditions are invariant under translation with constant velocity, in accordance with the Galilean principle of relativity.

From equations (54.09.2) and (54.08.2) we obtain

(55.03) $\quad (\tau_0 + \tau_1)(p_1 - p_0) = m(\tau_0 + \tau_1)(v_0 - v_1) = v_0^2 - v_1^2.$

Using this relation to eliminate v_0 and v_1 from (55.02) we find

(55.04) $$(p_1 - p_0)\frac{\tau_0 + \tau_1}{2} = i_1 - i_0,$$

and by $i = e + p\tau$,

(55.05) $$(\tau_0 - \tau_1)\frac{p_1 + p_0}{2} = e_1 - e_0.$$

The second of these two important shock relations could be interpreted to mean that the increase in internal energy across the shock front is due to the work done by the mean pressure in performing the compression. The first relation shows that the increase in enthalpy is due to the work done by the pressure difference on the mean volume.

Relations (55.04) and (55.05) are particularly noteworthy since they refer only to thermodynamical quantities. They were first introduced by Hugoniot; relation (55.05) is therefore called the *Hugoniot relation*.

56. Contact discontinuities

The discontinuity conditions (54.08–.10) admit a "trivial" or degenerate solution. If the flux m through the surface of discontinuity is zero, i.e., if no gas crosses it, then we have $v_0 = v_1 = 0$. Hence $u_0 = u_1 = U$, and from (54.09) we infer that $p_0 = p_1$, while (54.10.1) is automatically satisfied. However (55.01) can no longer be deduced from (54.10.3). Such a discontinuity surface is called a *contact surface*. A contact surface moves with the gas and separates two zones of different density (and temperature); but the pressure and flow velocity are the same on both sides. A contact discontinuity may separate not only parts of the same gas but also two different gases.

It is obvious that in reality such a contact surface cannot be maintained for an appreciable length of time; heat conduction between the permanently adjacent particles on either side of the discontinuity would soon make our idealized assumption unrealistic. While gas particles crossing a shock front are exposed to heat conduction for only a very short time, those that remain adjacent on either side of a contact surface are exposed to heat conduction all the time. Hence it is clear that a contact layer will gradually fade out.

The flow velocity is continuous across contact surfaces in one-dimensional flow. However, in flows in more than one dimension, as we shall show in Chapter IV, Section 118, the tangential component of flow velocity may suffer a discontinuity across a contact surface, while the normal component relative to the surface is always zero, as in the case under discussion.

57. Description of shocks

We recall the following definitions given in Section 52. The side of the shock front through which the gas enters the shock front was called the *front side* or the side ahead of the shock front. The other side was called the *back side*. In other words, the particles cross the shock front from the front toward the back side. This definition is independent of the choice of coordinate system. Usually we shall denote the front side of the shock front with the subscript ($_0$) and the back side with ($_1$). We also say that the shock front *faces* the front side or is *directed toward* the front side.

It should be clearly understood that the direction in which the shock front *moves*, given by the sign of U, has nothing to do with the direction toward which it *faces*, i.e., with the distinction between the front and back side of the shock which depends only on the relative velocity v. Whether the front advances, is stationary, or recedes, depends on the absolute velocity.

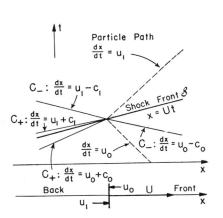
Fig. 31. Shock front.

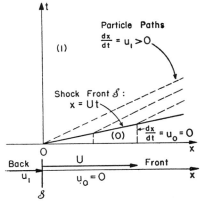
Fig. 32. Impinging shock front.

Pressure, density, temperature, and entropy are, as we shall see in Sections 65 and 67, always greater behind the shock front than ahead of it, and the degree of this increase can be used in various ways to measure the intensity of the shock, see Section 71.

We shall further see, in Sections 65 and 66, that the speed of the gas relative to the shock front, $|u - U|$, is always less behind the shock front than ahead of it. A consequence of this fact is that the velocity at the left side of the shock front is always greater than at the right side, irrespective of which side is the front side and which the back side.

(57.01) $\qquad u_{\text{left}} > u_{\text{right}} \quad \text{or} \quad v_{\text{left}} > v_{\text{right}}$.

We now discuss *three different interpretations of a shock front, all of which are equivalent* by the Galilean principle of relativity.

First, suppose that the velocity u_0 on the front side is zero. Then *the shock front advances into a zone* (0) *of rest* with the velocity U when observed from the front side, which will be shown to be supersonic, while the velocity $U - u_1 = -v_1$ of the shock front observed

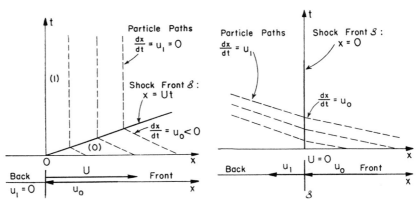

FIG. 33. Receding shock front.   FIG. 34. Stationary shock front.

from the high pressure zone on the back side is subsonic. The shock front moves rapidly into the zone of quiet, enveloping more and more of the gas which, after being overtaken, follows at a speed less than that of the shock front. At the same time the density and pressure are suddenly increased. We have already explained that such an advancing shock wave is generated when a piston is moved into the quiet gas.

Secondly, suppose the velocity u_1 on the back side is zero. Then the shock front may be interpreted as receding with the velocity U leaving behind a high pressure zone of quiet. Such *receding shock waves* will be encountered as shock waves reflected from a wall, see Section 70.

Finally, suppose that the velocity of the shock front is zero, in other words, that the shock front is *stationary*. (Any shock front is stationary if observed from a coordinate system moving with the instantaneous shock front velocity U.) Such a stationary shock front is simply described by a fixed point $x = \xi$ in the tube into which the gas flows at supersonic speed and behind which it is slowed down (to subsonic speed) while pressure and density are increased. The *discontinuity conditions that hold for stationary shocks* ($U = 0$) can be found immediately by putting $v_i = u_i$ in (54.08–.10):

(57.02) $$\rho_0 u_0 = \rho_1 u_1 = m,$$

(57.03) $$\rho_0 u_0^2 + p_0 = \rho_1 u_1^2 + p_1 = P,$$

(57.04) $$\tfrac{1}{2} u_0^2 + i_0 = \tfrac{1}{2} u_1^2 + i_1 = \tfrac{1}{2} \dot{q}^2.$$

58. *Models of shock motion*

Shocks in their different aspects can be visualized through *analogy with the motion of particles* such as a stream of fast automobiles on a highway. A shock can be produced as follows. We assume a steady flow of traffic at high speed. In such a flow there is a "sound speed," i.e. a speed at which small disturbances occurring in the traffic will spread. If the speed of the traveling cars exceeds this sound speed, then a shock occurs when the velocity is suddenly reduced, for example when one driver sees a slow-down sign. The driver of any following car will suddenly see the car in front reduce speed. He cannot transmit a warning signal to the driver in the rear before he also reduces speed. The resulting shock faces backwards. The increase in density is obvious; increase in pressure is also immediately represented in our model if we imagine the row of cars separated by springs or buffers with a nonlinear law of repulsion. An increase in temperature can perhaps be interpreted by means of such models if the energy of small "excitations" is considered as representing heat.

A strictly receding shock wave can be pictured as an extreme case. Let us assume, as before, a long column of equally spaced cars, traveling at supersonic speed, which strikes an unanticipated obstacle that suddenly brings the first car to a full stop. The second will press close to the first and stop; then the third will be abruptly stopped by the second, and so on. The point separating the stopped cars from the moving cars obviously represents a receding shock front.

A forward-facing shock front impinging on a zone of rest is represented by the phenomenon of a column of fast moving cars pounding against a row of widely spaced parked cars and setting them in motion.

Models of one-dimensional wave motion by means of individual particles connected by nonlinear law of repulsion are not only suggestive, but may even be used as approximations to actual situations and thus as a basis for numerical computation, see [58], in cases in which only the first two shock conditions need be considered, see Section 61.

59. Discussion of the mechanical shock conditions

Only the third condition explicitly introduces the thermodynamical nature of the substance represented by the energy e or the enthalpy i as a known function of p and ρ. Hence all conclusions drawn solely from the "mechanical conditions," the first two shock conditions (54.08–.09),

$$\rho_0 v_0 = \rho_1 v_1 = m$$
$$\rho_0 v_0^2 + p_0 = \rho_1 v_1^2 + p_1 = P,$$

are valid for any medium irrespective of its equation of state. This is true of the relations:

(59.01) $$m(v_1 - v_0) = p_0 - p_1$$

(59.02) $$m^2 = -\frac{p_0 - p_1}{\tau_0 - \tau_1}$$

(59.03) $$v_1 v_0 = \frac{p_0 - p_1}{\rho_0 - \rho_1}.$$

Relation (59.01) follows directly from (54.09). Relation (59.02) follows from (59.01) by setting $v_1 = m\tau_1$ and $v_0 = m\tau_0$; relation (59.03) follows from (59.01) by setting $mv_1 = \rho_0 v_0 v_1$ and $mv_0 = \rho_1 v_1 v_0$.

The velocities v_0, v_1 and the mass flux m obviously have the same sign. Relation (59.01) then shows that the pressure p changes in the sense opposite to that in which the relative speed $|v|$ changes.

Relation (59.02) shows that the density changes in the same sense as the pressure.

Anticipating the fact that the shock is compressive, i.e., $\rho_1 > \rho_0$ (a general proof will be given in Section 65; for polytropic gases the fact is established by explicit formulas, as shown in Section 67), we see that the pressure increases and the relative speed $|v|$ decreases as the gas crosses the shock front.

A different symmetric form for the mechanical shock conditions is

(59.04)
$$\tau_0(p_1 - p_0) = v_0(v_0 - v_1),$$
$$\tau_1(p_0 - p_1) = v_1(v_1 - v_0),$$

from which the equivalent relations

(59.05) $\qquad (\tau_0 - \tau_1)(p_1 - p_0) = (v_0 - v_1)^2,$

(59.06) $\qquad (\tau_0 + \tau_1)(p_1 - p_0) = v_0^2 - v_1^2,$

follow.

All the conditions (59.03–.06) clearly separate thermodynamic quantities from velocities.

60. Sound waves as limits of weak shocks

Let us consider a sequence of shocks for a fixed state (0) such that the shock strength measured by $p_1 - p_0$ tends to zero. If we anticipate the result, to be proved in Section 65, that, across a shock front, the entropy change is only of third order in $p_1 - p_0$, then equation (59.03) yields for $p_1 \to p_0$ the relation $v_0 v_1 \to p_\rho = c^2$. Since relation (59.01) entails $v_1 \to v_0$ we see that in the limit the flow speed relative to the shock front is the sound speed. Since we have recognized this fact, at the end of Section 34, as the condition for the propagation of sound waves, given by characteristics in the (x, t)-plane, we have shown: *A sound wave can be interpreted as an infinitely weak shock.*

61. Cases in which the mechanical shock conditions are sufficient to determine the shock

Certain further remarks should be made about the role of the first two shock conditions, the mechanical conditions, in contrast to the thermodynamical condition. There are cases of great practical importance in which the first two conditions alone are sufficient to determine the shock process, namely in the flow of fluids in which *the pressure depends on the density alone* and not, or not noticeably, on the entropy.

Water, for example, is approximately such a fluid, inasmuch as in its pressure-density relation, $p = A\rho^\gamma - B$, the coefficients A and B are approximately independent of the entropy, see Section 3. The same applies to the determination of shocks, or rather *bores, in shallow water* which are characterized by the relation $\bar{p} = A\bar{\rho}^2$, see (19.14).

The third shock relation remains, of course, valid in all these cases but it may be considered merely as a means of determining the energy balance after the problem has been solved.

The internal energy of such fluids splits into two parts, $e = e^{(1)}(\rho) + e^{(2)}(S)$, one depending only on the density, the other only on the entropy, see Section 3. The third shock condition can then be written in the form

$$[e_1^{(2)} - e_0^{(2)}] = -[\tfrac{1}{2}v_1^2 - \tfrac{1}{2}v_0^2 + e_1^{(1)} - e_0^{(1)} + p_1\tau_1 - p_0\tau_0].$$

Since the right-hand side is already determined by the first two shock conditions, one can calculate the increase of $e^{(2)}$, which may be interpreted as energy transformed into heat (or into energy of turbulence in shallow water).

These remarks also apply in general to *weak shocks*, i.e., to shocks for which the excess pressure ratio $\dfrac{p_1 - p_0}{p_0}$ is small. For such weak shocks, as we shall see, the entropy rise is very small, in fact of third order in $\dfrac{p_1 - p_0}{p_0}$, and can therefore be safely neglected, see Section 72.

62. Shock conditions in Lagrangian representation

Some remarks concerning the Lagrangian form of the shock relations will be useful later. If $x(t)$ is the coordinate of a moving particle, $x_0(t)$ referring to a specific "zero"-particle, see Section 18, then any particle is fixed (irrespective of the time) by the Lagrangian coordinate

$h = \int_{x_0}^{x} \rho dx$. With h and t as independent, u and $\tau = \rho^{-1}$ as dependent variables, the differential equations are, see (18.09–.10),

$$\tau_t = u_h, \quad u_t = k^2 \tau_h, \quad \text{with } k = \tau^{-1}c = \rho c$$

and $x_h = \tau$, $x_t = u$. Now let us consider a shock front S moving relative to the gas, enveloping at the time t a particle with the La-

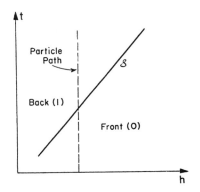

FIG. 35. Motion of a shock front in Lagrangian representation.

grangian coordinate $h = h(t)$. Then if $x(h, t)$ is the position of the particle with the coordinates h and t, the position of the shock front is given by

$$\xi = x(h(t), t),$$

and thus the shock velocity is

$$U = \tau \dot{h} + u.$$

With the symbol $[f]$ for $f_1 - f_0$ we immediately obtain the "kinematic" shock condition

(62.01) $$\dot{h}[\tau] + [u] = 0,$$

which replaces the automatically satisfied condition of conservation of mass. We note that $-\dot{h}$ is the mass crossing the shock front in unit time from the front side to the back side (the cross section is assumed to have unit area). Consequently, the conservation of momentum is expressed by the relation

(62.02) $$[p] - \dot{h}[u] = 0,$$

which by (62.01), in a form invariant under translatory motion, is

(62.03) $$[p] + \dot{h}^2[\tau] = 0,$$

while, from $v = u - U = -\tau\dot{h}$, the conservation of energy is expressed by

(62.04) $$[\tfrac{1}{2}(u - U)^2 + i] = 0$$

or

(62.05) $$\tfrac{1}{2}\dot{h}^2[\tau^2] + [i] = 0.$$

63. Shock relations derived from the differential equations for viscous and heat-conducting fluids

It seems appropriate to supplement the introductory remarks in Section 50 by a brief and somewhat more subtle analysis of how the shock conditions may be obtained by letting the coefficients μ of viscosity and λ of heat conduction approach zero. (The notation μ, λ for these two coefficients is limited to this section only; the factor μ used here is 4/3 times that usually introduced.) The differential equations involving these factors,[*] generalizing equations (17.01–.03) are

(63.01) $$\rho_t + (\rho u)_x = 0$$

(Conservation of mass),

(63.02) $$(\rho u)_t + (\rho u^2 + p - \mu u_x)_x = 0$$

(Conservation of momentum with viscous friction),

(63.03) $$[\rho(\tfrac{1}{2}u^2 + e)]_t + [\rho u(\tfrac{1}{2}u^2 + i) - \mu u u_x - \lambda T_x]_x = 0$$

(Conservation of energy),

(63.04) $$\rho T S_t + \rho u T S_x = \mu u_x^2 + (\lambda T_x)_x$$

(Heat balance).

[*] For a derivation of these equations see Goldstein [19, Volume II, Chapter 14].

The heat balance equation (63.04) *can be derived as a combination of the three conservation laws.* The left-hand side in equation (63.04) is the heat acquired by a unit volume in unit time. The second term on the right-hand side measures the contribution due to heat conduction, while the first term measures the contribution due to viscous friction, which is essentially positive in accordance with the second law of thermodynamics.

Objections have been raised to using the notions of viscosity and heat conduction in describing the internal mechanism of a shock process since the changes of all quantities in the narrow shock zone are so great that these notions are meaningless. It has been proposed instead to use the Boltzmann equation of the kinetic theory of gases. However, whether the notions of viscosity and heat conduction can or cannot be used at least for weak shocks does not seem to have been decided. In any case one may expect that the use of these notions leads to a picture of the situation which is qualitatively correct.

Using viscosity and heat conduction we ought to show that under initial and boundary conditions corresponding to a physical situation the system (63.01–.03) possesses a unique and continuous solution which, as $\lambda \to 0$ and $\mu \to 0$, converges to a solution of the differential equations of non-viscous and non-heat-conducting flow except along discrete lines in the (x, t)-plane. In the neighborhood of these lines the convergence is then non-uniform and the limit solution becomes discontinuous across them. It should be shown, moreover, that the conditions for shocks or contact discontinuities hold across these lines. A proof of these facts would support the idea that the previous theory is an adequate approximation to the physical state; such a proof has not yet been given.

It is nevertheless possible to analyze this passage to a limit in parts by simplifying the problem. We assume our statement concerning convergence and on this assumption we deduce the shock conditions.

We consider a sudden transition in the neighborhood of the point $x = 0$ at the time $t = 0$; with no restriction of generality we can refer the process to a moving coordinate system so that this point is at rest at the time $t = 0$. For simplicity, we assume further that in the neighborhood of $x = 0$, $t = 0$, the process can be considered steady, so that we may set $u_t = \rho_t = S_t = 0$ at $t = 0$ near $x = 0$, and write v instead of u.

The four laws (63.01–.04) then reduce to

(63.05) $$(\rho v)_x = 0,$$

(63.06) $$(\rho v^2 + p - \mu v_x)_x = 0,$$

(63.07) $$[\rho v(\tfrac{1}{2}v^2 + i) - \mu v v_x - \lambda T_x]_x = 0,$$

(63.08) $$\rho v T S_x = \mu v_x^2 + (\lambda T_x)_x.$$

The three conservation laws (63.05–.07) can evidently be integrated. Then they express the constancy of mass, momentum, and energy in the flow process. This possibility leads to the shock conditions in the following way. We integrate the equations (63.05–.07) between $-\epsilon$ and ϵ, where ϵ is arbitrarily small, with the result

(63.09) $$[\rho v]_{-\epsilon}^{\epsilon} = 0,$$

(63.10) $$[\rho v^2 + p - \mu v_x]_{-\epsilon}^{\epsilon} = 0,$$

(63.11) $$[\rho v(\tfrac{1}{2}v^2 + i) - \mu v v_x - \lambda T_x]_{-\epsilon}^{\epsilon} = 0,$$

in which $[f]_{-\epsilon}^{\epsilon}$ denotes the difference $f(\epsilon) - f(-\epsilon)$. For varying values of λ and μ with the limit $\lambda \to 0$, $\mu \to 0$, we consider a sequence of flows which are assumed to converge to a limit flow, except possibly at the point $x = 0$. Relations (63.09–.11), not involving quantities at the point $x = 0$, remain valid in the limit. Thus we obtain for the limit flow

(63.12) $$[\rho v]_{-\epsilon}^{\epsilon} = 0,$$

(63.13) $$[\rho v^2 + p]_{-\epsilon}^{\epsilon} = 0,$$

(63.14) $$[\rho v(\tfrac{1}{2}v^2 + i)]_{-\epsilon}^{\epsilon} = 0.$$

When we now let ϵ approach zero, we obtain the same shock conditions that we found earlier.

Likewise the fourth shock condition of increasing entropy is a consequence of our limiting process. Setting $\rho v = m =$ constant, in accordance with (63.05), and integrating equation (63.08) between $-\epsilon$ and ϵ, we find

(63.15) $$m[S]_{-\epsilon}^{\epsilon} = \int_{-\epsilon}^{\epsilon} \mu \frac{v_x^2}{T} dx + \int_{-\epsilon}^{\epsilon} \lambda \frac{T_x^2}{T^2} dx + \left[\frac{\lambda T_x}{T}\right]_{-\epsilon}^{\epsilon}.$$

For a fixed value of ϵ, the last term on the right-hand side tends to zero as $\mu \to 0$, $\lambda \to 0$; this is not necessarily true of the two other terms. For, these terms are integrals over an interval within which T_x^2 and v_x^2 become very large in the limiting process. Thus the positive contributions on the right-hand side eventually dominate, and we have in the limit

$$[S]_{-\epsilon}^{\epsilon} \geq 0,$$

which is the fourth shock condition. (We have already seen that the equality sign is excluded for an actual discontinuity.) It is remarkable that *this last shock condition, which is independent of the three conservation laws for shocks, results in the limit from a heat balance equation which is dependent on the three conservation laws for continuous flow.*

It should be emphasized again that the approximate description of flow in fluids with almost no viscosity and heat transfer by an idealized flow involving shock fronts, but no viscosity and heat conductivity, is necessarily inadequate in the region near the shock front, where the derivatives of v, ρ, and T become large. A closer analysis of the sudden transition for very small, but not vanishing, values of λ and μ is therefore desirable. It is in particular desirable to determine the *width of the shock zone* from such an analysis.

Let us consider the quantities ρ_0, τ_0, p_0 and ρ_1, τ_1, p_1 on both sides of the shock zone $-\epsilon \leq x \leq \epsilon$ as known and equal to those occurring in the limit case of a shock front. Then solutions (ρ, τ, p) of the three equations (63.05–.07) are to be found which assume the prescribed boundary values. The width 2ϵ of the shock zone is to be determined from the condition that such solutions exist for which the derivatives ρ_x, p_x, v_x vanish at both ends, $x = \pm \epsilon$.

It is doubtful whether or not such solutions exist. Instead, a different procedure can be tried which has been used successfully in other fields, in particular in Prandtl's boundary layer theory for viscous fluid flow, see [19]. The boundary conditions are imposed at $x = -\infty$ and $x = \infty$ instead of at $x = -\epsilon$ and $x = \epsilon$. The shock zone is then somewhat arbitrarily defined as an interval over which noticeable changes of ρ, p, and v take place.

This boundary value problem for the infinite interval has been treated by various authors, see [4, 17, 45], assuming constant coefficients μ and λ, and by L. H. Thomas [46], accounting for the variation of μ and λ with temperature. As a result of this theory, the shock zone has been found to be so narrow that its width is comparable to the mean free path of the gas molecules. This indicates that for a study of such transition zones the theory considering the gas as a continuum is not quite adequate and that a relevant theory will have to fall back on notions of the kinetic theory of gases.

64. Hugoniot relation. Determinacy of the shock transition

For the following considerations it will prove useful to introduce τ and p instead of τ and S as independent variables and to consider the energy as a function $e(\tau, p)$ of τ and p. That this is possible follows from the assumption $g_s > 0$, see (2.07), for the function $p = g(\tau, S)$. Using the *Hugoniot function* with the center (τ_0, p_0),

$$(64.01) \qquad H(\tau, p) = e(\tau, p) - e(\tau_0, p_0) + (\tau - \tau_0)\frac{p + p_0}{2},$$

we can write the Hugoniot relation (55.05) simply in the form

$$(64.02) \qquad H(\tau, p) = 0.$$

It characterizes all pairs of values (τ, p) for the state on one side of the shock front that are compatible with the three shock relations (54.08–.10) when the values (τ_0, p_0) on the other side are given. The graph of the Hugoniot relation in the (τ, p)-plane is called the *Hugoniot curve*, see Figure 36.

For polytropic gases with

$$(64.03) \qquad e = \frac{1}{\gamma - 1} p\tau = \frac{1 - \mu^2}{2\mu^2} p\tau,$$

see (3.04) and (14.06), the Hugoniot function is given by

$$(64.04) \qquad 2\mu^2 H(\tau, p) = (\tau - \mu^2\tau_0)p - (\tau_0 - \mu^2\tau)p_0 ;$$

the Hugoniot curve is therefore a rectangular hyperbola.

The Hugoniot function is very useful in investigating which data determine a shock transition. The three shock relations (54.08–.10) represent three relations between seven quantities τ_0, τ_1, p_0, p_1, u_0, u_1, and U, since e may be considered a given function of τ and p. Hence if three of these quantities are fixed, there is still a one-parameter family of shocks possible.

While the shock relations between the seven quantities are non-linear and thus do not necessarily define the shock if one more quantity

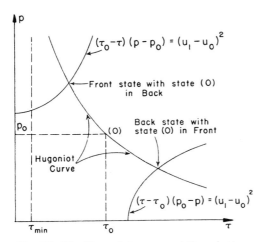

FIG. 36. The Hugoniot curve and the solution of problem (C).

is prescribed, we shall see that under wide conditions the following theorems hold:

(A) *The state (0) on one side of the shock front and the shock velocity U determine the complete state (1) on the other side of the shock front.*

(B) *The state (0) and the pressure p_1 determine the velocity of the shock front and the complete state (1).*

(C) *The state (0) and the velocity u_1 determine the speed of the shock front and the complete state (1) if it is specified whether the state (0) should be ahead of or behind the shock front.*

Note that in cases (A) and (B) the data already determine whether the state (0) is ahead of or behind the shock front. The condition for the former situation, as we shall see, is $|u_0 - U| > c_0$ in case (A), and $p_1 > p_0$ in case (B). In case (C), on the other hand, the data determine already which state is at the left and which at the right-hand side of the shock front, according to the relation, $u_{\text{left}} > u_{\text{right}}$, see (57.01).

The conditions under which the theorems will be proved* may be expressed with reference to the Hugoniot curve $H(\tau, p) = 0$ with the center (τ_0, p_0). They are :

1) Along the Hugoniot curve the pressure p varies from zero to infinity. The values of τ may vary between finite limits, τ_{min} and τ_{max}.

This is, for example, the case for polytropic gases for which τ varies along the Hugoniot curve between $\tau_{min} = \mu^2 \tau_0$ and $\tau_{max} = \mu^{-2} \tau_0$, as seen from (64.04).

2) Along the Hugoniot curve $dp/d\tau < 0$.

3) Every ray through the center (τ_0, p_0) intersects the Hugoniot curve at exactly one point, provided this ray intersects the τ-axis at a point $\tau \leq \tau_{max}$.

Condition 3) is satisfied quite generally, as will be shown in the next section. All three conditions are immediately verified for polytropic gases from the expression (64.04).

If the state (0) and p_1 are given, conditions 1) and 2) insure that there is one and only one value τ_1 satisfying $H(\tau_1, p_1) = 0$. The values $v_0 = m\tau_0$, $v_1 = m\tau_1$ are then found from (59.02). The sign of m depends on whether the shock front is facing to the left $m > 0$, or to the right $m < 0$. The shock velocity is given by $U = u_0 - v_0$. Thus theorem (B) is proved.

For theorems (A) and (C) a further condition is to be imposed, namely

(64.05) $$p_0 < \rho_0^2 v_0^2 (\tau_{max} - \tau_0) \qquad \text{for (A)}$$

(64.06) $$(u_1 - u_0)^2 < p_0(\tau_{max} - \tau_0) \qquad \text{for (C).}$$

These conditions affect, however, only the case in which the state (0) is behind the shock front.

To prove theorem (A) we note that the quantity $-m^2 = -\rho_0^2 v_0^2$, given by the data, equals the ratio $(p_1 - p_0)/(\tau_1 - \tau_0)$ by (59.02). Hence to find τ_1 and p_1 we need only intersect the Hugoniot curve with the ray through (τ_0, p_0) of slope $-m^2$. By condition 3) and (64.05) there is just one such intersection. The value of v_1 is then found from $v_1 = m\tau_1$.

To prove theorem (C) we use relation (59.05) for which the data give the right member $(u_1 - u_0)^2$. Hence, to determine τ_1 and p_1,

* For polytropic gases a different approach is given in Section 68.

we need only intersect the hyperbola $(\tau - \tau_0)(p - p_0) = -(u_1 - u_0)^2$ with the Hugoniot curve. Since the slope of the hyperbola is positive, it follows from condition 2) and (64.06) that there are just two intersections, corresponding to the two possibilities that the state (0) is ahead of or behind the shock front. The flux m is found from (59.02), positive if the shock front faces to the left, negative if it faces to the right. The shock velocity is then found from $U = u_0 - \tau_0 m = u_1 - \tau_1 m$.

65. Basic properties of the shock transition

In the present section we shall establish the four basic properties of the states of the gas at both sides of a shock front. By *shock strength* we mean any of the differences $\rho_1 - \rho_0$, $p_1 - p_0$, or $|v_1 - v_0|$.

I. The increase of entropy across a shock front is of the third order in the shock strength.

II. The rise of pressure, density, and temperature across a shock front differs from reversible adiabatic changes of these quantities at most in terms of the third order in the shock strength. It is here assumed that the initial state and one quantity in the final state are the same for both processes.

III. Shocks are compressive. More precisely: density and pressure rise across the shock front.

IV. The flow velocity relative to the shock front is supersonic at the front side, subsonic at the back side.

For polytropic gases these properties are easily read off from the explicit transition formulas which we shall discuss in Section 67. It is remarkable that, for general ideal gases, these properties depend essentially only on the basic assumptions (2.04–.07) made about the function $p = g(\tau, S)$, namely

(65.01) $\qquad g_\tau < 0, \; g_\tau = -\rho^2 c^2,$

(65.02) $\qquad g_{\tau\tau} > 0,$

(65.03) $\qquad g_S > 0.$

Before establishing these facts* in all generality, we mention that the first part of property IV follows immediately from III if

* Recognized by Bethe [47] and Weyl [48].

instead of (65.02) the somewhat stronger condition $d^2p/d\rho^2 > 0$ is assumed. For, because $S_1 > S_0$ and $g_s > 0$, equation (59.03) yields the relation

$$v_0 v_1 = \frac{p(\rho_1, S_1) - p(\rho_0, S_0)}{\rho_1 - \rho_0} > \frac{p(\rho_1, S_0) - p(\rho_0, S_0)}{\rho_1 - \rho_0}$$

$$= p_\rho(\bar{\rho}, S_0) > p_\rho(\rho_0, S_0) = c^2(\rho_0, S_0) = c_0^2,$$

in which $\bar{\rho}$ is a properly chosen intermediate value between ρ_0 and ρ_1. Hence

(65.04) $$v_0 v_1 > c_0^2.$$

Since $\rho_0 v_0 = \rho_1 v_1$, statement III, $\rho_1 > \rho_0$, leads to $|v_0| > |v_1|$. From (65.04) we then conclude the desired relation $|v_0| > c_0$. It may be noted that this argument has not made any use of the third thermodynamic shock relation.

We now proceed to establish our four statements in general.*

In a rather simple way we first derive the important fact that the distinction between a shock transition and an adiabatic change is only of the third order in the shock strength and hence becomes noticeable only for "strong" shocks. Precisely we shall prove:

I. *The increase of entropy across a shock is of third order in the difference $\tau_1 - \tau_0$ of specific volume, or, what is equivalent, in the difference of the pressures.*

Furthermore: II. Suppose we consider a shock and a reversible adiabatic change from the same initial state to states with the same specific volume. Then *the pressure rise across the shock front agrees with the pressure rise in the adiabatic change up to terms of the second order in the difference $\tau_1 - \tau_0$ of specific volume.* The same is then also true for the rise of the temperature, which is a function of p and τ. Moreover, similar statements may be immediately concluded if instead of $\tau_1 - \tau_0$ the differences $p_1 - p_0$, $T_1 - T_0$, $|v_0 - v_1|$, are introduced as shock strength.

To prove these two statements we study the Hugoniot function

(65.05) $$H(\tau, p) = e(\tau, p) - e(\tau_0, p_0) + \tfrac{1}{2}(\tau - \tau_0)(p + p_0),$$

* Prandtl and Busemann have given an illuminating geometric interpretation of the shock relations using a (v, p)-plane, from which statement IV could also be derived for general fluids, see [3]. For the following compare Weyl [48].

BASIC PROPERTIES OF THE SHOCK TRANSITION 143

see (64.01), and the *Hugoniot curve* $H(\tau, p) = 0$ in the (τ, p)-plane, with "center" (τ_0, p_0). It characterizes the one-parametric family of states (τ, p) which can be reached from (τ_0, p_0) by a shock. We assume that this curve can be represented in the form $p = G(\tau)$ and that $G(\tau) \to \infty$ for $\tau \to \tau_{\min}$, see conditions 1) and 2) in Section 64. Along the Hugoniot curve we have $dH = 0$, hence by (2.01)

(65.06) $\quad 2TdS - (p - p_0)d\tau + (\tau - \tau_0)dp = 0;$

consequently $TdS = 0$ or

(65.07) $\quad\quad\quad\quad dS = 0 \quad$ at (τ_0, p_0).

Differentiating (65.06) once more along the Hugoniot curve, and considering τ as the independent variable we find

$$2d(TdS) + (\tau - \tau_0)d^2p = 0,$$

hence at the center

$$d(TdS) = dTdS + Td^2S = 0,$$

therefore also

(65.08) $\quad\quad\quad\quad d^2S = 0 \quad$ at (τ_0, p_0).

Relations (65.07) and (65.08) show that the change of entropy is at least of third order.

Differentiating once more and then setting $p = p_0$, $\tau = \tau_0$, we obtain

$$2d^2(TdS) + d\tau d^2p = 0 \quad \text{at } (\tau_0, p_0)$$

or, because of (65.06–.07),

$$2Td^3S + d\tau d^2p = 0 \quad \text{at } (\tau_0, p_0).$$

Because $g_{\tau\tau} > 0$, see (65.02), this relation yields

(65.09) $\quad\quad\quad\quad d^3S > 0$ when $d\tau < 0$ at (τ_0, p_0).

Therefore the increase of the entropy is precisely of third order. Thus statement I is proved.

The assertion about the pressure is an immediate consequence. The entropy S is a function of τ along the Hugoniot curve; hence we have

$$p = G(\tau) = g(\tau, S(\tau)).$$

Consequently, by (65.07–.08),

$$G(\tau_0) = g(\tau_0, S_0), \ G'(\tau_0) = g_\tau(\tau_0, S_0), \ G''(\tau_0) = g_{\tau\tau}(\tau_0, S_0),$$

or, geometrically speaking: the Hugoniot curve $p = G(\tau)$ and the adiabatic curve $p = g(\tau, S_0)$ through the center have there a contact of second order. This establishes our second statement.

From (65.07–.09) it follows that the function $S(\tau)$ is monotonic at the center. We now prove that this is true in the large: III. *Along the whole Hugoniot curve the entropy increases with decreasing specific volume.*

For the proof of this statement we make use of an elegant reasoning by H. Weyl [48].

We rewrite the conditions (65.01–.03) in terms of the function $S = S(\tau, p)$. The identity $S = S(\tau, g(\tau, S))$ entails $S_p g_S = 1$; hence by relation (65.03),

(65.10) $$S_p > 0.$$

Furthermore, $0 = S_\tau + S_p g_\tau$ implies, through (65.01),

(65.11) $$S_\tau > 0.$$

Moreover, the convexity condition (65.02) $g_{\tau\tau} > 0$, leads to

(65.12) $$S_{\tau\tau} S_p^2 - 2 S_{\tau p} S_p S_\tau + S_{pp} S_\tau^2 < 0,$$

as we conclude by differentiating the identity $S_p g_\tau + S_\tau = 0$ once more with respect to τ.

Now, to prove the monotone character of S along the Hugoniot curve $H(\tau, p) = 0$, it suffices to show $dS \neq 0$ along this curve except at the center (0), (τ_0, p_0). If S were stationary along the Hugoniot curve at a point (1), (τ_1, p_1), that means, if dS and dH vanish there, simultaneously, the straight chord (0–1) would by (65.06) by tangent to the Hugoniot curve at the point (1). Such tangency is impossible, however, as shown by the following argument. On a ray $\mathcal{R}$ in the (τ, p)-plane, represented in terms of the parameter s by

$$p = p_0 + as, \ \tau = \tau_0 + bs,$$

with

$$a = p_1 - p_0, \ b = \tau_1 - \tau_0,$$

we have $dp = ads$ and $d\tau = bds$; consequently by (65.06) $dH = TdS$.

BASIC PROPERTIES OF THE SHOCK TRANSITION   145

Therefore, if we consider both S and H as functions of s along $\mathcal{R}$, then $S(s)$ and $H(s)$ are simultaneously stationary if one of them is anywhere stationary.

The ray $\mathcal{R}$ cannot coincide with the Hugoniot curve since otherwise a contradiction to the convexity at the center (0) would arise. Hence the fact that $H(s)$ vanishes at the center (0) and at the end-point (1) of the chord implies that $H(s)$ possesses at least one extremum in between. At the point of the extremum $S(s)$ is likewise stationary. This stationary value of S is a maximum; for, at that point

$$S_s = S_\tau b + S_p a = 0$$

or $S_\tau/S_p = -a/b$ and hence S_{ss} is, except for a positive factor, equal to the quantity

$$S_{\tau\tau}S_p^2 - 2S_{\tau p}S_p S_\tau + S_{pp}S_\tau^2,$$

which is negative by (65.12). Therefore S and thus H possess one and only one single stationary point on $\mathcal{R}$ situated between (0) and (1).

From the fact that S has just one maximum between (0) and (1) we infer the inequalities

(65.13.1) $$\frac{dS}{ds} > 0 \quad \text{at (0)}$$

(65.13.2) $$\frac{dS}{ds} < 0 \quad \text{at (1)}.$$

The second inequality excludes the possibility that S could be stationary along the Hugoniot curve at the point (1) because, as we saw, the ray $\mathcal{R}$ would be tangent to the Hugoniot curve at such a point. The relation $dH = 0$ at this point would therefore imply $\frac{dS(s)}{ds} = 0$, in contradiction to (65.13.2). Thus we have proved that the entropy increases along the Hugoniot curve with decreasing specific volume.

Since the entropy increases across a shock by condition (54.11), it now follows that the same is true for the density τ^{-1} and, because of (65.01) and (65.03), also for the pressure. Thus statement III is established.

The *fourth statement* is an immediate consequence of the inequalities (65.13). Because of $S_\tau/S_p = -g_\tau = \rho^2 c^2$, $d\tau/ds = \tau_1 - \tau_0$, $dp/ds = p_1 - p_0$, and (65.10), the inequalities (65.13) assume the form

$$(p_1 - p_0) + \rho_0^2 c_0^2 (\tau_1 - \tau_0) > 0,$$
$$(p_1 - p_0) + \rho_1^2 c_1^2 (\tau_1 - \tau_0) < 0.$$

Suppose now $\tau_1 < \tau_0$; then the two statements can be combined to

$$\rho_0^2 c_0^2 < \frac{p_1 - p_0}{\tau_0 - \tau_1} < \rho_1^2 c_1^2.$$

By (59.02) and (54.08) this relation is equivalent to

(65.14) $$v_0^2 > c_0^2, \; v_1^2 < c_1^2.$$

Thus statement IV is also proved.

66. Critical speed and Prandtl's relation for polytropic gases

In the following sections we shall investigate the special forms of the shock relations for polytropic gases. The thermodynamical shock condition is then particularly simple. The enthalpy of a polytropic gas is

$$i = \frac{\gamma}{\gamma - 1} \frac{p}{\rho} = \frac{1 - \mu^2}{2\mu^2} c^2,$$

with

$$\mu^2 = \frac{\gamma - 1}{\gamma + 1},$$

see (9.06), (14.06); hence the condition (55.02) becomes

(66.01) $$\mu^2 v_0^2 + (1 - \mu^2) c_0^2 = \mu^2 v_1^2 + (1 - \mu^2) c_1^2 = c_*^2.$$

This relation agrees completely with *Bernoulli's law* in the form (14.08), and c_* is the critical speed discussed in Section 15. Owing to this algebraic form of the third shock condition, the relations between the various quantities on both sides of the shock front and the velocity U of the shock front are of a purely algebraic character.

The relation between the relative velocities v_0, v_1 on both sides of the shock front can be put in a very elegant and useful form, due to Prandtl, namely

(66.02) $$v_0 v_1 = c_*^2.$$

This fundamental relation involves velocities only and does not refer explicitly to thermodynamic quantities such as pressure or density.

To prove Prandtl's relation we may, for example, derive from (54.09), (66.01), and $\gamma p = \rho c^2$, see (3.06), the relations

$$\mu^2 P + p_1 = \mu^2 v_1^2 \rho_1 + (1 + \mu^2) p_1 = c_*^2 \rho_1,$$
$$\mu^2 P + p_0 = \mu^2 v_0^2 \rho_0 + (1 + \mu^2) p_0 = c_*^2 \rho_0.$$

Subtracting, we find

$$p_1 - p_0 = c_*^2 (\rho_1 - \rho_0)$$

or

$$c_*^2 = \frac{p_1 - p_0}{\rho_1 - \rho_0},$$

and relation (66.02) follows by (59.03).

Prandtl's relation is evidently equivalent to the transition formula

(66.03) $$\frac{c_*}{v_1} + \frac{v_1}{c_*} = \frac{c_*}{v_0} + \frac{v_0}{c_*},$$

provided v_1 and $v_0 \neq 0$.

Incidentally, Prandtl's relation exhibits the fact that a *shock front approaches a sound wave as its strength approaches zero.* For if $v_0 = v_1$, (66.02) illustrates that both v_0 and v_1 have the common value c_*; since $c = c_*$ a weak discontinuity therefore progresses approximately with sound speed. Of course this fact is in agreement with the principle that disturbances occurring not in the quantities u and ρ, but only in their derivatives, are propagated along characteristics.

As an immediate consequence of Prandtl's relation we recognize the fact that *the speed of the gas relative to the shock front is supersonic on the front side, subsonic on the back side of the shock front*, in agreement with the general results of Section 65.

Formula (66.02) shows that $|v_0| > |v_1|$ implies $|v_0| > c_*$ and $|v_1| < c_*$, and our assertion follows immediately from the basic properties of the critical speed c_*, which for the present case can be read off from (66.01) when written in the form

$$(1 - \mu^2)(v_0^2 - c_0^2) = v_0^2 - c_*^2, \quad (1 - \mu^2)(v_1^2 - c_1^2) = v_1^2 - c_*^2.$$

67. Shock relations for polytropic gases

From the three standard shock relations various other relations can be derived for the quantities on both sides of a shock.

The Hugoniot relation $H(\tau_1, p_1) = 0$ for polytropic gases assumes by (64.04) the form $(\tau_1 - \mu^2 \tau_0)p_1 - (\tau_0 - \mu^2 \tau_1)p_0 = 0$, and thus yields the important formula

$$(67.01) \qquad \frac{p_1}{p_0} = \frac{\tau_0 - \mu^2 \tau_1}{\tau_1 - \mu^2 \tau_0} = \frac{\rho_1 - \mu^2 \rho_0}{\rho_0 - \mu^2 \rho_1}.$$

Inverting this relation we have

$$(67.02) \qquad \frac{\tau_0}{\tau_1} = \frac{\rho_1}{\rho_0} = \frac{p_1 + \mu^2 p_0}{p_0 + \mu^2 p_1}.$$

Relation (67.02) shows, as was already mentioned in Section 64, that the compression $\frac{\rho_1}{\rho_0}$ is always restricted to the range

$$(67.03) \qquad \mu^2 < \frac{\rho_1}{\rho_0} < \frac{1}{\mu^2};$$

so that the compression is never more than μ^{-2}-fold. For $\gamma = 1.4$ the density compression is therefore always less than 6-fold and for $\gamma = 1.2$ the limit of the compression ratio is 11.

The change of temperature and entropy across the shock is immediately read off from (67.02) by using $T_1/T_0 = p_1\tau_1/p_0\tau_0$ and $S_1 - S_0 = c_v \log p_1\tau_1^\gamma/p_0\tau_0^\gamma$, see (3.01) and (3.23). It is easy to verify that both S and T increase through the shock.

From the mechanical shock relations in Section 59 we can derive relations among the velocities, the pressures, and the density on only one side. From (59.02) we find by (67.01)

SHOCK RELATIONS FOR POLYTROPIC GASES

(67.04) $$m^2 = \frac{p_1 + \mu^2 p_0}{(1 - \mu^2)\tau_0} = \frac{p_0 + \mu^2 p_1}{(1 - \mu^2)\tau_1},$$

whence from (59.01)

(67.05)
$$(u_1 - u_0)^2 = (p_1 - p_0)^2 \frac{(1 - \mu^2)\tau_0}{p_1 + \mu^2 p_0}$$
$$= (p_1 - p_0)^2 \frac{(1 - \mu^2)\tau_1}{p_0 + \mu^2 p_1}.$$

A particularly simple relation holds between the pressure ratio $\frac{p_1}{p_0}$ and the *Mach number*

(67.06) $$M_0 = \frac{|v_0|}{c_0}$$

of the gas flow, see (10.01). From (67.04) and $\rho_0 c_0^2 = \gamma p_0$ we have

$$p_1 + \mu^2 p_0 = (1 - \mu^2)\rho_0 v_0^2 = \gamma(1 - \mu^2)p_0 M_0^2$$

or

(67.07) $$\frac{p_1}{p_0} = (1 + \mu^2)M_0^2 - \mu^2.$$

Shock relations which involve only the particle and sound velocities can most easily be derived from Prandtl's relation (66.02). By substituting $v_i = u_i - U$ in (66.02) and using (66.01) we find

(67.08)
$$(u_0 - U)(u_1 - U) = \mu^2(u_0 - U)^2 + (1 - \mu^2)c_0^2$$
$$= \mu^2(u_1 - U)^2 + (1 - \mu^2)c_1^2.$$

Hence we obtain the relation

(67.09) $$(1 - \mu^2)(U - u_0)^2 - (u_1 - u_0)(U - u_0) = (1 - \mu^2)c_0^2,$$

which represents a *quadratic relation* for $U - u_0$ if u_1 and c_0 are given; it is equivalent to

(67.10) $$\frac{u_1 - u_0}{c_0} = (1 - \mu^2)\left(\frac{U - u_0}{c_0} - \frac{c_0}{U - u_0}\right),$$

a form which is occasionally useful.

68. The state on one side of the shock front in a polytropic gas determined by the state on the other side

The various formulas already derived can be used to determine the shock transition completely if the state on one side is given, and in addition one other quantity such as the shock velocity U, or the pressure, or the velocity, on the other side. We thus obtain for polytropic gases a confirmation of theorems (A), (B), and (C) stated in Section 64. Instead of following the steps of Section 64 we use an alternative procedure involving the sound speeds instead of the densities.

A) Given p_0, ρ_0, u_0, U. First calculate $v_0 = u_0 - U$, next $M_0 = |v_0|/c_0$, and then p_1 from (67.07). Further determine $c_*^2 = \mu^2 v_0^2 + (1 - \mu^2) c_0^2$, and thereby

$$v_1 = \frac{c_*^2}{v_0}.$$

Finally find c_1^2 from

$$\mu^2 v_1^2 + (1 - \mu^2) c_1^2 = c_*^2.$$

Relation (67.10)

$$\frac{u_1 - u_0}{c_0} = (1 - \mu^2)\left(\frac{U - u_0}{c_0} - \frac{c_0}{U - u_0}\right)$$

serves as a check.

B) Given p_0, ρ_0, u_0 at the front of a backward-facing shock (which implies $v_0 > 0$) and $p_1 > p_0$. First find M_0^2 from (67.07), then $v_0 = M_0 c_0$, $U = u_0 - v_0$, $c_*^2 = \mu^2 v_0^2 + (1 - \mu^2) c_0^2$ and continue as for A).

C) Given p_0, ρ_0, u_0 at the front side of a backward-facing shock and $u_1 < u_0$. First find U by solving the quadratic equation (67.09) for $v_0 = u_0 - U > 0$. Then continue as for A). The relation $u_1 - U = c_*^2/v_0$ serves as a check.

69. Shock resulting from a uniform compressive motion

We have seen earlier, in Section 45, that the flow produced by a piston in a tube that is suddenly withdrawn with constant velocity, can be described in terms of a centered simple rarefaction wave.

SHOCK FROM A UNIFORM COMPRESSIVE MOTION 151

If, on the other hand, the piston is suddenly pushed into the tube with constant velocity, the resulting flow involves a shock, see Section 48. Mathematically speaking, we are faced with a mixed boundary initial value problem, see Section 38: $u = 0$, $p = p_0$, $\rho = \rho_0$ are prescribed for $t = 0$ on the semi-axis $x > 0$, while the condition $u = u_P$ is imposed on the line $x = u_P t$ representing the motion of the piston with the constant velocity u_P. A solution of the differential equations of flow (34.01) satisfing these conditions should be found; however, if $u_P > 0$ no such solutions exist unless discontinuities are admitted. A flow involving a shock discontinuity and satisfying all conditions can now be easily described as in Section 53 and Figure 30.

A shock of constant strength moves with constant velocity into the gas at rest and the gas behind the shock front is in a steady state. The gas velocity u behind the shock front then equals the given piston velocity, $u = u_P$. Theorem (C), see Sections 64 and 68, shows that under these circumstances the shock and the state behind it are completely determined for an arbitrary value of the piston velocity $u_P > 0$. For non-polytropic gases the same is true under very general conditions. That the preceding solution of the problem is the only one mathematically possible requires a mathematical proof, which is omitted here.

Since $u_1 = u_P$, $u_0 = 0$, we find U for polytropic gases from equation (67.09),

$$(69.01) \qquad U = \frac{1}{2}\frac{u_P}{1-\mu^2} + \sqrt{c_0^2 + \frac{1}{4}\left(\frac{u_P}{1-\mu^2}\right)^2}.$$

Clearly, the shock velocity U is greater than c_0 and greater than $u_P/(1 - \mu^2)$. The latter observation shows, for example, that for air with $\gamma = 1.4$, $\mu^2 = 1/6$, the shock is at least 20% faster than the piston.

With the shock velocity thus determined, the description of the basic compressive piston motion is shown to be consistent. Although we have given no proof of uniquenes, i.e., we have not mathematically excluded the possibility of other flow patterns, we accept the preceding reasoning as a satisfactory theory for the interpretation of actual phenomena observed under circumstances resembling our idealized model. Having obtained U, we find by the procedure A) of Sections 64 and 68 the pressure p_1, the sound speed c_1, and the density ρ_1 in the zone adjacent to the piston.

For a *high speed* u_P of the oncoming piston, i.e., for $u_P/c_0 \gg 1$, we have, by (69.01), (67.07), and (67.02),

$$(69.02) \qquad U \sim \frac{u_P}{1 - \mu^2},$$

$$(69.03) \qquad \frac{p_1}{p_0} \sim \frac{1 + \mu^2}{(1 - \mu^2)^2} \frac{u_P^2}{c_0^2},$$

$$(69.04) \qquad \frac{\rho_1}{\rho_0} \sim \frac{1}{\mu^2}.$$

70. Reflection of a shock on a rigid wall

We shall now discuss a point of great importance, the reflection of a shock. Suppose an oncoming column of gas of constant velocity u_1 behind a shock front impinges on a zone of quiet bounded by a rigid wall. Then the ensuing physical phenomenon can be described as a reflection of the shock wave from the wall, and can be represented mathematically by piecewise constant solutions of the differential

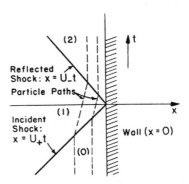

Fig. 37. Reflection of a shock wave on a rigid wall.

equations, satisfying the shock conditions across the *incident* shock wave and the *reflected* shock wave. Under the impact of the incident shock wave the zone (0) of quiet next to the wall shrinks to zero, say at $t = 0$; then a reflected shock starts in the opposite direction and in turn leaves a growing zone of quiet between itself and the wall. The situation can best be grasped from a diagram in the (x, t)-plane. State (0) is a zone of quiet characterized by the quantities $u_0 = 0$, ρ_0, p_0, c_0. In the state (1) following the incident shock we have $u = u_1$; in the state (2) adjacent to the wall we again have rest, $u = u_2 = 0$, but new values ρ_2, p_2, c_2. Our aim is to find the state (2) from the data ρ_0, p_0, $u_0 = 0$, u_1.

To this end we note that the pattern tentatively assumed in Figure 37 shows a state (1) with flow velocity u_1 and sound speed c_1 connected through a shock with a zone of rest (0) and through another shock with a zone of rest (2). U_+ is the velocity of the incident, U_- the velocity of the reflected shock; then according to equation (67.08), both these velocities satisfy the same quadratic equation

$$(U - u_1)^2 + \frac{(U - u_1)u_1}{1 - \mu^2} - c_1^2 = 0,$$

or the two numbers $M_+ = (u_1 - U_+)/c_1 < 0$ and $M_- = (u_1 - U_-)/c_1 > 0$ satisfy the quadratic equation

(70.01) $$M^2 - (1 - \mu^2)^{-1} c_1^{-1} u_1 M - 1 = 0,$$

so that

(70.02) $$M_+ M_- = -1.$$

Moreover, the pressure relations following from (67.07) are

(70.03) $$\frac{p_0}{p_1} = (1 + \mu^2) M_+^2 - \mu^2, \qquad \frac{p_2}{p_1} = (1 + \mu^2) M_-^2 - \mu^2;$$

using (70.02), we obtain for the *reflected pressure ratio*

(70.04) $$\frac{p_2}{p_1} = \frac{(2\mu^2 + 1)\frac{p_1}{p_0} - \mu^2}{\mu^2 \frac{p_1}{p_0} + 1}$$

and for the ratio of excess pressures,

(70.05) $$\frac{p_2 - p_0}{p_1 - p_0} = 1 + \frac{1 + \mu^2}{\frac{p_0}{p_1} + \mu^2}.$$

This is the basic relation for the important phenomenon of *reflection*. In a "sonic reflection" resulting from linear wave motion this ratio has the value 2, showing that the excess pressure after reflection is simply doubled. Here we find a totally different situation; in particular, if the incident shock is *strong*, i.e., one for which the ratio $\frac{p_1}{p_0}$ is large, we find

$$(70.06) \qquad \frac{p_2 - p_0}{p_1 - p_0} \sim 2 + \frac{1}{\mu^2} = \begin{cases} 8 \text{ for } \gamma = 1.4, \\ 13 \text{ for } \gamma = 1.2, \\ 23 \text{ for } \gamma = 1.1. \end{cases}$$

Thus, *reflection of strong shocks results in a considerable increase of the pressure at the wall*, a fact obviously of major importance.

For a *weak* incident shock, $\frac{p_1}{p_0} - 1$ is small, and we find from relation (70.05),

$$\frac{p_2 - p_0}{p_1 - p_0} \sim 2$$

in agreement with the case of sonic reflection.

71. Shock strength for polytropic gases

It is convenient for various considerations to introduce the notion of *shock strength*. Several parameters can be offered as measures for the strength of a shock:

the *excess pressure ratio* $\quad \dfrac{p_1 - p_0}{p_0},$

the *condensation* $\quad \dfrac{\rho_1 - \rho_0}{\rho_0},$

the parameters $\quad \dfrac{|u_1 - u_0|}{c_0} \quad$ or $\quad M_0^2 - 1.$

$M_0 = |v_0|/c_0$ is the *Mach number* of the incoming flow relative to the shock front, see (10.01). We write down the relations between these quantities for polytropic gases.

The relationship between these various interpretations of strength

SHOCK STRENGTH FOR POLYTROPIC GASES

is contained in the formulas of Section 67. In particular we have from (67.01)

(71.01) $$\frac{p_1 - p_0}{p_0} = \frac{1 + \mu^2}{1 - \mu^2 \frac{p_1}{p_0}} \frac{\rho_1 - \rho_0}{\rho_0} = \frac{\gamma}{\left(\frac{\rho_1 - \rho_0}{\rho_0}\right)^{-1} - \frac{\gamma - 1}{2}},$$

and from (67.02)

(71.02) $$\frac{\rho_1 - \rho_0}{\rho_0} = \frac{1 - \mu^2}{1 + \mu^2 \frac{p_1}{p_0}} \frac{p_1 - p_0}{p_0} = \frac{1}{\gamma\left(\frac{p_1 - p_0}{p_0}\right)^{-1} + \frac{\gamma - 1}{2}}$$

and

(71.03) $$\frac{\tau_0 - \tau_1}{\tau_0} = \frac{1 - \mu^2}{\frac{p_1}{p_0} + \mu^2} \frac{p_1 - p_0}{p_0}$$

From (67.07) we have the particularly simple formula

(71.04) $$\frac{p_1 - p_0}{p_0} = (1 + \mu^2)(M_0^2 - 1),$$

and from (67.05) and (67.07),

(71.05) $$\frac{|u_1 - u_0|}{c_0} = \frac{1 - \mu^2}{\sqrt{1 + \mu^2}} \frac{p_1 - p_0}{p_0} \sqrt{\frac{p_0}{p_1 + \mu^2 p_0}}$$

$$= (1 - \mu^2) \frac{M_0^2 - 1}{M_0}.$$

Of special importance are the extreme cases of *strong* and *weak* shocks. *Strong shocks* may be characterized by the condition that p_1/p_0 or M_0 is very large. From (67.02) or (71.02) we see that the density ratio ρ_1/ρ_0 approaches a finite limit as $p_1/p_0 \to \infty$, that is,

(71.06) $$\rho_1/\rho_0 \to 1/\mu^2 = 6 \text{ for } \gamma = 1.4.$$

The pressure ratio p_1/p_0 increases as the square of the Mach number according to (67.07), or (71.04), that is,

(71.07) $$(p_1/p_0)/M_0^2 \to 1 + \mu^2.$$

For the ratio of the sound velocities we then find

(71.08) $$(c_1/c_0)/M_0 \to \mu \sqrt{1 + \mu^2},$$

since $c_1^2/c_0^2 = p_1\rho_0/p_0\rho_1$. For the velocity difference $u_1 - u_0$ we find from (71.05)

(71.09) $$\frac{|u_1 - u_0|}{c_0} \Big/ M_0 \to (1 - \mu^2).$$

72. Weak shocks. Comparison with transitions through simple waves

Here is the place for an important theorem comparing the discontinuous transition through a shock with a continuous transition through a simple wave:

Suppose a shock and a simple wave transform a gas in the initial state given by (τ_0, p_0, u_0) into the states given by (τ, p, u) and (τ^*, p^*, u^*) respectively. We measure the shock strength by any of the three differences $\tau - \tau_0$, $p - p_0$, or $u - u_0$ and accordingly speak of a simple wave and a shock wave as being of equal strength if $\tau^* = \tau$, or $p^* = p$, or $u^* = u$. We then state the *Theorem*: *For a shock transition and a simple wave transition with the same initial state and the same strength the quantities τ^* and τ, p^* and p, u^* and u agree up to second order in the shock strength and differ in third order.*

To establish the theorem it is obviously sufficient to prove it for one of the three definitions of the strength. For this purpose, we choose the difference $\tau - \tau_0$.

This theorem implies two facts: both the *change of the entropy* and the *change of one of the Riemann invariants through a shock are of the third order*, since both quantities are constant across a simple wave (specifically the appropriate Riemann invariant is s for a forward-facing and r for a backward-facing shock front). The first of these facts was proved already in Section 65 and we can use it now to prove immediately the part of our theorem concerning pressures. However, the part concerning velocities does not follow from the condition on the entropy, and for it we need a more refined argument.

To determine the change in pressure we insert in the expansion

$$p = p_0 + g_\tau(\tau_0, S_0)(\tau - \tau_0) + g_S(\tau_0, S_0)(S - S_0)$$
$$+ \tfrac{1}{2} g_{\tau\tau}(\tau_0, S_0)(\tau - \tau_0)^2 + \cdots$$

the appropriate expansion of $S - S_0$ in terms of $\tau - \tau_0$; the terms of first and second order in $\tau - \tau_0$ are not affected and hence are the same as for a simple wave, across which $S = S_0$ is constant. Thus the pressures agree up to second order, see statement II in Section 65.

The part of the theorem concerning velocity is found from the mechanical shock relation (59.05)

$$(p - p_0)(\tau - \tau_0) = (u - u_c)^2.$$

Differentiating this relation successively three times with respect to τ and then setting $\tau = \tau_0$, $p = p_0$, $u = u_0$, we obtain the relations

(72.01) $\qquad dpd\tau = (du)^2,$

(72.02) $\qquad d^2pd\tau = 2d^2udu,$

in which the differentials are to be taken for $\tau = \tau_0$. Since $dp = g_\tau(\tau_0, S_0)d\tau = -\rho_0^2 c_0^2 d\tau$ and $d^2p = \frac{1}{2}g_{\tau\tau}(\tau_0, S_0)d\tau^2$ are already determined, du and d^2u can be calculated from (72.01) and (72.02). The sign of du depends on whether the shock is forward- or backward-facing (in the former case $du/d\tau$ is positive and in the latter negative).

Now consider a set of simple waves with the same state (τ_0, p_0, u_0) ahead of it, and the value τ at its end. Such a set is simply furnished by the sections of one simple wave between the straight head characteristic carrying τ_0, p_0, u_0 and the characteristic carrying the value τ. The dependence of p and u on τ is then the same as along a cross-characteristic. Consequently, relation $dpd\tau = (du)^2$ holds, see (34.05), and by differentiation with respect to τ also $d^2pd\tau = 2d^2udu$. These two relations taken for $\tau = \tau_0$ agree with the relations (72.01–.02) for shocks. Therefore du and d^2u at $\tau = \tau_0$ for simple waves also agree with those for shocks. In other words, the expansions of u in powers of $\tau - \tau_0$ for shock and simple wave are the same up to and including terms of second order.

It can easily be verified by calculation that the pressure and velocity changes for shock and simple wave actually do differ in the terms of third order.

Our theorem remains valid, of course, if we compare shock and simple wave transitions which end up with the same value of the pressure p or the velocity u, instead of τ, see Figure 38. The expansions of all quantities mentioned in powers of $p - p_0$ or $u - u_0$ respectively agree up to terms of second order.

We consider now expansions in powers of $u - u_0$ instead of $\tau - \tau_0$. For a shock wave, these are the same up to and including terms of

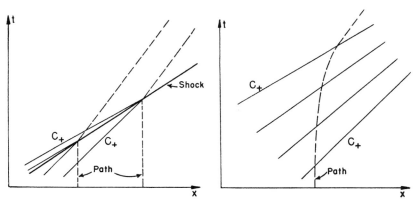

Fig. 38. Shock and simple compression wave starting with the same state ahead and producing the same increase in velocity.

second order as those given for simple waves in Section 40, see (40.10–.11), in particular, for a forward-facing shock front in a polytropic gas,

$$c = c_0 + \frac{\gamma - 1}{2}(u - u_0) + \cdots,$$

$$u + c = u_0 + c_0 + \frac{\gamma + 1}{2}(u - u_0) + \cdots,$$

$$u - \frac{2}{\gamma - 1}c = u_0 - \frac{2}{\gamma - 1}c_0 + \cdots,$$

(72.03)
$$p = p_0 + \rho_0 c_0 (u - u_0)$$
$$\qquad + \frac{\gamma + 1}{4}\rho_0(u - u_0)^2 + \cdots,$$

$$\tau = \tau_0 - \tau_0 c_0^{-1}(u - u_0)$$
$$\qquad + \frac{\gamma + 1}{4}\tau_0 c_0^{-2}(u - u_0)^2 + \cdots,$$

the dots indicating terms of third order.

There is one important quantity connected with a shock which is not affected by our theorem, the shock velocity U. By inserting the expansion of p in the mechanical shock relation (59.02) written in the form

$$(72.04) \qquad (u_0 - U)^2 = -\tau_0^2 \frac{p - p_0}{\tau - \tau_0}$$

one could easily obtain the expansion of U in powers of $\tau - \tau_0$. Here we give instead only the expansion of U for polytropic gases in powers of $u - u_0$. The first order term could be obtained by inserting in (72.04) the last two expansions (72.03). To derive the full expansion it is more convenient to use relation (67.10) which gives

$$(72.05) \qquad \begin{aligned} U = u_0 + c_0 + \frac{\gamma + 1}{4}(u - u_0) \\ + \frac{(\gamma + 1)^2}{32} \frac{(u - u_0)^2}{c_0} + \cdots \end{aligned}$$

for a forward-facing shock front. For later purposes the expansion of U in powers of $u + c - u_0 - c_0$ is useful; using (72.03) we obtain

$$(72.06) \qquad \begin{aligned} U = u_0 + c_0 + \tfrac{1}{2}(u + c - u_0 - c_0) \\ + \frac{1}{8}\frac{(u + c - u_0 - c_0)^2}{c_0} + \cdots. \end{aligned}$$

We see from this formula that *the velocity of a forward-facing shock front is in first order just the mean value $\tfrac{1}{2}(u_0 + c_0 + u + c)$ of the velocities of the forward-facing sound waves ahead of and behind the shock front.*

The inaccuracy resulting from substituting the simple wave relations for the shock relations is very small even for shocks with an excess pressure ratio $(p_1 - p_0)/p_0$ as large as 1.5, see [54]. For a shock of this strength, in a polytropic gas with $\gamma = 1.4$,

$$\frac{u_1 - u_0}{c_0} = .71, \qquad \frac{c_1 - c_0}{c_0} = .15,$$

$$\frac{c_1 - c_0}{c_0} - \frac{\gamma - 1}{2}\frac{u_1 - u_0}{c_0} = .01,$$

while for a simple wave of the same strength

$$\frac{u_1 - u_0}{c_0} = .70, \qquad \frac{c_1 - c_0}{c_0} = .14,$$

$$\frac{c_1 - c_0}{c_0} - \frac{\gamma - 1}{2}\frac{u_1 - u_0}{c_0} = 0.$$

The shock velocity U is given by

$$\frac{U - u_0}{c_0} = 1.51,$$

while formula (72.06) gives

$$\frac{U - u_0}{c_0} = 1.52.$$

This indicates that even for shocks with a strength somewhat greater than $(p_1 - p_0)/p_0 = 1.5$ the method of substituting the simple wave relations for the shock relations and using formula (72.06) for the shock velocity would be accurate enough for most purposes.

73. Non-uniform shocks

In the shock wave motions discussed earlier in Sections 53 and 57 the situation is very simple since the states on both sides of the shock front are constant. This implies constant speed and strength of the shock front. In the (x, t)-plane such a shock front is represented by a straight "shock path" whose slope with respect to the t-axis is the constant shock velocity U. Frequently, however, the states on the two sides are not both constant, but must be described by non-constant solutions of the flow equations. Then the shock front does not have a constant velocity, that is, the shock path in the (x, t)-plane is *curved*. In general the entropy change through such shocks also varies. Thus, even if the state ahead of the shock front is of uniform entropy, the state behind it is not and we are forced to use the differential equations (34.06) for non-isentropic flow. This mathematical complication precludes an explicit theory although calculations in specific problems are quite feasible. Fortunately, in many cases of practical importance, the shock is weak or of moderate strength so that changes in entropy may justifiably be neglected; then numerical treatment of the problems is much easier. In such cases we could use the simpler differential equations for isentropic flow, operating solely with the first two mechanical shock conditions and disregarding the third. A still simpler approximate treatment will be discussed in the next section.

74. Approximate treatment of non-uniform shocks of moderate strength

For non-uniform shocks which are weak, or of moderate strength, the results of Section 72 suggest replacing the shock transition relations by those for a transition through a corresponding simple compression wave. In other words, we might assume the entropy and the appropriate Riemann invariant to remain unchanged through

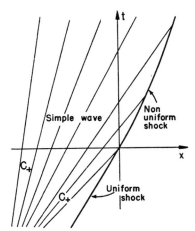

Fig. 39. Shock modified by a simple wave overtaking it.

the shock. An approximate treatment based on this procedure was first used by Chandrasekhar [54], and developed quite generally later on, see [55]. It is especially easy to handle if the flow ahead of the curved shock front is constant, e.g. in a state of rest. For the most part we shall confine ourselves to this case.

The most natural situation involving a non-uniform shock arises if a forward-facing uniform shock front with constant states ahead of and behind it is overtaken from behind by a forward-facing simple wave and subsequently modified, see Figure 39. (This is one of the problems of wave interaction which will be discussed from a different point of view in Part D.) Assuming the shock to be of moderate strength and adopting the procedure indicated, we assign to the wave flow behind the modified shock front a constant entropy and a constant Riemann invariant s. A flow with constant entropy and constant s is a forward-facing simple wave according to the theory of Chapter II, see Section 29. The C_+-characteristics, along which r is also constant, are straight. Consequently, this simple wave is just the continuation of the incident simple wave. Thus, within the approximation the shock does not influence the simple wave. It is for this reason, as we shall see, that the influence of the simple wave on the shock is so easily determined.

It is sufficient to describe the state of the gas by giving the particle velocity u and the sound speed c. We assume $u = u_0 = 0$, $c = c_0$ ahead of the shock front and suppose that it is overtaken at the time $t = 0$ at the place $x = 0$. Then the simple wave behind the shock front can be described by equations

(74.01) $$x = \xi + \omega(\xi)t,$$

(74.02) $$\omega(\xi) = u(\xi) + c(\xi),$$

for $\xi < 0$, in which $u(\xi)$ and $c(\xi)$ should satisfy the relation $u(\xi) - l(\xi) = -l_0$, see (40.01) and (49.01). Here l is a known function of c_0 since the entropy is assumed to be the same on both sides of the shock front, see Sections 40 and 49. From the simple form (40.09), which this relation has for polytropic gases, we derive

(74.03) $$u(\xi) = (1 - \mu^2)(\omega(\xi) - c_0),$$
$$c(\xi) - c_0 = \mu^2(\omega(\xi) - c_0),$$

see (40.06). The path of the shock front can now be described parametrically by giving t as a function of ξ and inserting this function in (74.01). Differentiating relation (74.01) with respect to ξ along the shock path and using $dx = U dt$ along this path we find the relation

(74.04) $$[U(\xi) - \omega(\xi)]\frac{dt}{d\xi} - \frac{d\omega(\xi)}{d\xi} t = 1,$$

which is a *linear differential equation* for the function $t = t(\xi)$, once U as a function of ξ is known. Now the shock velocity U depends on the state of the gas at both sides of the front. We use formula (72.06), writing it in the form

(74.05) $$U(\xi) = c_0 + \tfrac{1}{2}[\omega(\xi) - c_0] + \frac{1}{8c_0}[\omega(\xi) - c_0]^2.$$

Inserting this relation into (74.04) yields the differential equation

(74.06) $$\left\{\tfrac{1}{2}(\omega - c_0) - \frac{1}{8c_0}(\omega - c_0)^2\right\}\frac{dt}{d\xi} + \frac{d\omega}{d\xi}t = -1,$$

which is to be solved under the initial condition $t = 0$ for $\xi = 0$. Setting

(74.07) $$\omega(\xi) = c_0 + c_0\sigma(\xi),$$

we find the solution explicitly as

(74.08) $$c_0 t = 8\left(\frac{4 - \sigma(\xi)}{\sigma(\xi)}\right)^2 \int_\xi^0 \frac{\sigma(\eta)\,d\eta}{[4 - \sigma(\eta)]^3}.$$

The condition that the shock velocity is supersonic relative to the gas in front can be expressed by $0 < U - c_0 = \tfrac{1}{8}(\omega - c_0)(4 + \sigma)$, which shows that $\sigma > 0$ or $\omega > c_0$ behind the shock front. On the other hand, the condition that the shock velocity is subsonic relative to the gas behind it, as expressed by $0 < \omega - U = \tfrac{1}{8}(\omega - c_0)(4 - \sigma)$, shows that $\sigma < 4$ or $\omega - c_0 < 4c_0$. (As a matter of fact $|\sigma|$ is small compared with 4 whenever our approximation can be used; for otherwise the term of second order in (74.05) would not be small compared with the first order term.)

The function $t(\xi)$, defined as long as $\sigma > 0$, and with it the function $x(\xi)$ obtained from (74.01) describe the path of the shock front as it cuts through the simple wave region.

If the simple wave is compressive the shock front is accelerated and, consequently, increases in strength; if the simple wave is expansive the shock front is decelerated and, consequently, decreases in strength.

To prove this we first observe that $d\omega(\xi)/d\xi > 0$ in a forward-facing rarefaction wave, as follows from the statements made in Section 40; consequently $dU(\xi)/d\xi > 0$ by (74.05). Secondly we note that $d\xi/dt < 0$ as follows from (74.06) and $0 < \sigma < 4$ (for compressive waves $d\xi/dt$ might become infinite and change sign, but then our approximation can no longer be used).

75. Decaying shock wave. N-wave

Of particular importance is the decay of the shock wave produced when a shock impinging on a zone of rest is overtaken by a rarefaction wave which stays indefinitely in contact with the shock front. This is the case, as we shall see, if the gas velocity behind the rarefaction wave vanishes. From our assumption that the gas in the rarefaction wave has the same entropy and the same Riemann invariant s as the gas ahead of the shock front, it follows that the sound speed and the pressure behind the rarefaction wave are the same as those ahead of the shock front.

It is of great interest to study the *asymptotic* behavior of the position of the shock front and the distribution of u, c, and p in the wave zone behind it, i.e. their behavior for large times t. It is not at all obvious whether the *width of the shock wave*, i.e. the distance between the shock front and the tail of the rarefaction wave, increases, decreases, or approaches a finite value different from zero, as the time increases. We shall show that *the width of the shock wave increases like the square root of the time t*.

We observe from (74.08) that t approaches infinity when σ approaches 0, or ω approaches c_0. Thus the shock front gradually traverses the part of the simple wave in which the pressure and density are greater than ahead of the shock front. It will not penetrate into parts of the simple wave in which the pressure and density are less than ahead of it. Thus, if the pressure at the tail of the rarefaction wave is greater than the pressure ahead of the shock, the shock traverses the rarefaction wave and then continues on a straight path. In this case the width of the shock wave decreases to zero in a finite time.

To investigate the asymptotic behavior of the shock we consider the case in which the pressure at the tail of the simple rarefaction wave just equals the pressure ahead of the shock front. Within the accuracy of our approximation the gas is then at rest behind the rarefaction wave since it is at rest ahead of the shock front. The tail of this wave, given by the sound wave $\xi = \xi_0 < 0$, is characterized then by the condition $\omega(\xi_0) = c_0$ or $\sigma(\xi_0) = 0$.

With the notation

$$(75.01) \qquad A = 32 \int_{\xi_0}^{0} \frac{\sigma(\eta)\, d\eta}{[4 - \sigma(\eta)]^3}$$

we have from (74.08) the asymptotic expansion

DECAYING SHOCK WAVE

(75.02) $$\sqrt{c_0 t} = \left(\frac{1}{\sigma} - \frac{1}{4}\right)\sqrt{4A} + \cdots,$$

by which

(75.03) $$\sigma = 2\sqrt{\frac{A}{c_0 t}} - \frac{A}{c_0 t} + \cdots.$$

By inserting this expression in (74.01) we find

(75.04) $$x = \xi_0 + c_0 t + 2\sqrt{Ac_0 t} - A$$

as the asymptotic representation of the path of the shock front. Since the tail of the rarefaction wave is given by $x = \xi_0 + c_0 t$, we find for the *width of the shock wave* the asymptotic expression

(75.05) $$d(t) = 2\sqrt{Ac_0 t} - A.$$

This asymptotic expression is by no means to be understood in the sense that the deviation of the actual width from the expression $d(t)$ approaches zero as t increases indefinitely. The formula (75.05) is to be interpreted rather in the sense that the coefficient of $\sqrt{t}$ and the constant term in the genuine asymptotic expansion differ little from the terms $2\sqrt{Ac_0}$ and $-A$ given by (75.05).

To obtain the distribution of u and c in the wave zone we express σ in terms of x for fixed t. Setting

(75.06) $$\sigma = a(\xi - \xi_0) + b(\xi - \xi_0)^2 + \cdots,$$

and inserting in (74.01) we find

(75.07) $$x = \xi_0 + c_0 t + (ac_0 t + 1)(\xi - \xi_0) + bc_0 t(\xi - \xi_0)^2$$

asymptotically, and accordingly

(75.08) $$\xi - \xi_0 = \frac{x - \xi_0 - c_0 t}{ac_0 t + 1} - bc_0 t \frac{(x - \xi_0 - c_0 t)^2}{(ac_0 t + 1)^3},$$

and by (75.06) again

(75.09) $$\sigma = a\frac{x - \xi_0 - c_0 t}{ac_0 t + 1} + b\frac{(x - \xi_0 - c_0 t)^2}{(ac_0 t + 1)^3}$$

Inserting (75.09) in the expressions (74.02) and (74.03) for u and c written in the form

(75.10) $$u = (1 - \mu^2)c_0\sigma, \qquad c = c_0 + \mu^2 c_0\sigma,$$

gives *the asymptotic distribution of u and c in the wave zone* extending from $x = \xi_0 + c_0 t$ to $x = \xi_0 + c_0 t + d(t)$.

The pressure distribution is found from $p = p_0(c/c_0)^{2\gamma/\gamma-1}$ and (75.10),

(75.11) $\qquad p = p_0[1 + (1 + \mu^2)\sigma + \frac{1}{2}(1 + \mu^2)\sigma^2]$.

Immediately behind the shock front we have from (75.09)

(75.12) $\qquad \sigma = a \dfrac{d(t)}{ac_0 t + 1} + b \dfrac{d^2(t)}{(ac_0 t + 1)^3}$

which by (75.05) is equivalent to (75.03)

$$\sigma = 2\sqrt{\dfrac{A}{c_0 t}} - \dfrac{A}{c_0 t} + \cdots .$$

By substituting (75.03) in (75.11) we easily see that the pressure rise across the shock front or simply *the shock strength decreases in inverse proportion to the square root of the time.*

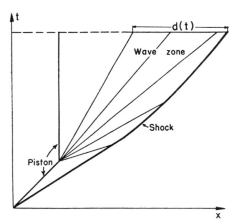

FIG. 40. Decaying shock wave produced by stopping a moving piston.

A decaying shock wave of this kind, produced by a rarefaction wave overtaking it, is found in a tube when a piston, after creating a shock of constant strength, is suddenly stopped. For, the head of the centered rarefaction wave sent out the moment the piston is stopped follows the shock front at sound speed, and catches up with it. For this case the function $\omega = \omega(\xi)$ and the constants A, a, and b are easily calculated, see [55]. The resulting flow pattern is shown in Figure 40.

If the piston is not suddenly stopped but suddenly retracted with constant speed and then suddenly stopped in its original position, a second shock front is sent out at the instant when the piston is finally stopped. The wave pattern, illustrated in Figure 41, consists then of a "head shock front," across which the gas is accelerated and compressed, a rarefaction wave across which gas is decelerated to a negative velocity and expanded to under-atmospheric pressure, and a "tail shock front," across which the gas comes to rest and acquires atmospheric pressure again.

A wave pattern of this type has been called an "N-wave." The tail shock of such an N-wave can be treated in the same manner as the head shock and its asymptotic behavior can be described by

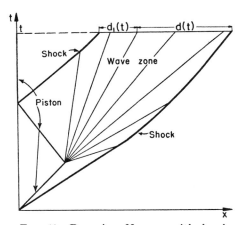

FIG. 41. Decaying N-wave with head and tail shock due to the retraction of a piston.

essentially the same formulas. The rarefaction wave may be divided into two parts separated by the sound wave, $\xi = \xi_0$, carrying the gas velocity $u = 0$ and the sound speed $c = c_0$. The head shock front traverses the forward part of the rarefaction wave, the tail shock front the backward part. If the original position of the tail shock front is at $c = \xi_1$, we find that the *width* of the N-wave zone is asymptotically,

(75.13) $\quad d(t) + d_1(t) = 2(\sqrt{A} + \sqrt{A_1})\sqrt{c_0 t} - (A + A_1)$

in which

(75.14) $\qquad\qquad A_1 = 32 \int_{\xi_1}^{\xi_0} \dfrac{\sigma(\eta)\, d\eta}{[4 - \sigma(\eta)]^3}.$

Again the width increases like the square root of t. The asymptotic distribution of u, c, p is given by the formulas (75.09–.11) as before.

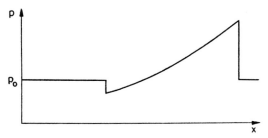

Fig. 42. Pressure distribution in a decaying N-wave with head and tail shock.

The asymptotic pressure distribution in the N-wave resulting from a retracting piston is shown in Figure 42.

76. Formation of a shock

As we have demonstrated in Part B, a discontinuity may arise in a compression wave. The wave form may gradually steepen so much that an infinite slope occurs at some point, see Section 41, and at this point a shock will develop. Described in an (x, t)-diagram the C_+-characteristics in a forward-facing compression wave may form an envelope starting at a cusp, see Section 48. At this cusp a shock path begins. The envelope (unless it falls entirely into one point) has two branches having the same direction at the cusp and the quantities u, c, p themselves are still continuous at the cusp

FORMATION OF A SHOCK

(only their derivatives become infinite there). Consequently, the shock begins with the strength zero and is weak during the early stages of its formation. The simplifying assumptions made in Section 74 may therefore also be made in describing the formation of a shock.

We confine ourselves to discussing the case in which the shock begins at the head of a forward-facing simple compression wave entering a zone of rest. This case arises, for example, in the compression wave produced by a uniformly accelerated piston, see the end of Section 49. The situation is then practically the same as if a simple wave overtakes a shock, see Section 75, except that in this case the initial shock vanishes.

Assuming that the shock begins at the time $t = 0$ at the place $x = 0$, and the zone $x \geq 0$ is at rest at $t = 0$, we require $\omega(\xi_0) = c_0$, $d\omega/d\xi < 0$ for $\xi < 0$, and in order that a shock be formed at $x = 0$, $t = 0$, $d\omega/d\xi = -\infty$ at $\xi = 0$. Specifically we assume

$$(76.01) \qquad \sigma = \sqrt{-\xi}(a + a_1\xi + \cdots), \qquad a > 0$$

with $\omega = c_0(1 + \sigma)$, see (74.08). Then we may use the formulas (74.08) and (74.01). Expanding the right members of these formulas with respect to powers of ξ we find

$$(76.02) \qquad \begin{aligned} c_0 t &= \frac{4}{3a}\sqrt{-\xi} - \tfrac{1}{12}\xi + \cdots, \\ x &= \frac{4}{3a}\sqrt{-\xi} - \tfrac{5}{12}\xi + \cdots, \end{aligned}$$

or

$$(76.03) \qquad x = c_0 t + \tfrac{3}{16}a^2 c_0^2 t^2 + \cdots$$

as the representation of the shock path. Thus the shock begins to move with the velocity c_0, as expected, and the acceleration $3a^2c_0^2/8$. The growth of the shock strength is seen from the formulas

$$(76.04) \qquad \begin{aligned} u &= \tfrac{3}{4}(1 - \mu^2)a^2 c_0^2 t + \cdots, \\ c - c_0 &= \tfrac{3}{4}\mu^2 a^2 c_0^2 t + \cdots, \\ p - p_0 &= \tfrac{3}{4}(1 + \mu^2)p_0 a^2 c_0 t + \cdots, \end{aligned}$$

which follow from (74.03) and (76.02).

We can specify these formulas easily for the formation of a shock in the compression wave produced by a uniformly accelerated piston moving according to $x = \frac{1}{2}bt^2$. The compression wave is given by

$$(76.05) \qquad x = \tfrac{1}{2}b\beta^2 + \left[c_0 + \frac{1}{1-\mu^2}b\beta\right](t-\beta),$$

see (49.06). The cusp is formed at

$$(76.06) \qquad t = t_c = (1-\mu^2)c_0/b, \qquad x = x_c = (1-\mu^2)c_0^2/b,$$

see (49.10). Shifting the origin of the coordinate system to this point and setting

$$\sigma = \beta/t_c$$
$$(76.07) \qquad \xi = -\frac{1}{2}\frac{1+\mu^2}{1-\mu^2}b\beta^2 = -\gamma b\beta^2/2,$$

we write (76.05) in the form

$$(76.08) \qquad x - x_c = \xi + c_0(1+\sigma)(t-t_c).$$

Hence the coefficient a in (76.01) is given by

$$(76.09) \qquad a^2 = \frac{2}{\gamma b t_c^2} = \frac{2b}{(1-\mu^4)c_0^2}.$$

After replacing x and t by $x - x_c$, $t - t_c$ and inserting (76.08) and (76.09) in formulas (76.02–.04), we obtain the description of the development of the shock. For a more detailed discussion, in particular of cases in which the shock develops in the interior of the wave region, see [55].

It should be emphasized again that all the preceding results and their numerical application in specific cases are valid only as long as the shock strength is sufficiently small. As soon as this assumption becomes inadmissible, we must not only consider non-isentropic flows but also realize that the shock modifies the simple wave behind it. This is an interaction effect which may be interpreted by a "reflected wave" of noticeable strength (while weak shocks do not noticeably reflect the flow behind them). In Part D we shall study interactions without any assumptions on the weakness of the shock.

77. Remarks on strong non-uniform shocks

It is a complex analytical problem to determine precisely the development of the shock that inevitably occurs within a compression wave beyond the early stages in which it is weak or moderately strong. The region $\mathcal{R}$ in the (x, t)-plane in which the flow is influenced by the shock is evidently bounded by the shock path beginning at the cusp A, a section of the cross-characteristic from A to a point B on the piston path and the arc of the piston path beyond B, see Figure 43. If the shock path $\mathcal{S}$ and its strength were known we could obtain along $\mathcal{S}$ definite initial values of u, ρ and the entropy S in the region $\mathcal{R}$ by means of the shock relations. With these initial values we could solve the differential equations (34.06) and determine u, ρ, S in $\mathcal{R}$. Now the shock path must be chosen in such a way that on the arc of the piston path beyond the point B the solution has the values u prescribed by the velocity of the piston.

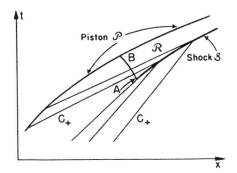

Fig. 43. The region R influenced by the shock originating in a compression wave.

This is an initial boundary value problem with an unknown boundary, and no direct theoretical treatment seems possible. The solution of each particular problem could be determined by numerical methods, for example by finite differences. The following reverse procedure may also be feasible. Assume a shock path $\mathcal{S}$ and determine the initial values on the other side of it according to the shock conditions. Then solve the initial value problem and find the corresponding piston motion as the motion of the flow through B. By carrying out such computations for a suitable variety of assumed shock paths $\mathfrak{S}$, an assortment of flow patterns can be obtained from which one can choose the one most closely representing a given piston motion.

To clarify the basic questions one must formulate in general terms the mathematical problem presented by the flow of a gas in a semi-infinite tube under the influence of a piston operating at the closed end: A solution of the differential equations (34.06) is sought in the domain of the (x, t)-plane bounded by the positive x-axis and an arbitrarily prescribed "piston path" $\mathcal{P}$: $x = X(t)$, $0 \leq t \leq \infty$. For $t = 0$ the state ρ, p, S, u is prescribed and along $\mathcal{P}$ the value of u is prescribed to equal $u_P(t) = \dot{X}(t)$.

We know that this initial boundary value problem does not in general possess a continuous solution. The question arises: If shock discontinuities across certain not a priori given lines are admitted, is the problem always solvable and is the solution uniquely determined?

The answer to such general questions is still beyond the present state of knowledge. Even more is this true of corresponding or similar questions for flow in two or three dimensions. But at least such questions serve to direct the groping efforts in the few special cases which alone are accessible at present.

D. Interactions

78. Typical interactions

As stated repeatedly before, a general solution of flow problems is not feasible. If the beginning of the flow is of a simple nature, we can, however, achieve a fairly complete analysis of the subsequent phases of the motion characterized by the "interaction" of initial "elementary waves," i.e. expansion waves, compression waves, shock waves originating from a state of rest and contact discontinuities. In many cases of major interest the gas moves initially in elementary waves, separated from each other. What happens shortly afterwards is that such waves are reflected, see Section 70, meet or overtake one another, so that by *interaction* a more general state of motion results. The motion thus produced is the subject of the following discussions. Unlike the case of linear wave motion *no principle of superposition is valid here*. Phenomena altogether different from those of linear wave motion occur. As a classical instance we mention *Lagrange's problem in interior ballistics* treated in detail by Love and Pidduck [50]. A tube is closed at a fixed point O by a rigid wall and at the other end by a piston of given mass with a variable position B. Up to time $t = 0$ there is atmospheric pressure

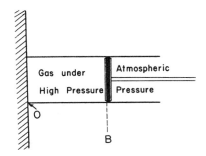

FIG. 44. Initial situation in Lagrange's problem.

in the tube. Then at $t = 0$ an explosion in the tube produces a gas *still at rest* with constant entropy S_0, density ρ_0, and a very high pressure p_0. The piston is accelerated by the pressure difference between its two sides; consequently the gas in the tube behind the piston is also accelerated and thinned out. This expansion spreads into the interior of the tube as a rarefaction wave. It travels from the piston to the wall at the point O, is reflected on the wall, meets and intersects the wave still traveling from the piston into the tube, is reflected on the piston, etc. The problem is to describe the motion of the gas as well as that of the piston.

Another typical problem of interior ballistics involving interactions occurs in the following case: We have the same situation as before, but in place of the piston we imagine that the exploded gas is separated from the atmospheric air in the tube by a membrane which is instantaneously removed. The initial discontinuity at the interface between the exploded gas and the air splits into two waves, a shock wave racing into the quiet air and compressing it, and a rarefaction wave sweeping backward over the exploded gas and expanding it. The rarefaction wave is reflected at the closed end. The reflected rarefaction wave catches up first with the interface between the exploded gas and the air, which has been pushed away from the wall by the expansion of the gas. Finally it catches up with the shock. Each "catching-up" produces a complicated interaction process involving more reflected waves. Some indication of the process is given in Figure 45.

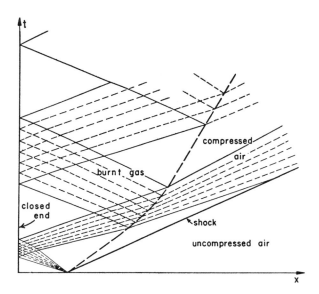

Fig. 45. Wave motion produced by removing a membrane between gas under high pressure and atmospheric air.

As stated, we shall content ourselves with the study of *elementary interactions*, between two waves colliding, or one wave overtaking another, or one wave meeting a discontinuity. We shall see that as the ultimate outcome of such interactions we can, in general, expect two waves moving in opposite directions away from the place of interaction.

TYPICAL INTERACTIONS 175

A general observation on the problem of one wave overtaking another should be made here. *Any two waves facing in the same direction, with the exception of two rarefaction waves, eventually overtake each other.* To demonstrate this important fact we must examine the four possible cases. (All velocities are observed relative to the gas in the region (1) between the two waves.)

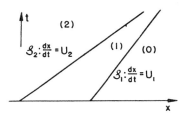

FIG. 46. Overtaking of one shock by another.

1) A shock front S_1 travels behind a shock front S_2 facing in the same direction, see Figure 46. Then, the first shock S_1 travels with subsonic speed and the second shock S_2 travels with supersonic speed, see Section 65. Therefore S_2 will catch up with S_1.

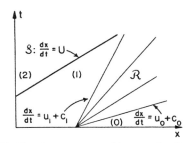

FIG. 47. Shock wave overtaking rarefaction wave.

2) A rarefaction wave $\mathcal{R}$ is followed by a shock wave S, see Figure 47. The tail of $\mathcal{R}$ travels with sound speed c_1 while the shock front travels with a speed greater than c_1.

3) A shock wave S is followed by a rarefaction wave $\mathcal{R}$, see Figure 48. The head of the rarefaction wave travels with sound speed c_1,

but the shock front travels with subsonic speed. Therefore $\mathcal{R}$ overtakes $\mathcal{S}$.

4) However, two rarefaction waves facing in the same direction never meet, see Figure 49. The tail of one travels with the same sound speed as the head of the other.

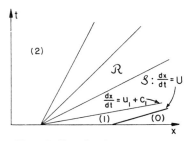

Fig. 48. Rarefaction wave overtaking shock wave.

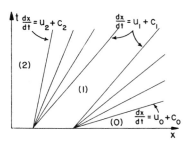

Fig. 49. Two rarefaction waves facing in the same direction.

The fact that two shock waves or a shock wave and a rarefaction wave, facing in the same direction, always overtake each other implies that two shock fronts or a shock front and a centered rarefaction wave may never start from the same point at the same time, and face in the same direction.

79. Survey of results

It is convenient to describe the results of interactions by symbolic equations. We denote a shock front facing in the direction of increasing x by $\mathcal{S}_{\rightarrow}$, an opposite shock front by $\mathcal{S}_{\leftarrow}$, and similarly, rarefaction waves by $\mathcal{R}_{\rightarrow}$ or $\mathcal{R}_{\leftarrow}$ according to whether the gas moves into the wave from the right or from the left.* Contact discontinuities, see Section 56, which frequently arise from interactions, are denoted by the symbol $\mathcal{T}$. Contact discontinuities may be distinguished as $\mathcal{T}_{>}$ or $\mathcal{T}_{<}$ according to whether the sound speed on the right side of $\mathcal{T}$ is greater or smaller than that on the left side of $\mathcal{T}$; for polytropic gases with the same value of γ, higher sound speed cor-

* Once more it should be stressed that the direction in which an elementary wave *faces* has nothing to do with the direction in which the wave front *moves*.

responds to higher density. By the symbol $\mathcal{JJ}$ we denote a zone in which the pressure and flow velocity are constant but in which the density, entropy, and temperature vary from one particle path to another.

In a head-on collision of two shock waves of different intensities the following situation results. After the two shock waves have penetrated (and thereby weakened and retarded) each other, they leave behind them an expanding zone of constant pressure and flow velocity. In this zone, however, the density is not uniform; instead, a point moving with the flow velocity of the zone separates two regions of different (uniform) density (and temperature). In other words, a *contact discontinuity* of the type envisaged in Section 56 appears.*
This fact, well established experimentally, shows that we must consider contact discontinuities together with shock and rarefaction waves in interactions.

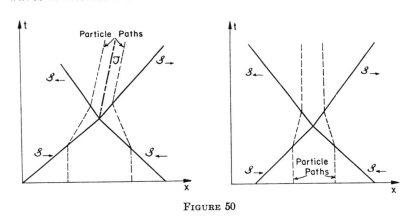

FIGURE 50

Head-on collision of two unequal shock waves.

Head-on collision of two equal shock waves.

The effect of the *collision of two shock waves* can therefore be described symbolically by the formula

$$\mathcal{S}_\rightarrow \mathcal{S}_\leftarrow \rightarrow \mathcal{S}_\leftarrow \mathcal{JS}_\rightarrow .$$

* This fact seems to have escaped many writers in the field and was brought to general attention by von Neumann, see [51].

In other words, head-on collisions of shock waves result in two shock fronts moving apart and separated by a contact discontinuity. In the case of two symmetric waves or what is equivalent, the reflection of a shock on a rigid wall along the line of symmetry, the contact discontinuity vanishes and we have, for reflection, the explicit results given in Section 70.

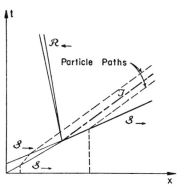

Fig. 51. Overtaking of one shock wave by another.

The overtaking of shock waves in gases with an adiabatic exponent $\gamma \leqq 5/3$ results in a transmitted shock, a reflected (in general weak) rarefaction wave, and a contact discontinuity between them:

$$\mathcal{S}_\leftarrow \mathcal{S}_\rightarrow \to \mathcal{R}_\leftarrow \mathcal{J} \mathcal{S}_\rightarrow \text{ for } \gamma \leqq \tfrac{5}{3},$$

a result first obtained by von Neumann [51]. For $\gamma > \tfrac{5}{3}$, which is not supposed to occur for an ideal gas, we may have the same situation, but there are also cases where a shock is reflected:

$$\mathcal{S}_\rightarrow \mathcal{S}_\rightarrow \to \begin{cases} \mathcal{R}_\leftarrow \mathcal{J} \mathcal{S}_\rightarrow \\ \quad \text{or} \\ \mathcal{S}_\leftarrow \mathcal{J} \mathcal{S}_\rightarrow \end{cases} \text{ for } \gamma > \tfrac{5}{3}.$$

Reflection and refraction of shock waves on contact surfaces (or between different media) occur according to the two formulas

$$\mathcal{S}_\rightarrow \mathcal{J}_< \to \mathcal{S}_\leftarrow \mathcal{J}_< \mathcal{S}_\rightarrow ,$$

$$\mathcal{S}_\rightarrow \mathcal{J}_> \to \mathcal{R}_\leftarrow \mathcal{J}_> \mathcal{S}_\rightarrow .$$

This means that if a shock wave in one gas hits a second gas with a higher sound speed, at a discontinuity surface, a reflected and a transmitted shock wave result. In case the second medium has a lower sound speed than the first there are two possibilities. First, if the second medium has a lower value of $c/(1 - \mu^2)$ or if the shock is sufficiently weak a rarefaction wave is reflected but a shock wave is still transmitted. Second, if the second medium has a higher value of $c/(1 - \mu^2)$ and the shock is sufficiently strong a rarefaction wave is transmitted and a shock wave is reflected, see Section 83.

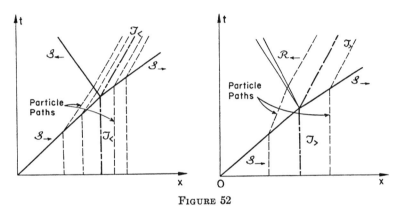

FIGURE 52

Interaction of a shock wave and a contact surface $\mathcal{J}_<$.

Interaction of a shock wave and a contact surface $\mathcal{J}_>$.

In interactions not involving rarefaction waves, a reflected and transmitted wave always emerge immediately after the collision. Interactions with rarefaction waves, however, lead at first to a period of *penetration* during which the flow cannot be described as made up of simple waves.

The waves emerging from the zone of penetration are simple waves according to our fundamental theorem since they are adjacent to constant states. If this penetration is completed in a finite time the zone between the two emerging simple waves is in a constant *terminal* state for similar reasons, see the fundamental lemma of Chapter II, Section 29. The following descriptions refer only to these terminal states. We imagine that the process starts with two simple waves separating zones of constant pressure and flow velocity. Then, interactions with rarefaction waves lead to terminal states as follows. The head-on collision of two rarefaction waves (in the case of symmetry this is equivalent to the reflection of a rarefaction wave on a rigid wall) again yields two rarefaction waves as the terminal state:

$$\mathcal{R}_\rightarrow \mathcal{R}_\leftarrow \rightarrow \mathcal{R}_\leftarrow \mathcal{R}_\rightarrow .$$

Likewise, the outcome of a *collision of a rarefaction wave with a zone with higher sound speed* is simply described by

$$\mathcal{R}_\rightarrow \mathcal{J}_< \rightarrow \mathcal{R}_\leftarrow \mathcal{J}_< \mathcal{R}_\rightarrow$$

while the interaction $\mathcal{R}_\rightarrow \mathcal{J}_>$ produces a compression wave in which a shock wave eventually forms, see [52].

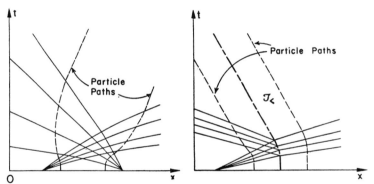

FIG. 53. Head-on collision of two rarefaction waves.   FIG. 54. Interaction of a rarefaction wave and a contact surface $\mathcal{J}_<$.

When a shock overtakes a rarefaction wave or a rarefaction wave overtakes a shock the process of interaction may go on indefinitely. We met such possibilities in the approximate treatment of the interaction of weak shock and simple wave, see Section 75. It is probable, however, that the interaction is completed in a finite time if the strength of the overtaking wave is definitely greater than that of the overtaken wave. Under these circumstances, we would have the following result:

$$\mathcal{S}_\leftarrow \mathcal{R}_\rightarrow \rightarrow \mathcal{S}_\leftarrow \mathcal{J}\mathcal{J}\mathcal{S}_\rightarrow ,$$

$$\mathcal{R}_\leftarrow \mathcal{S}_\rightarrow \rightarrow \mathcal{R}_\leftarrow \mathcal{J}\mathcal{J}\mathcal{S}_\rightarrow .$$

What is here denoted as a reflected shock is really a compression wave which leads to a shock—we should note that it was an essential part of our approximate treatment to neglect the reflected waves.

One further comment relating to these descriptions and figures is necessary. Zones $\mathcal{TT}$ result from shocks penetrating into a rarefaction wave (with the effect of mutual weakening); the shock path is bent and the particles crossing the shock undergo different changes in entropy. That ultimately all these particles emerge with the same velocity is a simplifying assumption which has been found true in the first approximation. This assumption, however, cannot be expected to represent the facts accurately.

80. Riemann's problem. Shock tubes

Interactions involving two shocks are related to a more general type of phenomena, studied by Riemann in his classical paper. Riemann's problem is to find the gas flow resulting from an initial state in which the gas at the right-hand side, $x > 0$, is in a constant state (r) given by u_0, p_0, ρ_0 and at the left-hand side, $x < 0$, in a constant state (l): u_1, p_1, ρ_1, τ_1. The difference between Riemann's problem and that discussed above is that the two given states are here considered independent, while only five of their data can be arbitrarily prescribed if they both are connected by shocks with a previous intermediate state (m) (which has shrunk to a point by the time $t = 0$).

Riemann's answer is: there are four possible types of subsequent flow, inasmuch as in both directions from the origin a shock or a centered rarefaction wave may proceed depending on inequalities prevailing for the initial quantities. The analysis of interaction processes discussed in the following section likewise applies to Riemann's problem.

We are particularly interested in the special case when in the initial state the gas is at rest, $u_l = u_r = 0$, and when $p_l > p_r$, $\rho_l > \rho_r$. This case represents an important experimental arrangement for the production of constant shocks in a long tube. Suppose the gas in the tube at the side $x < 0$ is of higher pressure and density than that at the side $x > 0$ and a membrane originally separating the two parts of the tube is suddenly removed at the time $t = 0$. Then a motion results as indicated in Figure 55 (in the neighborhood of the origin) and described by a shock racing into the quiet low pressure gas, raising the pressure p to a value p_{m^*} between p_l and p_r and followed by an expanding column of gas moving at a constant speed u_m towards the right. The end of this column is traversing the high pressure gas to the left at local sound speed and is followed by a rarefaction wave facing to the left, raising the pressure p_{m^*} to the pressure p_l of the high pressure zone, which remains at rest ahead of the expanding rarefaction zone.

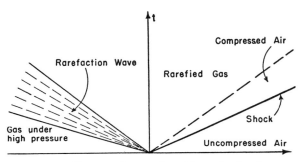

FIG. 55. Wave motion produced in a shock tube.

Devices of this type, introduced by Payman [62], have often been used since in experimental investigations.

81. Method of analysis

While reference must be made to a detailed report [52] for the justification of the statements of Section 79, the general method for obtaining these results in the case of polytropic gases can be indicated here. It amounts to an algebraic discussion of the transition relations for shock fronts and rarefaction waves, with the help of a graphical representation in the (u, p)-plane.

By subscripts a, b, m, l, r, k we denote zones with constant values of p and u, the letters l and r meaning "left" and "right" or smaller and larger values of x respectively.

From our previous analysis we recall the following results for shocks. For all shock waves we have $u_r < u_l$, see (57.01); in addition for forward-facing shocks we have $p_r < p_l$, and for backward-facing shocks $p_r > p_l$, see Section 65. Furthermore if a state (a) of the gas $u = u_a$, $p = p_a$, $\tau = \tau_a$, and another state (b), with the

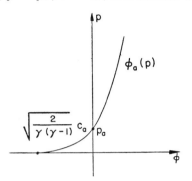

Fig. 56. The pressure behind or ahead of a shock front as a function of the velocity change.

quantities u_b, p_b, τ_b, are connected through a shock wave, then we have, see (67.04), (67.05)

$$(81.01) \qquad m = \frac{p_b - p_a}{u_b - u_a} = \pm\sqrt{\frac{p_b + \mu^2 p_a}{(1 - \mu^2)\tau_a}}.$$

From this we infer that

$$(81.02) \qquad u_b = u_a \pm \phi_a(p_b),$$

where

$$(81.03) \qquad \phi_a(p) = (p - p_a)\sqrt{\frac{(1 - \mu^2)\tau_a}{p + \mu^2 p_a}},$$

and that the plus sign is to be taken for shock fronts $\mathcal{S}_\to$, the minus sign for shock fronts $\mathcal{S}_\leftarrow$. The monotonic function $\phi_a(p)$ has the following simple properties:

$$(81.04) \qquad \phi_a(p_b) = -\phi_b(p_a),$$

$$(81.05) \qquad \begin{aligned} \phi_a(p) &\to \infty \quad \text{for} \quad p \to \infty, \\ \phi'_a(p) &\to 0 \quad \text{for} \quad p \to \infty, \end{aligned}$$

and the curve $\phi = \phi_a(p)$ meets the ϕ_a-axis at

$$(81.06) \qquad \phi = \phi_a(0) = -\sqrt{\frac{2}{\gamma(\gamma - 1)}}\, c_a,$$

where c_a is the sound speed in region (a). Relations (81.02) are shown graphically in Figure 57; the various branches of the curves

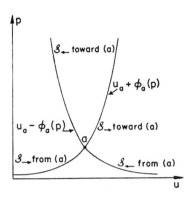

Fig. 57. Locus of all states which can be connected with a given state (a) through shock waves $\mathcal{S}_\rightarrow$ and $\mathcal{S}_\leftarrow$ facing toward and away from (a) as indicated.

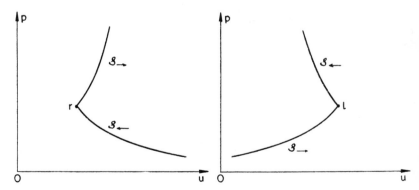

Fig. 58. (1) Locus of possible states (l) on the left of a shock front if the state (r) on the right is prescribed. (2) Locus of possible states (r) on the right of a shock front if the state (l) on the left is prescribed.

correspond to shock waves facing toward or away from (a) and are determined from $u_r < u_l$.

In Figure 58 the loci of all states which can be connected by shock waves $\mathcal{S}_\rightarrow$ and $\mathcal{S}_\leftarrow$ with a given state (r) to the right or (l) to the left are indicated.

A similar representation can be obtained for the one-parametric family of constant states which can be connected with a fixed

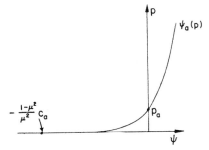

FIG. 59. The pressure behind or ahead of a simple wave as a function of the velocity change.

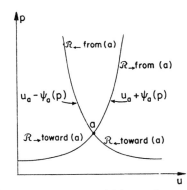

FIG. 60. Locus of all states which can be connected with a state (a) through rarefaction waves $\mathcal{R}_\rightarrow$ and $\mathcal{R}_\leftarrow$ facing toward or away from (a) as indicated.

state by a rarefaction wave. We saw in Section 40 that across a rarefaction wave $\mathcal{R}_\rightarrow$ or $\mathcal{R}_\leftarrow$, $u \mp \dfrac{2}{\gamma - 1} c = u \mp \dfrac{2}{\gamma - 1} \sqrt{\gamma \tau p} =$ constant. Now, since only adiabatic changes of state occur, we have

$\rho_a/\rho_b = (p_a/p_b)^{1/\gamma}$ or $\sqrt{\tau_b p_b}\, p_b^{-(\gamma-1)/2\gamma} = \sqrt{\tau_a p_a}\, p_a^{-(\gamma-1)/2\gamma}$. Thus

(81.07)
$$u_b - u_a = \pm \frac{2}{\gamma - 1}\left(\sqrt{\gamma \tau_a p_a} - \sqrt{\gamma \tau_b p_b}\right)$$
$$= \pm \frac{\sqrt{1-\mu^4}}{\mu^2}\, \tau_a^{1/2} p_a^{1/2\gamma}(p_a^{(\gamma-1)/2\gamma} - p_b^{(\gamma-1)/2\gamma})$$

so that, analogous to (81.02), we have

(81.08) $\qquad u_b = u_a \pm \psi_a(p_b),$

where

(81.09) $\quad \psi_a(p) = \dfrac{\sqrt{1-\mu^4}}{\mu^2}\, \tau_a^{1/2} p_a^{1/2\gamma}(p^{(\gamma-1)/2\gamma} - p_a^{(\gamma-1)/2\gamma}).$

The plus sign prevails for waves $\mathcal{R}_\rightarrow$ and the minus sign for waves $\mathcal{R}_\leftarrow$

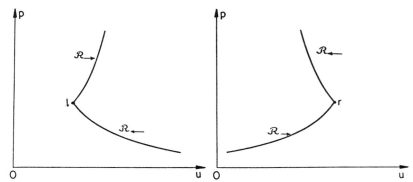

Fig. 61. (1) Locus of possible states (r) on the right of a rarefaction wave if the state (l) on the left is prescribed. (2) Locus of possible states (l) on the left of a rarefaction wave if the state (r) on the right is prescribed.

since u_l is now less than u_r, and hence $p_r > p_l$ for forward-facing waves, $p_r < p_l$ for backward-facing waves. The monotonic function $\psi_a(p)$ has the properties

(81.10) $\qquad\qquad \psi_a(p_b) = -\psi_b(p_a),$

(81.11) $\quad\begin{aligned}\psi_a(p) &\to \infty \quad \text{for} \quad p \to \infty, \\ \psi_a'(p) &\to 0 \quad \text{for} \quad p \to \infty,\end{aligned}$

and the curve $\psi = \psi_a(p)$ touches the ψ-axis tangentially at

(81.12)
$$\psi = \psi_a(0) = -\frac{\sqrt{1-\mu^4}}{\mu^2}\sqrt{\tau_a p_a}$$
$$= -\frac{1-\mu^2}{\mu^2} c_a.$$

As in the case of shock waves, and for similar reasons, we show graphically in Figure 60 the possible states (b), given by (81.08), which can be connected with a fixed state (a) through rarefaction waves facing toward and away from (a). Figure 61 shows the loci of all states which can be connected with a given state (l) or (r), on the left or right of the wave respectively, by rarefaction waves $\mathcal{R}_\rightarrow$ or $\mathcal{R}_\leftarrow$.

We note now that all states which can be reached from a state (r) by a forward wave may be represented in the (u, p)-plane by a curve

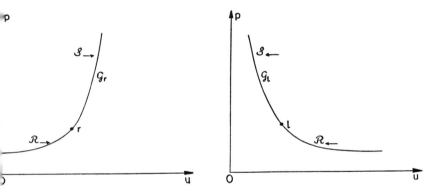

FIG. 62. (1) Locus $\mathcal{G}_r$ of all states which can be reached from a given state (r) on the right of a forward wave $\mathcal{S}_\rightarrow$ or $\mathcal{R}_\rightarrow$. (2) Locus $\mathcal{G}_l$ of all states which can be reached from a given state (l) on the left of a backward wave $\mathcal{S}_\leftarrow$ or $\mathcal{R}_\leftarrow$.

$\mathcal{G}_r$ expressed analytically by

(81.13)
$$\mathcal{G}_r \begin{cases} u = u_r + \phi_r(p), & p > p_r, & \mathcal{S}_\rightarrow, \\ u = u_r + \psi_r(p), & p < p_r, & \mathcal{R}_\rightarrow, \end{cases}$$

see Figure 62, while all points representing states connected with (l) by a backward wave are on a curve

(81.14) $\mathfrak{G}_l \begin{cases} u = u_l - \phi_l(p), & p > p_l, \quad \mathfrak{S}_\leftarrow, \\ u = u_l - \psi_l(p), & p < p_l, \quad \mathfrak{R}_\leftarrow. \end{cases}$

Now the problem of interactions can be attacked as follows. Suppose that before the interaction we have a state (l) to the left and a state (r) to the right, both joined to a middle zone (m) of constant velocity u_0 and constant pressure p_0 by certain given waves. At the instant the interaction begins, the middle zone (m) disappears and, either instantaneously or after a period of penetration, a forward wave moves into the state (r) and a backward wave moves into the zone of state (l). These two waves are now separated by a new middle zone (m^*) of constant velocity u^* and constant pressure p^*, see Figure 63. All that we have to find is the state (m^*) and the waves connecting (r) and (l) with (m^*).

If the two states (l) and (r) are known we need simply draw the two $\mathfrak{G}$ curves through them. The point of intersection then deter-

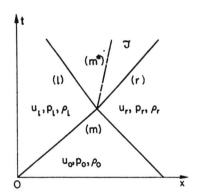

FIG. 63. The various zones of a gas before and after the collision of two shock fronts.

mines the state (m^*) and the waves from (m^*) to (l) and to (r). In practice, graphical construction often will not be sufficiently precise, but it indicates the proper arrangement for numerical calculation.

ANALYSIS OF INTERACTIONS

To illustrate the procedure we consider a few cases in more detail.

To study the *collision of two shock waves* $\mathcal{S}_{\rightarrow}$ and $\mathcal{S}_{\leftarrow}$ we observe that our states (l) and (r) must be represented in the (u, p)-diagram as in Figure 64 since they are obtained from (m) by a forward and backward shock respectively (and are therefore not entirely independent, a fact which is not essential to our procedure). The curves $\mathcal{G}_r$ and $\mathcal{G}_l$, according to our diagram, must intersect in a point m^*, and m^* is on the upper part of $\mathcal{G}_r$ as well as $\mathcal{G}_l$. Hence the two transitions from m^* are shocks as stated, see Figure 50.

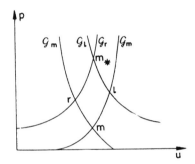

FIG. 64. Head-on collision of two shock waves.

In the state (m^*) we have constant values of p and u. The shock transition from (r) to (m^*), however, in general determines a value ρ_+^* different from the value ρ_-^* obtained by the shock transition from (l). Hence, in the zone (m^*) of the (x, t)-plane, there is a line of contact discontinuity coinciding with the particle path from the point of collision.

The *collision between two rarefaction waves* leads to a terminal state which can be determined just as easily. Here the relative positions of (m), (l), and (r) and consequently of (m^*) are immediately seen to be those of Figure 65, see also Figure 53. But this shows that (m^*) is on those parts of $\mathcal{G}_r$ and $\mathcal{G}_l$ to which rarefaction waves correspond. Hence our previous statement $\mathcal{R}_{\rightarrow}\mathcal{R}_{\leftarrow} \rightarrow \mathcal{R}_{\leftarrow}\mathcal{R}_{\rightarrow}$ is justified. (Of course, no contact discontinuities occur since no changes of entropy have taken place.) There are cases in which the two curves do not intersect. In such cases the process of penetration continues indefinitely, see [52].

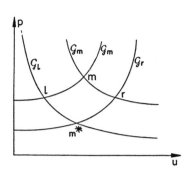 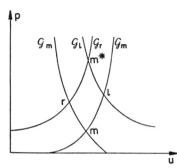

FIG. 65. Collision of two rarefaction waves.

FIG. 66. One shock wave overtaking another.

The problem of *one shock wave overtaking another* is slightly more delicate. Here the original states (l), (m), (r) are separated by two shock fronts facing to the right; we have

$$p_l > p_m > p_r,$$
$$u_l > u_m > u_r.$$

In the (u, p)-diagram the point m is evidently on the curve $\mathcal{G}_r$ since m is connected to r by a shock and the unknown point m^* must be situated on the same $\mathcal{G}_r$. On the other hand, m^* lies on a $\mathcal{G}_l$ which connects it with l, and hence a situation as in Figure 66 results if $\mathcal{G}_r$ passes on the right side of l. It is possible to show that this is the case from the algebraic form of our functions ϕ and ψ if $\gamma \leqq \frac{5}{3}$, see [51 and 52, A3]. Then our diagram indicates that an intensified forward shock $S_\rightarrow$ (between r and m^*) and a weak backward rarefaction wave result. For the same reason as in (a) we can expect a contact discontinuity in zone (m^*).

If $\gamma > \frac{5}{3}$, however, situations are possible in which $\mathcal{G}_r$ passes to the left of l, so that we obtain a (weak) reflected shock instead of a reflected rarefaction wave.

The method of analysis here discussed for elementary interactions, obviously applies also to the general problem of Riemann, see Section 80.

In the problem of the shock tube, mentioned in Section 80, the two initial states (l) and (r) are represented by two points $(0, p_l)$,

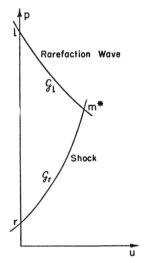

FIG. 67. (u, p) - diagram for shock and rarefaction wave occurring in a shock tube.

$(0, p_r)$ in the (u, p)-diagram as indicated in Figure 67 and it is immediately seen how the intermediate state (m^*) and other quantities pertaining to the phenomenon are represented.

82. *The process of penetration for rarefaction waves*

As emphasized before, the analysis of Section 81 yields only the elementary waves that are assumed to emerge ultimately from elementary interactions. However, when rarefaction waves or contact surfaces are involved in the interaction a more complicated flow occurs before the elementary waves move apart (the ultimate stage of such a motion may sometimes never be reached). This process of "penetration" requires the solution of boundary value problems for the differential equations of flow. As observed before, we must in general consider the equations (17.01–.03) in which non-uniform shocks, and thus variable entropy, are involved. Fortunately, in the case of the collision of two simple waves the problem of penetration depends only on the differential equations of isentropic flow and can be treated explicitly for polytropic gases on the basis of Riemann's theory, see Section 38.

Suppose two given rarefaction waves $\mathcal{R}_\rightarrow = \mathcal{R}_1$ and $\mathcal{R}_\leftarrow = \mathcal{R}_2$ collide, see Figure 68. After first meeting at the point $P(x_0, t_0)$ the two waves $\mathcal{R}_1$ and $\mathcal{R}_2$, consisting of families of straight line characteristics C_- and C_+, respectively, along which u and c are constant, interact to form the penetration region Q bounded by the characteristics C_+^0, C_+^1, C_-^1 and C_-^0. Upon entering Q the straight line characteristics no longer remain straight, and u and c vary along them. However, $\mu^2 u + (1 - \mu^2)c$ and $\mu^2 u - (1 - u^2)c$ remain constant along the C_+- and C_-- characteristics respectively. Finally upon emerging from the region Q, these characteristics are again straight lines and compose the two "transmitted" simple waves $\tilde{\mathcal{R}}_1$ and $\tilde{\mathcal{R}}_2$.

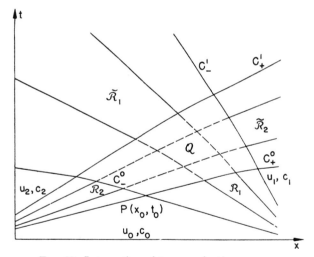

FIG. 68. Interaction of two rarefaction waves.

If the Riemann invariants r and s are introduced as characteristic parameters through equation (37.05), the problem becomes one of solving the differential equation,

(82.01) $$t_{rs} + \frac{\lambda}{r+s}(t_r + t_s) = 0;$$

here $\lambda = \dfrac{1}{2\mu^2}$. This equation is to be solved in a rectangle, since the region Q is mapped in a one-to-one manner into a rectangle $\mathcal{D}$ with sides parallel to the axes in the (r, s)-plane. Here the lines C_+^0 and C_-^0 are represented by

(82.02) $$r = r_0 = \frac{1}{2}\left(u_0 + \frac{2}{\gamma - 1} c_0\right)$$

and

(82.03) $$s = s_0 = -\frac{1}{2}\left(u_0 - \frac{2}{\gamma - 1} c_0\right),$$

respectively, while C_+^1 and C_-^0 are represented by

(82.04) $$r = r_1 = \frac{1}{2}\left(u_2 + \frac{2}{\gamma - 1} c_2\right)$$

and

(82.05) $$s = s_1 = -\frac{1}{2}\left(u_1 - \frac{2}{\gamma - 1} c_1\right),$$

respectively, see Figure 69.

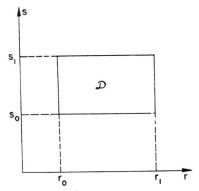

Fig. 69. Domain $\mathcal{D}$ in the plane of the Riemann invariants corresponding to the domain Q of penetration, see Figure 68.

Furthermore, since the characteristics C_+^0 and C_-^0, as well as the distribution of u and c along them, are explicitly given as soon as the waves $\mathcal{R}_1$ and $\mathcal{R}_2$ are known, $t(r, s_0)$ and $t(r_0, s)$ can be calculated. Hence, the problem of determining the penetration region

Q of two colliding simple waves is equivalent to solving a "characteristic initial value problem" for equation (82.01). For a specific case this will be carried out explicitly.

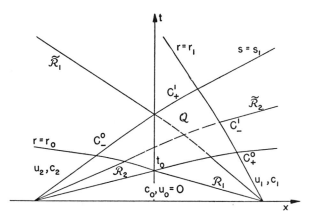

Fig. 70. Interaction of two centered rarefaction waves.

We consider, as an example, the case in which the two waves $\mathcal{R}_1$ and $\mathcal{R}_2$ are centered at the points $(x_1, 0)$ and $(x_2, 0)$ so that they first meet at $(0, t_0)$, see Figure 70. Furthermore, we may assume that $u_0 = 0$, and thus formulas (82.02–.03) simplify to

$$(82.06) \qquad r_0 = s_0 = \frac{1}{\gamma - 1} c_0.$$

The distribution of x and t along $r = r_0$ is obtained by substituting (82.01) in (38.02),

$$(82.07) \qquad x_s = \left(\frac{\gamma + 1}{2} r_0 + \frac{\gamma - 3}{2} s\right) t_s,$$

and from the equations of the straight line characteristics that sweep $\mathcal{R}_1$,

$$\frac{x - x_0}{t} = u - c = -\left(\frac{\gamma - 3}{2} r_0 + \frac{\gamma + 1}{2} s\right)$$

If x_s is eliminated from these two equations, we obtain, for $r = r_0$

(82.08) $$t_s + \frac{\lambda}{r_0 + s} t = 0.$$

Likewise for $s = s_0$,

(82.09) $$t_r + \frac{\lambda}{r + s_0} t = 0.$$

Thus we find for the initial conditions

(82.10) $$t(r, s_0) = t_0 \left(\frac{r_0 + s_0}{r + s_0}\right)^\lambda, \qquad r_0 \leqq r \leqq r_1;$$

(82.11) $$t(r_0, s) = t_0 \left(\frac{r_0 + s_0}{r_0 + s}\right)^\lambda, \qquad s_0 \leqq s \leqq s_1.$$

As Riemann [38] first found and as is easily verified, the solution of (82.01), satisfying the initial conditions (82.08–.11), can be given explicitly:

(82.12) $$\begin{aligned} t(r, s) &= t_0 \left(\frac{r_0 + s_0}{r + s}\right)^\lambda F\left(1 - \lambda, \lambda, 1, -\frac{(r_0 - r)(s_0 - s)}{(r_0 + s_0)(r + s)}\right) \\ &= t_0 \frac{(r_0 + s_0)^{2\lambda}}{(r_0 + s)^\lambda (r + s_0)^\lambda} F\left(\lambda, \lambda, 1, \frac{(r_0 - r)(s_0 - s)}{(r_0 + s)(r + s_0)}\right) \end{aligned}$$

in terms of the hypergeometric function $F(a, b, c, z)$.

Interesting formulas may be obtained from (82.12). For example, the duration of the interaction of the two waves is

(82.13) $$t(r, s) \begin{cases} = t_0 \left[\frac{c_0}{(c_1 + c_2 - c_0)}\right]^\lambda F\left(1 - \lambda, \lambda, 1, -\frac{(c_0 - c_1)(c_0 - c_2)}{c_0(c_1 + c_2 - c_0)}\right) \\ = t_0 \left[\frac{c_0^2}{c_1 c_2}\right]^\lambda F\left(\lambda, \lambda, 1, \frac{(c_0 - c_1)(c_0 - c_2)}{c_1 c_2}\right) \end{cases}$$

In accordance with our previous remarks, see Section 38, the function $F(1 - \lambda, \lambda, 1, z)$ reduces to a polynomial of degree $\lambda - 1$ if $\lambda = \frac{1}{2}\frac{\gamma + 1}{\gamma - 1}$ is a positive integer N or if $\gamma = \frac{2N + 1}{2N - 1}$, and consequently the solution of the problem can be given explicitly in those cases.

For arbitrary values of λ the following remark, due to C. De Prima*, is of mathematical interest. As known from the theory of hyperbolic differential equations, see e.g. [32, Ch. 5], the solution of initial value problems can always be expressed in terms of *Riemann's* function $R(\bar{r}, \bar{s}; r, s)$ of r, s and two parameters $\bar{r}, \bar{s}$, which as a function of r, s satisfies the differential equation (82.01), and as a function of $\bar{r}$, $\bar{s}$ the adjoint differential equation

$$(82.14) \qquad R_{\bar{r}\bar{s}} - \frac{\lambda}{\bar{r}+\bar{s}}(R_{\bar{r}} + R_{\bar{s}}) + \frac{2\lambda R}{(\bar{r}+\bar{s})^2} = 0$$

and the additional conditions

$$(82.15) \qquad \begin{aligned} R_{\bar{r}} - \frac{\lambda}{\bar{r}+\bar{s}} R &= 0 & \text{for } \bar{s} = s \\ R_{\bar{s}} - \frac{\lambda}{\bar{r}+\bar{s}} R &= 0 & \text{for } \bar{r} = r \\ R &= 1 & \text{for } \bar{r} = r, \bar{s} = s. \end{aligned}$$

These relations (82.14–.15) characterize Riemann's function. Moreover, Riemann's function $R^*(\bar{r}, \bar{s}; r, s)$ of the differential equation adjoint to (82.01) is given by the identity

$$(82.16) \qquad R^*(\bar{r}, \bar{s}; r, s) = R(r, s; \bar{r}, \bar{s})$$

and of course satisfies (82.01) as well as the corresponding boundary conditions.

From the properties it follows: *The solution $t(r, s)$ of the characteristic initial value problem for the penetration of two centered rarefaction waves is given explicitly by Riemann's function*:

$$(82.17) \qquad t(r, s) = t_0 R(r_0, s_0; r, s).$$

Another explicit expression,

$$(82.18) \qquad t(r, s) = t_0 \left(\frac{r_0 + s_0}{r + s}\right)^\lambda P_{\lambda-1}\left(\frac{1 + \alpha\beta}{\alpha + \beta}\right)$$

in which $\alpha = \dfrac{r}{r_0}$, $\beta = \dfrac{s}{s_0}$ and $P_\mu(z)$ denotes *Legendre's function* of order μ, was derived in an elegant way by R. Shaw**; it could be reduced to the preceding one by a known transformation formula connecting hypergeometric with Legendre functions.

* In an unpublished note.
** In a master's thesis at New York University.

These explicit solutions, particularly in cases in which they can be expressed by polynomials, are very suitable for numerical computations, and compare in this respect favorably with the approximate step-by-step procedure through finite differences discussed in the next section. However, when the rarefaction waves are not centered, or when the medium is not a polytropic gas, and even more when interaction of rarefaction waves with contact surfaces or shocks is considered, then the method of finite differences is apparently more tractable than any explicit representation of the solution.

83. *Interactions treated by the method of finite differences*

To find the flow in the zone of penetration of two elementary waves, the most suitable method in general is the use of finite differences. The following section taken from [53] explains the method. How it applies to the problem of penetration of two simple waves is obvious.

Next in simplicity is the problem of a rarefaction wave catching

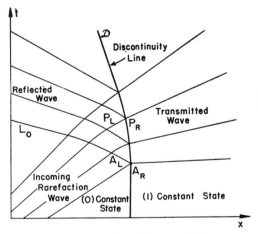

Fig. 71. Interaction of a rarefaction wave with a contact discontinuity.

up with a "contact discontinuity" or "separation line" $\mathfrak{D}$ (represented in the (x, t)-plane). The contact discontinuity originally separates two constant states (0) and (1), see Figure 71, with different values of entropy, temperature, and density, but with the same values of pressure and particle velocity. In the (x, t)-plane the discontinuity line $\mathfrak{D}$ is straight until it is met by the incoming rarefaction wave. The effect of the interaction is a gradual slowing down of the discontinuity point, or, in the (x, t)-representation, a bending of the line $\mathfrak{D}$ until $\mathfrak{D}$ has crossed the entire width of the wave, after which it again continues straight. The problem is then to determine from the given data the position of $\mathfrak{D}$ at any time and thereby the resulting flow during and after the penetration by the rarefaction wave. For simplicity the gas is taken to be polytropic.

The flow is isentropic on each side of $\mathfrak{D}$ and the differential equations can be written in the following characteristic form, see (34.02–.03),

$$
\begin{aligned}
x_\alpha - (u + c)t_\alpha &= 0, \\
x_\beta - (u - c)t_\beta &= 0, \\
u_\alpha + \frac{2}{\gamma - 1} c_\alpha &= 0, \\
u_\beta - \frac{2}{\gamma - 1} c_\beta &= 0;
\end{aligned}
$$

(83.01)

here x, t, u, c are the space, time, particle velocity, and local sound velocity respectively and γ is the adiabatic exponent, while the lines $\alpha = \alpha(x, t) = $ constant and $\beta = \beta(x, t) = $ constant are the characteristics. The initial data for the "zone of penetration," see Figure 71, in which the flow is not a simple wave are then obtained as follows: Since the incoming rarefaction wave generated by straight characteristics $\beta = $ constant is given, the first "reflected" characteristic L_0 which meets $\mathfrak{D}$ is known as well as the distribution of u and c along L_0. Although the position of $\mathfrak{D}$ is to be determined, we do know some data along $\mathfrak{D}$: firstly, $\mathfrak{D}$ is a particle path, i.e., along $\mathfrak{D}$ we have $\frac{dx}{dt} = u$; secondly, the portion of the rarefaction wave transmitted through $\mathfrak{D}$ is again a simple wave so that for any point P_R on the right hand side of $\mathfrak{D}$ the relation $u(P_R) - \frac{2}{\gamma_1 - 1} c(P_R) = u_1 - \frac{2}{\gamma_1 - 1} c_1$, holds in which the subscript 1 denotes quantities in the constant state (1). Thus, since $p(P_R) = p(P_L)$ (P_L is the point opposite to P_R on the left side of $\mathfrak{D}$) and $u(P_R) = u(P_L)$, we may express $u(P_L)$ as a function of $c(P_L)$ along $\mathfrak{D}$. Explicitly this relation is

$$(83.02) \quad u(P_L) - u(A_L) = \left[\left(\frac{c(P_L)}{c(A_L)}\right)^{\gamma_0(\gamma_1-1)/\gamma_1(\gamma_0-1)} - 1\right]\frac{2c(A_R)}{\gamma_1 - 1}.$$

As will be indicated, these initial data, together with the equations (83.01), are sufficient to determine $\mathfrak{D}$ as well as the flow uniquely. Figure 71 exhibits the region of penetration, the transmitted wave, and the wave reflected from $\mathfrak{D}$.

Problems in which the flow does not remain isentropic are of a more complex character. As an example we consider, by a method different from that of Section 75, the process of a straight shock overtaken by a rarefaction wave. During the penetration the shock gradually diminishes in intensity and speed and eventually emerges as a weakened constant shock. Because of the decay of the shock the isentropic character of the process is lost. Therefore, one has to consider the equations (17.01–.03) of one-dimensional non-steady flow involving entropy changes. Now we have three families of characteristics defined by equations of the form:

$$(83.03) \quad \begin{aligned} x_\alpha &= (u + c)t_\alpha, \\ x_\beta &= (u - c)t_\beta, \\ x_\omega &= ut_\omega. \end{aligned}$$

The first two are formally equivalent to those in the isentropic case whereas the third consists of the particle paths. Introducing the parameters $\alpha = \alpha(x, y) = $ constant and $\beta = \beta(x, y) = $ constant as new independent variables in the differential equation, we obtain, see Sections 3 and 4,

$$(83.04) \quad \begin{aligned} x_\alpha &= (u + c)t_\alpha, \\ x_\beta &= (u - c)t_\beta, \\ u_\alpha + \frac{2}{\gamma - 1}c_\alpha &= \frac{c\eta_\alpha}{\gamma(\gamma - 1)}, \\ u_\beta - \frac{2}{\gamma - 1}c_\beta &= -\frac{c\eta_\beta}{\gamma(\gamma - 1)}, \\ \eta_\alpha t_\beta + \eta_\beta t_\alpha &= 0; \end{aligned}$$

with $\theta = (\gamma - 1)c_v S$,

We arrive at the formulation of the boundary problem for the system (83.04) as follows, see Figure 72. From the given rarefaction wave we determine the first "reflected" characteristic L_0 as well as the distribution of u, c, η along L_0. The other data are given along the shock line $\mathcal{S}$ whose position is also to be found. These data now take

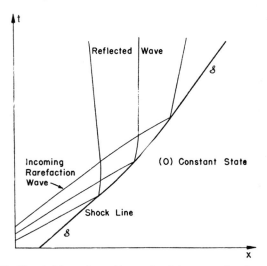

FIG. 72. Overtaking of a uniform shock by a rarefaction wave.

the form of the three shock conditions connecting the known constant state (0) in front of the shock with the state behind the shock. In these relations, we express the shock velocity $\dfrac{dx}{dt}$ in terms of the particle velocity u, sound speed c, and entropy $\eta/(\gamma - 1)c_r$ just behind the shock, i.e.,

$$\frac{dx}{dt} = h(u),$$

(83.05) $$\frac{dx}{dt} = g(c),$$

$$\frac{dx}{dt} = k(\eta).$$

The functions h, g, k are given by the Hugoniot relations. With these data it is then possible to determine the flow in the penetration region shown in Figure 72 as well as the position of the shock line $\mathfrak{S}$.

For both problems mentioned, a finite difference method for the approximate solutions, utilizing the characteristic equations (83.01) or (83.04), has proved to give reasonable results with relatively little labor as many calculations have shown. The procedure is as follows: In the case of a rarefaction wave overtaking a contact discontinuity, we consider in the (α, β)-plane the network of straight characteristics $\alpha = $ constant and $\beta = $ constant. The characteristic L_0 becomes a given line $\alpha = $ constant whereas the characteristics in the wave become straight lines $\beta = $ constant. The discontinuity line $\mathfrak{D}$ is mapped into a curve to be determined in the (α, β)-plane, and the corresponding characteristic lines are drawn as in Figure 73.

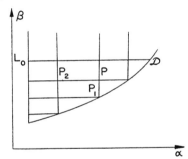

Fig. 73. Domain in the (α, β)-plane representing the interaction of a rarefaction wave with the discontinuity front $\mathfrak{D}$.

At the intersection of any two such characteristics we may then determine the values of u and c, since along $\alpha = $ constant and $\beta = $ constant we have from (83.01) the two relations $u - \dfrac{2}{\gamma - 1} c = $ constant and $u + \dfrac{2}{\gamma - 1} c = $ constant, respectively. The constants are determined by the initial distribution on L_0 and by the relation (83.02) between u and c on $\mathfrak{D}$. We then consider the first two equations of (83.01) written as

(83.06) $\qquad \Delta_\alpha x = (\tilde{u} + \tilde{c}) \Delta_\alpha t \qquad$ along $\beta = $ constant,

$\qquad \Delta_\beta x = (\tilde{u} - \tilde{c}) \Delta_\beta t \qquad$ along $\alpha = $ constant, where Δ_α or Δ_β means the difference between two successive points of intersection on β = constant or α = constant, respectively; $\tilde{u} + \tilde{c}$ or $\tilde{u} - \tilde{c}$ means the average between the values of $u + c$ or $u - c$ at the two successive net points. If one knows the value of x and t at two neighboring points such as P_1 and P_2 of an interior net point P, then by means of (83.05) the values of x and t are determined at P, see Figure 73. All that remains is to determine the values of x and t along $\mathfrak{D}$. This is accomplished by writing the relation $\dfrac{dx}{dt} = u$ on $\mathfrak{D}$ as a difference relation $\Delta x = u \Delta t$ on $\mathfrak{D}$. From this relation and the first equation of (83.05), the position of $\mathfrak{D}$ is determined. This completes the procedure.

In the case of the overtaking of an originally constant shock by a rarefaction wave the finite difference method is somewhat more complex. Again a similar net configuration is drawn in the (α, β)-plane where $\mathfrak{S}$ is to be determined. Now, however, the distribution of u, c at the net points is not immediately determined as before. Instead, knowing the values of x, t, u, c, η on L_0 one writes the system (83.04) as a system of difference equations:

$$\Delta_\alpha x = (\tilde{u} + \tilde{c})\Delta_\alpha t \qquad \text{along } \beta = \text{constant},$$

$$\Delta_\beta x = (\tilde{u} - \tilde{c})\Delta_\beta t \qquad \text{along } \alpha = \text{constant},$$

(83.07) $\quad \Delta_\alpha\left(u + \dfrac{2}{\gamma - 1}c\right) = \dfrac{\tilde{c}}{\gamma(\gamma - 1)}\Delta_\alpha \eta \qquad \text{along } \beta = \text{constant},$

$$\Delta_\beta\left(u - \dfrac{2}{\gamma - 1}c\right) = \dfrac{\tilde{c}}{\gamma(\gamma - 1)}\Delta_\beta \eta \qquad \text{along } \alpha = \text{constant},$$

$$\Delta_\alpha \eta \Delta_\beta t + \Delta_\beta \eta \Delta_\alpha t = 0.$$

The difference equations (83.07), together with the initial values of x, t, u, c, η along L_0 and the relation (83.05) along $\mathfrak{S}$, serve to determine the flow at all the net points of Figure 72 as well as the position of $\mathfrak{S}$. Specifically, suppose that P is an interior net point and that the values of x, t, u, c, η are known at the two neighboring points P_1 and P_2, see Figure 74. By utilizing the first and third equations of (83.07) between P_1 and P, the second and fourth between P_2 and P, and the fifth equation for the three points P_1, P, P_2, then the x, t, u, c, η at P may be determined. If, however, Q is a point on

S and if the solution is already determined at the neighboring points Q_0 (on S) and Q_2, then the second and fourth difference equations of (83.07) are taken between Q_2 and Q, while the three relations (83.05) along S (written as difference relations) are used between Q_1 and Q. These relations are sufficient to determine Q (in the (x, t)-plane) as well as u, c, and θ at Q.

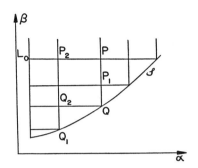

FIG. 74. Domain in the (α,β)-plane representing the interaction of a rarefaction wave and a shock.

For solving the difference equations (83.07) as well as the relations (83.05) an iteration process can be employed. First assume that $\Delta\eta = 0$ at the point in question and the first four equations of (83.07) give a first approximation for x, t, u, c. The fifth equation of (83.07) determines a new value for η, which can then be used to obtain new values of x, t, u, c. This iteration converges quite rapidly. Furthermore, it is convenient to tabulate the functions h, g, k of relation (83.05) numerically.

An interesting concrete result of such computations may be mentioned. In the overtaking of a shock wave by a rarefaction wave the "decay time" is quite long in general. As an example, consider a shock traveling at 2.6 times sound speed into atmosphere with a pressure of 7.3 atm. behind it; it is overtaken by a rarefaction wave which needs one unit of time to pass the point on the x-axis at which the rarefaction first meets the shock wave. Then a computation shows that after 230 units of time, the speed of the weakened shock drops to 1.5 times sound speed while the pressure behind it falls to 2.4 atm.

It should be mentioned that a satisfactory theoretical treatment of boundary value problems, as formulated, seems well within reach but has not yet been carried out.

E. Detonation and Deflagration Waves

84. Reaction processes

About 1880 a number of French physicists, chiefly Vielle, Mallard, Le Chatelier and Berthelot, began experimenting with the propagation of flames. They found that under normal circumstances the flame in a tube, filled with combustible gas ignited at one end, is propagated with a low velocity of a few meters per second. Under certain circumstances, however, the slow *combustion process* changes over into a very fast process which races ahead with a tremendous speed, 2000 meters per second and more. This second type of combustion process was called *detonation*. Naturally, the strange fact that there are two such completely different types of propagation of combustion (occurring time and again not only in gases but also in solid explosives) called for a theoretical explanation. A very simple and convincing explanation was given in 1899 by Chapman and independently in 1905 by Jouguet. In this explanation it is assumed that the chemical reaction takes place instantaneously; in other words, that there is a sharply defined front sweeping over the unburnt gas and changing it instantaneously into burnt gas. Evidently the transition across such a front is analogous to the transition of uncompressed to compressed gas across a shock front. The only difference from a shock transition is that the chemical nature of the burnt gas differs from that of the unburnt gas and that the reaction influences the energy balance.

It is true that the assumption of sharp flame fronts is an extreme idealization, generally acceptable for detonation but rarely satisfied for combustion processes. Combustion processes are in general strongly influenced by the viscous boundary layer which develops at the walls of the tube. They are also strongly dependent on gravity and frequently the resulting flow is turbulent. Nevertheless, valuable insight of a general character is gained on the basis of the preceding assumption.

In the present part we shall discuss the theory of gas flow involving chemical reaction processes which, by assumption, take place across sharply defined fronts. We shall see that these reaction processes have many features in common with shocks but that there are striking differences, particularly as to the unique determinacy of the flow process. Nevertheless, on the basis of very simple arguments we can explain why, in spite of their great similarities to shock processes, detonations and combustion flames are propagated in a different manner. One of the differences between shocks and reaction processes is the fact that the internal energy function $e(\tau, p)$ for the burnt gas is different from that for the unburnt gas. In addition we must take into account the energy liberated during the chemical reaction and transformed partly into kinetic and partly into internal energy of the burnt gas. The liberated energy is taken from the molecular binding energy of the unburnt gas. Here we denote as the molecular binding energy the potential energy of the atoms in the molecule, contrary to common usage. We assume that the process is exothermic, i.e. that the energy used up in recombining the atoms to form the new molecules is smaller for the burnt gas than the binding energy of the unburnt gas. We denote the binding energy or "*energy of formation*" per unit mass by g and introduce the complete energy

$$E = e + g.$$

We assume that this complete energy is a known function $E = E^{(0)}(\tau, p)$, of specific volume τ and pressure p for the unburnt gas, and $E = E^{(1)}(\tau, p)$ for the burnt gas, even though the chemical composition of the burnt gas may actually vary with pressure and specific volume. After these stipulations the relations governing the transition from unburnt to burnt gas are derived from the three conservation laws of mass, momentum, and energy in the same way as were the relations governing a shock transition.

We suppose that the flow of the burnt and unburnt gases takes place in a cylindrical tube, and that across each cross section the state of the gases, characterized by specific volume τ, pressure p, and velocity u is uniform. As before we denote the velocity of the reaction front by U and the velocity of the gases relative to the front by $v = u - U$. If we observe the process from a frame moving with the reaction front and indicate the state in the unburnt gas by the subscript 0 and the state in the burnt gas by the subscript 1, the two laws of conservation of mass and momentum are identical with the corresponding laws for shock fronts

(84.01) $$\rho_0 v_0 = \rho_1 v_1 = m,$$

(84.02) $$p_0 + \rho_0 v_0^2 = p_1 + \rho_1 v_1^2.$$

The law of conservation of energy takes the form

(84.03) $$E^{(0)}(\tau_0, p_0) + p_0 \tau_0 + \tfrac{1}{2} v_0^2 = E^{(1)}(\tau_1, p_1) + p_1 \tau_1 + \tfrac{1}{2} v_1^2.$$

The latter formula is essentially different from the corresponding formula for shocks; for instead of the internal energy we have the complete energy which is, moreover, a different function of τ and p on the two sides of the front. We shall see that this difference has important consequences.

From the first two (mechanical) conservation laws (84.01) and (84.02) the relation

(84.04) $$\frac{p_1 - p_0}{\tau_1 - \tau_0} = -\rho^2 v^2 = -\rho_0^2 v_0^2$$

follows, as for shocks, see (54.08–.09). Formula (84.04) implies that the sign of $p_1 - p_0$ is opposite to that of $\tau_1 - \tau_0$; in other words, that pressure and specific volume decrease or increase in opposite directions or, what is equivalent, that both *pressure and density increase and decrease in the same direction*. Thus two different types of processes are compatible with the conservation laws; those in which both pressure and density increase and those in which both pressure and density decrease. Processes of the first kind are called *detonations;* processes of the second kind correspond to slow combustions. Combustions are sometimes called *deflagrations* in contrast to detonations. In non-reacting gases, processes in which pressure and specific volume decrease, and which satisfy the shock transition relations, were excluded because they would involve a decrease of entropy. However, deflagration processes cannot, as we shall see, be excluded on such grounds. A characteristic difference between detonations and deflagrations is implied in the formula

(84.05) $$\frac{p_1 - p_0}{u_1 - u_0} = -\rho_0 v_0 = -\rho_1 v_1,$$

which follows from the conservation of momentum (84.02). For simplicity, take a specific case in which the reaction front is facing backward and accordingly v_0 and v_1 are positive; the unburnt gas (0) is to the left of the front and the burnt gas (1) is to the right. Relation (84.05) then shows that for a detonation, in which the pressure

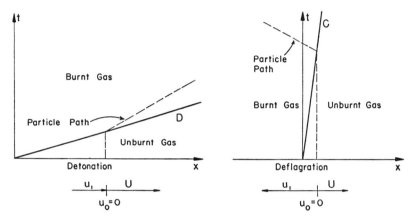

Fig. 75. Change of gas velocity through a detonation (D) and a deflagration (C).

increases, the velocity of the gas decreases when the reaction front sweeps over it. In other words, *a detonation retards the burnt gas relative to the front*. On the other hand, *in a deflagration*, in which the pressure decreases, *the gas is accelerated away from the reaction front*, when the reaction front sweeps over the unburnt gas.

85. Assumptions

For the following arguments we shall make a few assumptions about the nature of the energy function $E(\tau, p)$ usually satisfied in physical reality. First we shall assume that the partial derivatives of E with respect to p, and of the complete enthalpy

(85.01) $$I = E + p\tau$$

with respect to τ, are positive, i.e.,

(85.02) $$E_p > 0, \quad I_\tau > 0.$$

A second assumption is that the relation

(85.03) $$dE = de = -p\,d\tau + T\,dS$$

holds. It is valid if we can disregard the dependence of the energy of formation g on p and τ.

Incidentally, assumptions (85.02) follow from (85.03) if we make the basic assumptions $g_\tau < 0$, $g_S > 0$, see (2.04) and (2.07), for the function $p = g(\tau, S)$. In that case $e_p g_S = T > 0$ yields $e_p > 0$ or $E_p > 0$; further, $i_\tau = e_\tau + p = -e_p g_\tau$ is then positive, and thus $I_\tau = i_\tau > 0$.

A further requirement expressing the exothermic character of the process is that for the same pressure and density the total energy and enthalpy of unburnt gas is always greater than the total energy and enthalpy of burnt gas. Specifically we require this for the values of (τ_0, p_0) and (τ_1, p_1) in the unburnt and burnt gas respectively, just before and just after the process, i.e.

(85.04)
$$E^{(0)}(\tau_0, p_0) > E^{(1)}(\tau_0, p_0),$$
$$I^{(0)}(\tau_0, p_0) > I^{(1)}(\tau_0, p_0),$$

(85.05)
$$E^{(0)}(\tau_1, p_1) > E^{(1)}(\tau_1, p_1),$$
$$I^{(0)}(\tau_1, p_1) > I^{(1)}(\tau_1, p_1).$$

86. *Various types of processes*

It is useful to eliminate the velocities from the three relations, (84.01), (84.02), and (84.03), as we did for shocks in Section 64. We then obtain the *Hugoniot relation*

(86.01) $$E^{(1)}(\tau_1, p_1) - E^{(0)}(\tau_0, p_0) = -\tfrac{1}{2}(\tau_1 - \tau_0)(p_1 + p_0),$$

which by (85.01) is equivalent to

(86.02) $$I^{(1)}(\tau_1, p_1) - I^{(0)}(\tau_0, p_0) = \tfrac{1}{2}(\tau_1 + \tau_0)(p_1 - p_0).$$

Suppose we consider a process in which the specific volume remains unchanged, that is $\tau_1 = \tau_0$. Relation (86.01) then implies

(86.03) $$E^{(1)}(\tau_1, p_1) = E^{(0)}(\tau_0, p_0).$$

From assumptions (85.04) and (85.02) the inequality $p_1 > p_0$ follows immediately. In other words, the process is a detonation, a so called *"constant volume detonation."* From relation (84.04) it is seen that the *speed* $|v_0|$ of the front relative to the unburnt gas is infinite in such a constant volume detonation. That indicates that such a process should be considered only as a limiting case of detonations involving small volume increases and racing ahead with tremendous speeds.

Consider, on the other hand, a process in which the pressure does not change, $p_1 = p_0$. By relation (86.02) this implies

(86.04) $\quad\quad I^{(1)}(\tau_1, p_1) = I^{(0)}(\tau_0, p_0).$

By assumptions (85.04) and (85.02) this implies $\tau_1 > \tau_0$. Therefore such a process is a deflagration, a so called *"constant pressure deflagration."* From relation (84.04) we see that in such a deflagration the flame front would be at rest relative to the unburnt gas. Therefore, such a process should be considered as the limiting case of deflagrations involving small pressure drops and progressing very slowly.

For discussing reaction processes it is useful to employ the *Hugoniot function* for the burnt gas

(86.05) $\quad H^{(1)}(\tau, p) = E^{(1)}(\tau, p) - E^{(1)}(\tau_0, p_0) + \frac{1}{2}(\tau - \tau_0)(p + p_0),$

see Section 64. Omitting the subscript 1 for the burnt gas, relation (86.01) can be written in the form

(86.06) $\quad\quad H^{(1)}(\tau, p) = E^{(0)}(\tau_0, p_0) - E^{(1)}(\tau_0, p_0),$

in which the right member is positive according to assumption (85.04). Suppose the specific volume τ_0 and pressure p_0 of the unburnt gas are given, but not the velocity of the reaction front. Then the pressure and specific volume of the burnt gas satisfy relation (86.06) in all reaction processes compatible with the three conservation laws (84.01–.03). However, not all values of τ and p satisfying this relation actually correspond to a reaction process compatible with (84.01–.03) because of the condition

(86.07) $\quad\quad\quad\quad \dfrac{p_1 - p_0}{\tau_1 - \tau_0} < 0,$

derived from (84.04).

A graph of those points in a (τ, p)-plane which satisfy equation

(86.06) and inequality (86.07) is called the *Hugoniot curve*, see Figure 76. It is remarkable that this graph consists of two separate branches, exhibiting the fact that the conservation laws are compatible with two quite distinct types of processes. The two branches will be referred to as "detonation" and "deflagration branches", according as $\tau \leq \tau_0$ or $p \leq p_0$.

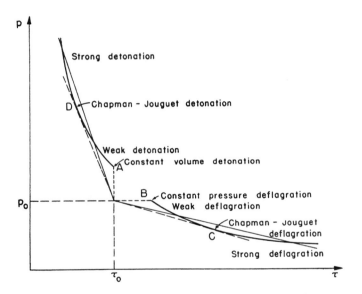

Fig. 76. Hugoniot curve for detonations and deflagrations.

Among the detonations and deflagrations we shall again distinguish several types. For this discussion we make the additional assumption that the pressure increases indefinitely along the detonation branch of the Hugoniot curve.

We consider in the (τ, p)-plane a straight line through the point (τ_0, p_0) and its intersections with the Hugoniot curve. If the slope $\dfrac{p - p_0}{\tau - \tau_0}$ is a large negative number there will be an intersection near the point A, corresponding to a constant volume detonation. From the assumption that the pressure increases indefinitely along the detonation branch of the Hugoniot curve it follows that this straight line has another intersection with the detonation branch. We shall show later, in Section 88, that there are only these two points of intersection, at least if assumption (85.03) is valid. If we increase the slope $(p - p_0)/(\tau - \tau_0)$, the two points of intersection eventually coalesce at a point D because, when p reaches the value p_0, the only intersection lies on the deflagration branch as we have shown before. Thus there is a point D on the detonation branch which separates it into two parts. In other words, as we note for later use, any ray through (τ_0, p_0) with a slope somewhat less than that of the ray through D intersects the Hugoniot curve in two points. Detonations represented by points on the lower part of this branch, i.e. the part with smaller values of p, will be called *weak detonations;* those represented by points on the upper part of this branch will be called *strong detonations*. The detonation corresponding to the point D separating these two points is called a *Chapman-Jouguet detonation*.

Similarly, if the slope $(p - p_0)/(\tau - \tau_0)$ of the straight line is a small negative number, it will intersect the deflagration branch near the point B which corresponds to a constant pressure deflagration. On decreasing this slope a second intersection and, as we shall show, at most a second intersection, may occur. Deflagrations represented by the first intersection points will be called weak, those represented by the second points will be called strong. *Weak and strong deflagrations* are separated by a so called *Chapman-Jouguet deflagration* represented by the point C. For later use we note that any ray through (τ_0, p_0) with a slope somewhat greater than that of the ray through C intersects the Hugoniot curve in two points.

87. Chapman-Jouguet processes

The Chapman-Jouguet reaction processes have a variety of peculiar properties which we now discuss using assumption (85.03) throughout. By the definition of the points D and C the straight line through the point (τ_0, p_0) and D or C is a tangent to the Hugoniot curve. In other words, for Chapman-Jouguet reactions the relation

(87.01) $$\frac{dp}{d\tau} = \frac{p - p_0}{\tau - \tau_0}$$

holds, in which the differentiation refers to the Hugoniot curve. From (87.01) and from the identity

(87.02) $\quad dH^{(1)}(\tau, p) = TdS + \tfrac{1}{2}\{(\tau - \tau_0)dp - (p - p_0)d\tau\}$,

which follows from (86.05) by assumption (85.03), we have the relation

(87.03) $\quad\quad\quad\quad\quad dS = 0 \quad \text{at } D \text{ and } C.$

In other words, if the state of the unburnt gas is fixed, *the entropy of the burnt gas behind the various possible reaction processes assumes a stationary value for a Chapman-Jouguet process.* Furthermore since an adiabatic process is characterized by $dS = 0$, the derivative $dp/d\tau$ along the Hugoniot curve at D or C agrees with the derivative $dp/d\tau$ along the adiabatic curve through D or C.

The sound speed c of the burnt gas satisfies the relation

(87.04) $\quad\quad\quad\quad\quad \rho^2 c^2 = \dfrac{dp}{d\tau},$

in which the differentiation $dp/d\tau$ refers to the adiabatic curve; hence the same relation holds at the point D or C when the differentiation refers to the Hugoniot curve. Combining relations (87.04), (87.01), and (84.04) we obtain

(87.05) $\quad\quad\quad\quad\quad c = |v| \quad \text{at } D \text{ or } C.$

That is to say, in a Chapman-Jouguet process the speed $|v|$ of the burnt gas relative to the front equals the sound speed of the burnt gas, or *a Chapman-Jouguet reaction front, when observed from the burnt gas behind it, moves with sound speed.* This is the famous statement made by Jouguet in 1905. A further property is derived from the relation

(87.06) $\quad\quad\quad\quad (u_0 - U)^2 = v_0^2 = -\tau_0^2 \dfrac{p - p_0}{\tau - \tau_0},$

which follows from (84.04). Differentiating this relation along the Hugoniot curve, keeping τ_0 and p_0 fixed, we obtain the relation

(87.07) $\quad\quad dv_0^2 = \dfrac{-\tau_0^2}{(\tau - \tau_0)^2} \{(\tau - \tau_0)dp - (p - p_0)\,d\tau\},$

from which

(87.08) $\quad d(u_0 - U) = dv_0 = 0 \quad$ at D and C.

Relation (87.08) states: *Among the various possible reaction processes starting from the fixed state (0), the Chapman-Jouguet process yields a stationary value for the velocity of the reaction front relative to the unburnt gas.* This is the property formulated by Chapman in 1899.

More specifically the character of the stationary values of the entropy and the reaction front velocity in Chapman-Jouguet reactions can be described as follows: *for Chapman-Jouguet detonations the speed $|v_0|$ and the entropy of the burnt gas are relative minima, while for Chapman-Jouguet deflagrations the speed $|v_0|$ and the entropy are relative maxima.*

Significantly, for a Chapman-Jouguet detonation, which is the one that occurs under normal circumstances as we shall see, the entropy is a minimum. One might, on the contrary, have expected a detonation for which the entropy is a maximum. (This expectation has misled some writers into claiming that S has a maximum at D.) It is also interesting that the Chapman-Jouguet detonation is the slowest of all detonations.

We proceed to justify these statements by first deriving the relation

$$\frac{d^2 p}{d\tau^2} \geq 0 \quad \text{at } D \text{ and } C,$$

in which the differentiation is to be taken along the Hugoniot curve. This follows immediately from the statement made earlier, at the end of Section 86, that the ray through (τ_0, p_0) with a slope somewhat less or greater than the slope of the tangent ray through D or C respectively intersects the Hugoniot curve at two points near D or C.

Next we differentiate the relation

$$2T \frac{dS}{d\tau} = (p - p_0) - (\tau - \tau_0) \frac{dp}{d\tau}$$

along the Hugoniot curve, see (87.02). The result is

(87.09) $\quad 2T \dfrac{d^2 S}{d\tau^2} = -(\tau - \tau_0) \dfrac{d^2 p}{d\tau^2} \quad$ at D and C

since $\frac{dS}{d\tau} = 0$ at these points. Consequently,

$$\frac{d^2 S}{d\tau^2} \geq 0 \quad \text{at } D,$$

$$\leq 0 \quad \text{at } C.$$

We now exclude the equality sign by the following argument. By differentiating the relation $p = g(\tau, S)$ along the Hugoniot curve, we find

$$\frac{dp}{d\tau} = g_\tau + g_S S_\tau \quad \text{and}$$

$$\frac{d^2 p}{d\tau^2} = g_{\tau\tau} + g_S S_{\tau\tau} \quad \text{at } D \text{ or } C,$$

since $S_\tau = 0$ there. Through the basic assumption $\varrho_{\tau\tau} > 0$, the last relation shows that $p_{\tau\tau}$ and $S_{\tau\tau}$ cannot vanish simultaneously at D or C. Relation (87.09) then shows that these quantities cannot vanish at all at D or C. Hence we have

(87.10) $$\frac{d^2 p}{d\tau^2} > 0 \quad \text{at } D \text{ or } C,$$

(87.11) $$\frac{d^2 S}{d\tau^2} > 0 \quad \text{at } D, \quad \frac{d^2 S}{d\tau^2} < 0 \quad \text{at } C.$$

The latter relation implies that S has a minimum at D and a maximum at C.

Differentiating relation (87.07) along the Hugoniot curve we obtain, through (87.01),

$$\frac{d^2 v_0^2}{d\tau^2} = -\frac{\tau_0^2}{\tau - \tau_0}\frac{d^2 p}{d\tau^2} \quad \text{at } D \text{ and } C,$$

from which by (87.10)

(87.12) $$\frac{d^2 v_0^2}{d\tau^2} > 0 \quad \text{at } D, \quad \frac{d^2 v_0^2}{d\tau^2} < 0 \quad \text{at } C.$$

It then follows that $|v_0|$ is a minimum at D and a maximum at C.

88. Jouguet's rule

We shall now formulate characteristic properties by which we can distinguish weak from strong reaction processes. These properties are embodied in the following statement which we shall refer to as *Jouguet's Rule: The gas flow relative to the reaction front is*

supersonic	*ahead of a detonation front*
supersonic	*behind a weak detonation front*
subsonic	*behind a strong detonation front*
subsonic	*ahead of a deflagration front*
subsonic	*behind a weak deflagration front*
supersonic	*behind a strong deflagration front.*

Jouguet's rule is best derived on the basis of the considerations introduced by H. Weyl for the discussion of shocks and explained in Section 65, see also [73]. The following facts were shown to hold for any ideal gas and hence for the burnt gas in a reaction process.

On any straight ray in the (τ, p)-plane through the point (τ_0, p_0), any stationary value of the entropy S is a maximum. Consequently, there is at most one such stationary value. Further, because of the relation

$$(88.01) \qquad dH^{(1)} = TdS,$$

which by (87.02) holds along any ray, the Hugoniot function $H^{(1)}$ also has at most one stationary value on that ray, and this value is also a maximum.

An immediate consequence of this fact is the important statement made previously in Section 87: *A straight line through the point* (τ_0, p_0) *intersects the Hugoniot curve in at most two points.* Otherwise the function $H^{(1)}$ would have at least two stationary values on this line.

We now consider the two intersections which the straight ray through the point (τ_0, p_0) makes with any of the two branches of the Hugoniot curve characterized by equation (86.06),

$$(86.06) \qquad H^{(1)}(\tau, p) = E^{(0)}(\tau_0, p_0) - E^{(1)}(\tau_0, p_0),$$

and the inequality (86.07). The "first" point of intersection corresponds to a weak process, the "second" one to a strong process. At these two points of intersection the Hugoniot function $H^{(1)}$ has the same value, and hence it has a maximum at some intermediate point. Consequently, the entropy S also has a maximum at this point, and since S has no other stationary value along the ray, S increases at the first point of intersection and decreases at the second.

If the ray intersects the detonation branch, τ decreases along it from the point (τ_0, p_0) on, and thus the last statement is expressed by

(88.02) $$\frac{dS}{d\tau} < 0, \quad \text{and} \, > 0,$$

respectively, at the first and second points of intersection with the detonation branch. Similarly

(88.03) $$\frac{dS}{d\tau} > 0, \quad \text{and} \, < 0,$$

respectively, at the first and second points of intersection with the deflagration branch. Along a straight line $dp/d\tau = (p - p_0),'(\tau - \tau_0)$; hence $dp/d\tau = -\rho^2 v^2$ by (84.04). Further, along an adiabatic $S_\tau/S_p = -dp/d\tau$; hence $S_\tau/S_p = \rho^2 c^2$ by (2.05). Consequently

(88.04) $$\frac{dS}{d\tau} = \rho^2(c^2 - v^2)S_p$$

since $S_p = 1/p_S > 0$ by (2.07).

Thus conditions (88.02) and (88.03) are equivalent to

(88.05) $$c < |v|, \quad \text{and} \quad c > |v|,$$

respectively, at the first and second points of intersection with the detonation branch,

(88.06) $$c > |v|, \quad \text{and} \quad c < |v|,$$

respectively, at the first and second points of intersection with the deflagration branch.

Thus the part of *Jouguet's rule referring to the state behind the detonation or deflagration front is proved.*

In order to establish *Jouguet's rule for the state ahead of a reaction front* we consider a state behind the front as fixed and vary the state ahead of it. Equation (86.01) or (86.06) now takes the form

(88.07) $$H^{(0)}(\tau, p) = E^{(1)}(\tau_1, p_1) - E^{(0)}(\tau_1, p_1),$$

the right member of which is now negative by assumption (85.05).

Again it remains true that $H^{(0)}$ and S have at most one stationary value along the ray and that this value is a maximum. In contrast to the situation for the Hugoniot curve given by (86.06), any ray through the point (τ_1, p_1) can intersect the Hugoniot curve given by (88.07) in at most one point. For, suppose that there were more than one point of intersection; then the quantity $H^{(0)}$, which vanishes at the point (τ_1, p_1) and assumes the same value at the points of intersection, would have a minimum between two such points. For the

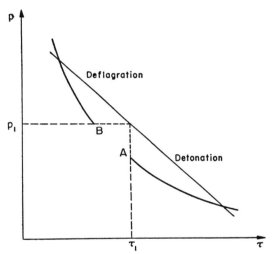

FIG. 77. Hugoniot curve for the state ahead of the reaction front if the state (τ_1, p_1) behind it is given.

same reasons it is clear that $H^{(0)}$ decreases along the ray at a point of intersection. This is also true for the entropy S, of (88.01). At a point of intersection with the detonation branch we have $\tau > \tau_1$; hence $dS/d\tau < 0$ there, a result which by (88.04) is equivalent to $v^2 > c^2$. Thus we have shown that the flow ahead of a detonation wave is supersonic. In a similar manner we find that ahead of a deflagration wave the flow is subsonic $v^2 < c^2$. Thus Jouguet's rule is completely proved.

This rule, it should be emphasized, simply enumerates the various combinations mathematically compatible with the conservation laws across a discontinuity front. Physical reality, however, does not present all the possibilities suggested by this mathematical pattern. As a further discussion in Sections 93 and 94 will show, *strong deflagrations can never actually occur, and weak detonations are possible only under extreme and rare circumstances.* Postponing this deeper analysis, we first draw conclusions within our present mathematical framework.

89. Determinacy in gas flow involving a reaction front

We assume that at an initial time, $t = 0$, a cylindrical tube extending from $x = 0$ to $x = \infty$ is filled with combustible gas, and that at the time $t = 0$ at the cross section $x = 0$ a reaction starts which proceeds in a forward direction. Usually one is concerned with the case in which the end $x = 0$ remains closed or is completely open. We shall, however, give a slightly more general analysis by considering the case in which a piston starts to move at the time $t = 0$ from the end $x = 0$ with the prescribed constant velocity u_P. This analysis not only contributes to the clarification of the problems of determinacy but also is helpful for undersanding more complicated phenomena.

Our object is to solve the problem: To what extent is the gas flow determined by the initial state and the piston motion, provided that there occurs one sharp reaction front moving at constant speed into the combustible gas? Of course, these discontinuities must satisfy the conservation conditions (84.01–.03).

With this provision we can now discuss the existence and uniqueness of a solution for the flow. If no chemical reaction takes place, the flow problem has a uniquely determined solution. This solution involves a uniquely determined shock for positive piston velocity, and a centered simple rarefaction wave for negative piston velocity. The question arises whether the same kind of *determinacy* prevails when a chemical reaction takes place. In other words, given a state of rest for $x > 0$, a reaction starting at $x = 0, t = 0$, and a constant piston velocity $u = u_P$, is there one and only one solution for the flow in the tube? As we shall see, this is not necessarily the case. The explanation for this difference is found in Jouguet's rule.

According to the principles of determinacy formulated in Section 36 the determinacy depends on whether or not the curves in the (x, t)-plane on which data are prescribed are time-like or space-like.

When such a curve is considered, time-like character corresponds to subsonic, space-like character to supersonic velocity relative to the gas flow. Time-like or space-like character in the (x, t) representation means that the two characteristic directions with $dt > 0$ point respectively toward different or toward the same sides of the curve. We assume for simplicity that the flow of unburnt and burnt gas is isentropic, so that there are only two characteristic directions. The statements made in Section 36 then imply that the solution of a flow problem in a region between two curves is unique in two cases:

1) Both curves are time-like and one quantity is prescribed on each, and at the point of intersection two quantities are prescribed.

2) One curve is space-like and has two quantities prescribed on it, and the other curve is time-like and carries one prescribed quantity.

Without proof we further state:

If in case 1) the data are continuous at the intersection of the two lines and the two lines are time-like there, the solutions exist as long as these two lines remain time-like.

The solution exists in case 2) if shocks are admitted as part of it.

In our present case the piston path is always time-like since it is identical with the path of the adjacent gas particles; it carries one quantity, the velocity u_P. The x-axis is evidently space-like; it carries two quantities, the velocity $u = 0$ and the sound speed $c = c_0$ (or p_0 or τ_0) corresponding to the assumption that the unburnt gas is originally at rest. However, the statement on uniqueness cannot be applied directly to these two lines because of the interference of the unknown path of the reaction front. The statement must therefore be applied separately to the sector between the x-axis and the reaction path $\mathcal{W}$, and the sector between the reaction path and the piston path $\mathcal{P}$.

We distinguish four cases AA, AB, BA, BB according as the flow relative to the reaction path $\mathcal{W}$ is supersonic (A) or subsonic (B), before or behind the front. The determining elements are the relative positions of the characteristics C_+, C_-, and of the particle path C_0 and accordingly there are the following four possibilities as represented in Figure 78:

Suppose that the slope of $\mathcal{W}$, that is, the velocity of the reaction front, is prescribed arbitrarily. If this reaction is supersonic relative to the gas in front of it, the reaction path $\mathcal{W}$ is space-like observed from the region ahead of it. The initial values, $u_0 = 0$ and c_0, along

220 ONE-DIMENSIONAL FLOW CHAP. III

the x-axis at $t = 0$ determine the flow uniquely up to $\mathfrak{W}$, namely $u = u_0$, $c = c_0$ throughout. By the transition conditions (84.01–.03), the values of u and c immediately behind the reaction path are

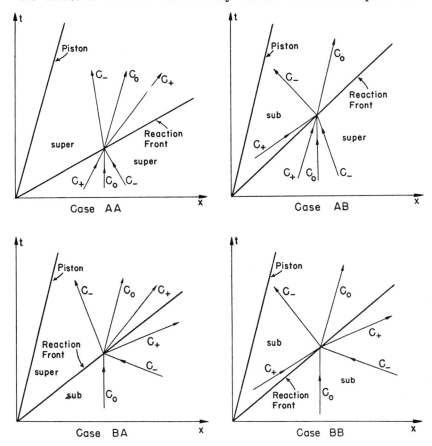

Fig. 78. Possible relations between characteristic directions and reaction front.

determined. For the continuation of the flow there are now two possible cases.

In the first case, AA, the flow behind $\mathfrak{W}$ is also supersonic relative to $\mathfrak{W}$. Then $\mathfrak{W}$ is space-like for the region behind it and carries two data, u and c. By Jouguet's rule, the reaction then is a *weak detonation*. Since the piston path $\mathcal{P}$ is always time-like, the flow in the domain between $\mathcal{W}$ and $\mathcal{P}$ is uniquely determined according to the general principle, see Section 28. Since the velocity of the reaction front could be chosen arbitrarily (subject only to the condition of being supersonic), there remains one *degree of indeterminacy*.

In the second case, AB, however, the path $\mathcal{W}$ is time-like with respect to the region behind it. Accordingly, immediately behind $\mathcal{W}$ considered as initial curve for the region between $\mathcal{W}$ and $\mathcal{P}$, only one quantity can be prescribed arbitrarily in order that a unique solution exist. The other quantity behind $\mathcal{W}$ is then determined. Since this quantity is also determined by the slope of $\mathcal{W}$, it is clear that this slope, the velocity of the reaction front, is no longer arbitrary but is to be adjusted properly. (We shall show in Section 90 that this can be done if the piston velocity u_P is constant and sufficiently large.) While thus in the case AA the wave front $\mathcal{W}$ could be chosen arbitrarily, there is no such leeway in the choice of $\mathcal{W}$ in the case AB, in which by Jouguet's rule the reaction is a *strong detonation*. The situation is similar to that for a flow in a non-reactive fluid which involves a *shock*, where the velocity of the shock front is determined in such a way that adjustment to the prescribed piston motion results. Similarly, in case AB, the flow is determined by the initial conditions and the piston velocity. The degree of indeterminacy is then zero.

In the case BA the reaction path $\mathcal{W}$ observed from the state ahead of it is time-like, so that in front of $\mathcal{W}$ one boundary condition can be arbitrarily imposed. Then behind the reaction front the state is determined by the transition conditions; $\mathcal{W}$ is space-like with respect to this region, $\mathcal{P}$ is time-like, and according to the general principle, the flow between $\mathcal{W}$ and $\mathcal{P}$ is determined. Thus there are two degrees of indeterminacy in the case BA, in which the reaction is a *strong deflagration*. In addition to the initial state and the piston motion two more quantities, the reaction path and one quantity ahead of it, can and must be prescribed to determine the flow.

In case BB, the reaction path $\mathcal{W}$ is time-like with respect to both adjacent zones. The quantity prescribed in front of $\mathcal{W}$ must then be chosen so that the resulting values of the three quantities behind $\mathcal{W}$ are compatible with the boundary condition at the piston. Accordingly, there is one degree of indeterminacy in this case, in which the reaction is a *weak deflagration;* a single quantity, such as the wave velocity of the reaction front, can and must be prescribed to determine the flow.

It should be emphasized that *in both cases BA and BB* the state ahead of the reaction front is not necessarily constant; in other words, *the state ahead of the reaction front is influenced by the reaction.*

The higher degree of indeterminacy of flows involving a detonation or deflagration front as compared with flows involving merely shock discontinuities is due to the fact that all cases AA, AB, BA, BB are compatible with the conservation laws in reaction processes, while shocks always belong to the case AB.

90. Solution of flow problems involving a detonation process

We now supplement the preceding general remarks by showing in detail, for the case in which the piston velocity u_P is constant, how the adjustment of the flow to the boundary conditions at the piston is effected.

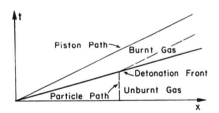

FIG. 79. Strong detonation supported by a piston moving with velocity $u_P > u_D$.

Let us first consider a *detonation*, beginning at the point $x = 0$ at the time $t = 0$, with a state of rest, $u_0 = 0$, ahead of it. As we saw at the end of Section 84, the velocity u behind the front is positive. The velocity of the reaction front is then determined from the three conservation laws. By $u_D = c_D$, see (87.05), we denote the particular value of the gas velocity behind the detonation front which satisfies the Chapman-Jouguet condition.

We start with the case

(90.01) $$u_P > u_D$$

and maintain that then just one flow, involving a *strong detonation front*, is possible. The state behind the front in this flow is constant; in particular the velocity u equals u_P. The situation is similar to that in a flow involving a shock front. Since the strong detonation front is time-like when observed from behind, the remarks made at the beginning of Section 89 make it likely that no other solution exists.

Next we show that whenever

(90.02) $$u_P \leq u_D$$

a flow involving a *Chapman-Jouguet detonation* is possible. The gas flow $u = u_D$ immediately behind the Chapman-Jouguet front need not be equal to the piston velocity u_P. Adjustment in this

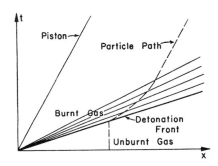

FIG. 80. Chapman-Jouguet detonation. The flow of the burnt gas comes to rest through a centered rarefaction wave.

case *is effected by a centered simple wave which follows the detonation front immediately*, since both the front and the first sound wave in the simple wave move with sound velocity relative to the gas behind the front, see Figure 80. Only if $u_P = u_D$ does this rarefaction wave drop out.

If the end of the tube is fixed, $u_P = 0$, or withdrawn, $u = u_P < 0$, the present case applies. A Chapman-Jouguet detonation is then possible. That this detonation is the one that actually occurs is essentially the famous *Chapman-Jouguet hypothesis*.

On the basis of the present analysis a flow involving a *weak detonation* would be possible for any piston velocity u_P. If the velocity u immediately behind the front lies between u_P and u_D, the flow would also involve a centered rarefaction wave which, however, would lag behind the front. This wave would drop out if $u = u_P$. If $u < u_P$, adjustment would be effected by a shock which also would lag behind the detonation. In the latter case u is restricted to a value below a certain value $u_* < u_P$. For $u = u_*$ the shock and detonation fronts would coalesce and together have the same effect as a strong detonation. We do not discuss these matters in greater detail for the simple reason that weak detonations do not exist, as we shall show in Section 94; we mention them only in order to make it clear that weak detonations cannot be excluded on the basis of the three conservation laws alone.

91. Solution of flow problems involving deflagrations

In *deflagration* processes the situation is in many respects quite different from that encountered with detonation processes. Suppose a weak deflagration wave begins at the piston, $x = 0$, and moves into the unburnt gas at rest in the interior of the tube, $x > 0$. Then by (84.05) the velocity u of the burnt gas behind the deflagration front is negative, $u < 0$. This is compatible with the conditions of the problem only if the piston is withdrawn with a speed at least equal to u. Otherwise it is impossible for a deflagration front to move into the quiet explosive gas. We have here the case BB; the line $\mathcal{W}$ is time-like with respect to the region behind it; an adjustment to a forward moving piston is impossible. Therefore the flow must necessarily take place according to a pattern different from that assumed in the preceding analysis. What actually happens is that *a pre-compression wave is sent out into the explosive*. It pushes the explosive gas ahead with a velocity just sufficient to ensure that it may come to rest when it is swept over and burnt by the deflagration front.

The occurrence of a pre-compression wave is in complete agreement with, or rather a consequence of, the theoretical considerations of Section 88. Indeed, according to Jouguet's rule, the flow ahead of a deflagration is subsonic (case B) and consequently the deflagration influences the state of the gas ahead of it.

If the piston velocity is constant, for example, if the piston is kept fixed, $u_P = 0$, a flow exists involving a constant deflagration and a centered pre-compression wave. (It is likely that this is the only solution of the flow problem.) A centered compression wave is nothing but a constant shock wave. The solution of the problem accordingly involves *a constant pre-compression shock sent out into the unburnt gas compressing and accelerating the unburnt gas so that, through the constant deflagration wave following the shock wave, the gas burns and comes to rest or acquires the prescribed piston velocity*, see Figure 81.

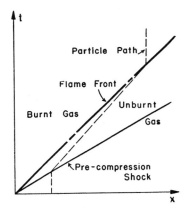

FIG. 81. Weak deflagration with pre-compression shock and closed end $u_P = 0$.

To each assumed pre-compression shock there is still a great variety of possible adjustments, either by a uniquely determined weak deflagration wave (case BB) or by a strong deflagration wave (case BA) (which may be arbitrarily chosen within a one-parametric variety and is followed again by rarefaction or shock waves). These possibilities are, of course, limited by the condition that the velocity of each wave is less than that of the preceding wave.

The main point in the discussion of deflagrations is that deflagration processes have a higher degree of indeterminacy than detonation processes. Again it is necessary to exclude certain deflagration processes on the basis of a discussion of their internal mechanisms, although such processes are compatible with the conservation laws. We shall indicate such a discussion later on, see Sections 93 and 94.

The result will be that *strong deflagrations are never possible and that weak deflagrations are possible only with a well determined speed v_0 depending on the state of the unburnt gas and also on the heat conductivity of the gases.* On the basis of our previous discussion it could then be shown that, again, to every piston velocity u_P *a flow involving a constant pre-compression shock and a constant deflagration wave exists*, under certain limitations for u_P. The limiting case is that in which the pre-compression shock coalesces with the deflagration.

92. Detonation as a deflagration initiated by a shock

A deflagration wave coalescing with its pre-compression shock wave is equivalent to a detonation wave, see Jouguet [64], v. Neumann [71]. Thus, by increasing the strength of the pre-compression wave so much that its speed equals that of a possible detonation, one may envisage a continuous transition from a deflagration process to a detonation process. We shall see in Section 94 that the interpretation of a detonation as a shock followed by a deflagration describes to a certain extent the internal mechanism of a detonation process.

The formal equivalence of these two processes is easily verified. Consider a shock with the velocity U through which the pressure p_0 and the density ρ_0 are raised to p_* and ρ_* and which produces a flux $m = \rho_0(u_0 - U) = \rho_*(u_* - U)$. Let this shock be followed by a deflagration with the velocity U_* through which the pressure p_* and the density ρ_* drop to p_1 and ρ_1 in such a way that the relative flux is the same as through the shock, $m = \rho_*(u_* - U_*) = \rho_1(u_1 - U_*)$. It follows that both processes have the same velocity, $U_* = U$, and hence coalesce if they did so at some initial time. From the mechanical conditions (84.01–.02) the relation

$$\frac{p_* - p_0}{\tau_* - \tau_0} = \frac{p_1 - p_*}{\tau_1 - \tau_*} = \frac{p_1 - p_0}{\tau_1 - \tau_0} = -m^2$$

follows. Since the energy law takes the form $v_0^2 + E_0 = v_*^2 + E_* = v_1^2 + E_1$, it is clear that the transition from the state (0) to the state (1) is equivalent to a single reaction process. That this process is a detonation is seen from Figure 82 in which the Hugoniot curves for shocks and reaction processes are shown. There are two processes beginning with (τ_0, p_0) having the same flux m. Since the pressure drop in a weak deflagration is less than in a strong one we

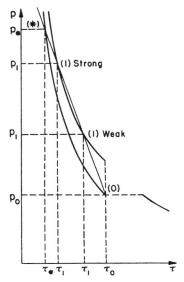

Fig. 82. A detonation as a deflagration coalescing with its pre-compression shock. The two curves are the Hugoniot curves with and without reaction.

see: *a strong detonation is equivalent to a shock followed by a weak deflagration, and a weak detonation is equivalent to a shock followed by a strong deflagration.*

93. Deflagration zones of finite width

We have seen that reaction processes are not determined by the conservation laws, as shocks are. To determine a reaction process its internal mechanism must be considered. A first step in this direction consists in assuming that the change of chemical composition does not take place instantaneously as we have assumed so far, but occurs gradually over a zone of finite width. In this way v. Neumann [71] has arrived at a justification of the Chapman-Jouguet hypothesis for detonations, see also [68]. Before presenting his argument we shall first employ similar reasoning to exclude strong deflagrations.

We assume a steady deflagration process takes place in a finite zone $x_0 \leq x \leq x_1$, unburnt gas (0) being at the left, $x \leq x_0$, burnt gas (1) at the right, $x \geq x_1$. The intermediate compositions of the gas are assumed mixtures of burnt and unburnt gas. We denote by ϵ and $(1 - \epsilon)$ respectively the fraction of burnt and unburnt gas in the mixture. The thermodynamic nature of the mixtures may be characterized by the total energy $E^{(\epsilon)}$ as a function of τ and p. We assume specifically that this total energy is the sum of the total energies that the burnt and unburnt gas contained in the mixture would have for the same values of τ and p:

$$(93.01) \qquad E^{(\epsilon)}(\tau, p) = (1 - \epsilon)E^{(0)}(\tau, p) + \epsilon E^{(1)}(\tau, p).$$

This assumption can be justified if the gases are ideal and accordingly $E^{(\epsilon)}$ depends only on the temperature or, ignoring the variation of the molecular weight, only on the product τp.

The reaction is to take place according to a definite reaction rate K which is a given function of $T = R p \tau$, p, and ϵ:

$$(93.02) \qquad \frac{d\epsilon}{dt} = K(T, p, \epsilon).$$

All that we need know about the reaction rate is that it is not negative:

$$(93.03) \qquad K(T, p, \epsilon) > 0.$$

(Later on we shall permit K to vanish at the ends of the reaction zone.) Since we are dealing with a steady process the reaction equation (93.02) takes the form

$$(93.04) \qquad v \frac{d\epsilon}{dx} = K;$$

v, the velocity of the flow, is positive since the flow is to take place from the unburnt gas at x_0 to the burnt gas at $x_1 > x_0$. Therefore we have

$$(93.05) \qquad \frac{d\epsilon}{dx} > 0.$$

The argument now is that the conservation laws should hold throughout the process, see [70], and that, therefore, the three laws (84.01–.03) in the form

$$\rho_0 v_0 = \rho_\epsilon v_\epsilon = m,$$

(93.06) $$p_0 + \rho_0 v_0^2 = p_\epsilon + \rho_\epsilon v_\epsilon^2,$$

$$E^{(0)}(\tau_0, p_0) + p_0 \tau_0 + \tfrac{1}{2} v_0^2 = E^{(\epsilon)}(\tau_\epsilon, p_\epsilon) + p_\epsilon \tau_\epsilon + \tfrac{1}{2} v_\epsilon^2,$$

should hold; here the subscript ϵ indicates the values of quantities at the place x_ϵ at which the mixture ratio has the value ϵ. Through elimination we obtain relations corresponding to (59.02) and (55.05),

(93.07) $$\frac{p_\epsilon - p_0}{\tau_\epsilon - \tau_0} = -m^2,$$

(93.08) $$E^{(\epsilon)}(\tau_\epsilon, p_\epsilon) - E^{(0)}(\tau_0, p_0) = -\tfrac{1}{2}(\tau_\epsilon - \tau_0)(p_\epsilon + p_0).$$

Equation (93.07) means that in the (τ, p)-plane the process is represented by a straight line, or

(93.09) $$\frac{dp_\epsilon}{d\tau_\epsilon} = \frac{p_\epsilon - p_0}{\tau_\epsilon - \tau_0}.$$

Differentiating relation (93.08) we obtain by (93.01), (93.09), and (85.03) the relation

(93.10) $$T_\epsilon \frac{dS_\epsilon}{d\tau_\epsilon} \frac{d\tau_\epsilon}{dx} = [E^{(0)}(\tau_\epsilon, p_\epsilon) - E^{(1)}(\tau_\epsilon, p_\epsilon)] \frac{d\epsilon}{dx}.$$

We use this relation in particular for $\epsilon = 1$. The right-hand member is then positive because of (93.05) and assumption (85.05). Since we are dealing with a deflagration we have $\dfrac{d\tau_\epsilon}{dx} > 0$. Hence (93.10) implies

(93.11) $$\frac{dS}{d\tau} > 0 \quad \text{for} \quad \epsilon = 1.$$

From inequality (87.11) we now see that the point (τ_1, p_1) is a "first" point of intersection of the ray through (τ_0, p_0) with the Hugoniot curve for $\epsilon = 1$, see Section 88. Hence the process is a weak deflagration. Thus it is shown that *strong deflagrations are impossible*.

230   ONE-DIMENSIONAL FLOW   CHAP. III

The basis of the argument will become clearer if we consider the set of Hugoniot curves expressed, according to (86.06), by

(93.12) $\quad H^{(\epsilon)}(\tau, p) = E^{(0)}(\tau_0, p_0) - E^{(\epsilon)}(\tau_0, p_0)$

$\qquad\qquad\qquad = \epsilon[E^{(0)}(\tau_0, p_0) - E^{(1)}(\tau_0, p_0)]$

(in obvious notation). If the set of curves is as indicated in Figure 83, a transition from (τ_0, p_0) to a "second" point of intersection

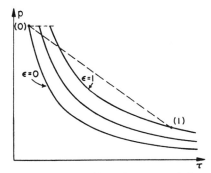

FIG. 83. The impossibility of a strong deflagration.

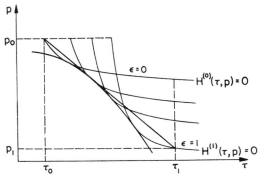

FIG. 84. A situation permitting a strong deflagration but excluded by conditions on the Hugoniot function.

(τ_1, p_1) is not possible along the straight ray without passing through a region which would correspond to $\epsilon > 1$, which is meaningless. A situation, as indicated in Figure 84, in which it would be possible to reach such a point, see [72], is excluded by condition (85.05), because from the definition of $H^{(0)}$ and (85.05) we have $H^{(0)}(\tau_1, p_1) > H^{(1)}(\tau_1, p_1) = 0$, while the figure indicates $H^{(0)}(\tau_1, p_1) < 0$.

The assumptions on which the argument was based are unrealistic in some respects. It has been assumed that no reaction takes place in the gas until it reaches the cross section $x = x_0$, when suddenly the reaction rate assumes a positive value. A more adequate description would be as follows: After ignition has taken place, the unburnt gas is heated through conduction from the parts already burnt; only when the unburnt gas has acquired a sufficiently high temperature will the reaction rate assume noticeable values increasing rapidly with increasing temperature. Thus a proper description of a deflagration process should necessarily take heat conduction into account. *Only by accounting for heat conduction is it possible to understand why, among the set of weak deflagrations compatible with the conservation laws, only one with a particular velocity is possible.* A discussion of this question will be given in Section 96.

94. Detonation zones of finite width. Chapman-Jouguet hypothesis

By analyzing the beginning of the reaction one could also show that a *detonation cannot be described*, in the same manner as we have described a deflagration, *as a gradual change of state guided by the reaction rate*. (Otherwise the preceding argument which excluded strong deflagrations would also exclude strong detonations in contradiction to the facts). The proper description of a detonation is: *a detonation is a combustion process initiated by a shock.* Through the shock, pressure, density, temperature, and entropy are raised instantaneously; through the subsequent combustion process pressure and density are decreased again, while temperature and entropy rise further. Clearly, the combustion process in a detonation is of the same character as a deflagration except that it is initiated differently. Since strong deflagrations are not possible we see that those detonations are impossible in which the combustion process corresponds to a strong deflagration; these are the weak detonations as was shown in Section 92. Consequently, *weak detonations are impossible.* Thus only strong and Chapman-Jouguet detonations remain possible; we have seen in Section 90 that it is plausible that then the flow process is uniquely determined.

Consider the flow in which the detonation starts at the end of a tube where a piston is moved with a velocity $u_P < u_D$, see (90.02); for example, assume that the end is closed, $u_P = 0$, or withdrawn, $u_P < 0$. It was found that in this case no strong detonation is possible. Now we see that *in this case only a Chapman-Jouguet detonation is possible*. Thus we have justified the *Chapman-Jouguet hypothesis*, see Section 90. This justification is largely due to v. Neumann [71].

95. The width of the reaction zone

The width l of a deflagration or detonation zone can be calculated when the reaction rate K is known, by

$$(95.01) \qquad l = x_1 - x_0 = \int_0^l K^{-1} v \, d\epsilon;$$

here K is a given function of p, τ, and ϵ, and v, p, τ are known functions of ϵ through relations (93.06). If, as is more appropriate to assume, the reaction rate vanishes for $\epsilon = 1$, the width l becomes infinite. In that case the definition of the width l is to be modified. It may be defined in a somewhat arbitrary way as the width of a zone over which the changes of the quantities T and ϵ are noticeable.

Often the reaction rate K is assumed to have the form

$$(95.02) \qquad K = (1 - \epsilon) K_\infty(p) e^{-A/T}$$

Other semi-empirical expressions for the reaction rate, depending on the type of chemical reaction, have also been used.

It turns out that the value of l is extremely sensitive to variations of the coefficient A in the exponent in (95.02), which in general is not very accurately known. Therefore formula (95.01) has been used in the reverse direction to determine the value of A from the experimentally measured width of the reaction zone.

96. The internal mechanism of a reaction process. Burning velocity

Assuming that the chemical reaction takes place over a zone of finite width is only the first step in analyzing the mechanism of the reaction process. A more refined discussion would have to account for the way in which the state of the gas changes in addition to the change of composition. It is natural to proceed in the same way as with shocks, see Section 63, and to introduce the effects of heat conduction and viscosity. The same objections to using these notions that were raised in the case of shocks could be raised here. But again it is probable that the picture developed on that basis is qualitatively correct.

The differential equations, in a one-dimensional steady process, are then

(96.01) $$\rho v = m = \text{constant},$$

(96.02) $$\rho v^2 + p - \mu v_x = P = \text{constant},$$

(96.03) $$\rho v(\tfrac{1}{2}v^2 + I^{(\epsilon)}) - \mu v v_x - \lambda T_x = Q = \text{constant},$$

(96.04) $$v \epsilon_x = K.$$

The first two equations are the same as equations (63.05–.07), except for the presence of the total enthalpy $I^{(\epsilon)}$ which depends on the mixture ratio ϵ and includes the energy of formation, see Section 84 and (93.01). Assuming the gas mixture is ideal, we have $T = R\rho\tau$. The last equation (96.04) is the same as (93.04); K depends on T, p, and ϵ.

Once the three constants m, P, Q are known, equations (96.02–.04) are three differential equations of the first order for the quantities v, T, and ϵ through which all other quantities can be expressed. As in the case of a shock we prescribe boundary conditions at $x = -\infty$ and $x = +\infty$. We assume the process to be backward-facing, so that the values $v = v_0$, $\tau = \tau_0$, $p = p_0$, $\epsilon = 0$ for the unburnt gas are to be prescribed at $x = -\infty$. The values $v = v_1$, $\tau = \tau_1$, $p = p_1$, $\epsilon = 1$ for the burnt gas are then prescribed at $x = +\infty$. We seek a solution existing in the whole range $-\infty < x < \infty$ and approaching values at $x = \pm\infty$ in such a way that the derivatives v_x, T_x, and ϵ_x approach zero. To make this possible we require that the reaction rate K vanish for $\epsilon = 1$ and for $\tau = \tau_0$, $p = p_0$, $\epsilon = 0$. The constants m, P, Q can then be expressed in terms of the boundary values. The requirement that these expressions have the same values for the states (0) and (1) imposes three conditions on the boundary values, which of course are nothing but the three *conservation laws*, (84.01–.03). We assume that these three conditions are satisfied.

The question then is *whether any two such states* (0) *and* (1) *obeying the conservation laws can be connected by solutions of the equations* (96.02–.04), or whether additional conditions must be satisfied by the data.

A preliminary investigation shows that the *number of these conditions depends just on whether the flow in the states* (0) *or* (1) *is supersonic or subsonic*, in other words, it depends on *whether the process is a weak or strong detonation or deflagration*, see [72]. More specifically it turns out that *the number of conditions to be satisfied in the four cases* just agrees with *the degree of indeterminacy* found in Section 89 for flows involving any of these four processes. In other words, the "indeterminacy" of what might be called the "external flow problem", found by taking only the conservation laws into account, is balanced by the "over-determinacy of the internal flow problem."

A closer analysis of the solutions of equations (96.01–.04) gives the following results, see [72]. No conditions are to be imposed for strong detonations; one condition is to be imposed on weak detonations, namely that it reduces to the limiting case of a Chapman-Jouguet detonation (except if the reaction rate K assumes extremely high values which are probably excluded by additional physical conditions). On the basis of such an analysis, strong deflagrations should be subjected to two conditions. A more refined analysis, however, shows that they do not exist at all, in agreement with the results of Section 93.

For weak deflagrations the result is that the data at the two ends are subjected to one condition which depends essentially on the coefficient λ *of heat conduction*. In other words only one point on the deflagration branch of the Hugoniot curve, Figure 76, is possible. Here we have a materially new result which could not have been derived by the considerations of Section 93. This condition does not apply to the combustion process occurring in a detonation because this process is initiated differently, by a shock.

To illustrate these statements we consider the limiting case of an "almost constant pressure" deflagration, see Section 86. We assume that the velocity v is small of the same order as the heat conductivity λ and the reaction rate K; further, that the variation of the pressure and the viscous stress μv_x is of the same order of magnitude as v^2, consistent with (96.02). We introduce ϵ and $T = Rp\tau$ as dependent variables. Equations (96.03–.04) then assume the form

(96.05) $\quad \lambda T_x - mI^{(e)} = -mI^{(0)} = -mI^{(1)} = $ constant,

(96.06) $\quad m\epsilon_x = RpK(T, p, \epsilon)/T;$

here the pressure p is a given constant. From these relations we obtain the equation

(96.07) $$\frac{dT}{d\epsilon} = \frac{m^2}{\lambda} \frac{I^{(e)}(T, p) - I^{(0)}}{K(T, p, \epsilon)} \frac{T}{Rp}.$$

A solution of this equation is to be found which satisfies $T = T_0$ for $\epsilon = 0$, $T = T_1$ for $\epsilon = 1$. The values T_0 and T_1 are restricted by the condition

$$I^{(0)}(T_0, p) = I^{(1)}(T_1, p) = I^{(0)} = I^{(1)}.$$

If m is nearly zero, the solution $T = T(\epsilon)$ beginning with $T(0) = T_0$ remains approximately constant. If m is very large, this solution will reach the value T_1 before ϵ has become equal to one. Consequently, there is one particular value of m for which this solution just assumes the value T_1 for $\epsilon = 1$. That m should have just this value is the condition to be imposed on the data. Thus, the relative *burning velocity* $v_0 = m\tau_0$ is determined in its dependence on λ, p, T_0, τ_0, and the functions $I^{(e)}(T, p)$ and $K(T, p, \epsilon)$.

APPENDIX

Wave Propagation in Elastic-Plastic Material

97. *The medium*

Solid matter is capable of *elastic* deformations under certain conditions, and of plastic changes in shape under others. The property of matter characterizing it as elastic or plastic can be expressed mathematically by the relation existing between the *stress* and the *strain*, and will be defined in the following paragraphs, see also Section 5.

In such elastic-plastic materials an important variation of wave propagation occurs which differs in many respects from wave motion in gases. The decisive new feature (see the bibliography) is that shock waves and continuous simple waves occur in both expansive and compressive motion. It is also interesting that there is always a sonic discontinuity at the head of a rarefaction wave entering a zone in which the material is unstrained. In contrast to a gas, which expands indefinitely under zero pressure, an elastic-plastic material assumes a well-defined original state when it suffers no stress.

The Lagrangian representation seems the natural one to employ for the treatment of motion in such material. Let us consider an elastic-plastic cylindrical bar of uniform cross section in its original (unstrained) state. When the bar is deformed in the direction of the axis, the axial coordinate x of a particle depends on its "original" abscissa a and on the time t: $x = x(a, t)$. The *strain* is then defined in terms of the rate of change $x_a = \partial x/\partial a$ by

(97.01) $$\epsilon = x_a - 1.$$

When ρ is the density of mass and ρ_0 the "original" density, we clearly have $\rho_0 \, da = \rho \, dx$ or

(97.02) $$\frac{\rho_0}{\rho} = (1 + \epsilon).$$

The *stress* is the force per unit area acting in normal direction against a cross section; for the following considerations, however, a somewhat different quantity is to be used. Actually, the motion of the bar does not take place solely in the axial direction since an extension in the axial direction is always connected with a contraction in the perpendicular direction. Thus for the desired approximate one-dimensional treatment, the significant quantity is not the stress, but the total force acting in the normal direction against a cross section. This total force divided by the constant area of the original cross section, the so-called *engineering stress*, is the one denoted in the following by σ and simply called the *stress*. This stress is then assumed to be a known function of the strain

(97.03) $$\sigma = \sigma(\epsilon),$$

this function depending only on the nature of the material. One always has

(97.04) $$\sigma \gtreqless 0 \text{ for } \epsilon \gtreqless 0,$$

that is, the stress is positive in tension and negative in compression

($\sigma = 0$ for $\epsilon = 0$ is true by definition). For most materials the further inequality

(97.05) $$\frac{d\sigma}{d\epsilon} > 0$$

is satisfied throughout; that is, increasing strain implies increasing stress. In the following discussion we assume relation (97.05) to hold.

A material is called *elastic* when the stress depends *linearly* on the strain. Most materials are elastic when the strain does not exceed a certain limit, the critical strain ϵ_*. The stress-strain relation is then

(97.06) $$\sigma = E\epsilon, \quad |\epsilon| \leq \epsilon_*,$$

the constant E being Young's modulus.

A material is here called *plastic* when the stress is a *nonlinear* function of the strain, the latter being greater than the critical strain. For the plastic region we assume

(97.07) $$0 < \frac{d\sigma}{d\epsilon} < E, \quad |\epsilon| > \epsilon_*$$

and

(97.08) $$\frac{d^2\sigma}{d\epsilon^2} \begin{cases} <0 \\ >0 \end{cases} \text{ for } \begin{matrix} \epsilon > \epsilon_* \\ \epsilon < -\epsilon_* \end{matrix}.$$

It should be noted that for some materials there is a certain range of values of the strain where the stress is independent of the strain but depends on the rate of strain. Some authors reserve the notion "plastic" for such a state of the material. We have excluded these cases by condition (97.05). A typical function $\sigma = \sigma(\epsilon)$ is indicated in the accompanying graph, see Figure 85.

It is interesting to compare the stress-strain relation for an elastic-plastic material with the adiabatic pressure-density relation for a polytropic gas. To this end we identify the pressure with the negative stress, $p = -\sigma$ (although this is not quite proper since σ is the "engineering stress"). We further set $\rho = \rho_0/(1 + \epsilon)$ in accordance with (97.02). Then the adiabatic relation for a gas becomes

$$\sigma = -\frac{p_0}{(1 + \epsilon)^\gamma},$$

the graph of which is given in Figure 86. We observe that for tensile strain, $\epsilon > 0$, the trend of the two (ϵ, σ)-curves is the same, in that

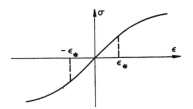

Fig. 85. Graph of the stress-strain relationship for an elastic-plastic material.

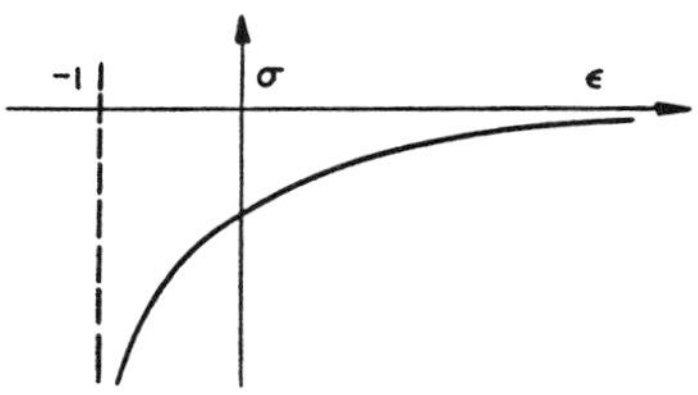

Fig. 86. Graph of the "stress-strain" relationship for a gas, $\sigma = -p$.

$d\sigma/d\epsilon$ decreases for increasing ϵ. However, for compressive strain, $\epsilon < 0$, the slope $d\sigma/d\epsilon$ decreases as ϵ decreases for elastic-plastic material, while for gas it increases as ϵ decreases. The significance of this fact will become apparent in the following sections.

98. The equations of motion

The motion of a particle with the original abscissa a is given by a function $x = x(a, t)$; its velocity is therefore given by

$$(98.01) \qquad u = x_t = \frac{\partial x}{\partial t}.$$

The equation of motion, $\rho u_t = \sigma_a/x_a$, becomes by (97.01) and (97.02)

$$(98.02) \qquad u_t = g^2 \epsilon_a,$$

with

$$(98.03) \qquad g = \sqrt{\frac{d\sigma}{d\epsilon} \bigg/ \rho_0}.$$

As a second equation we have from (97.01) and (98.01)

$$(98.04) \qquad \epsilon_t = u_a.$$

The difference between the present equations and the form of the Lagrangian equations which we have employed for gases, see Section 18, is that ϵ is used instead of $\tau = 1/\rho$, and a instead of $h = \rho_0 a$.

The quantity g is clearly the rate of change da/dt with which a disturbance shifts from particle to particle. We call a rate of change da/dt a *shift rate* and $g(\epsilon)$ is in particular called the *characteristic shift rate*. The shift rate g is connected with the sound speed c and the impedance $k = \rho c$, previously defined for gases, by the relations

(98.05) $$g = \frac{k}{\rho_0} = \frac{\rho}{\rho_0} c,$$

where c is defined by

(98.06) $$c = \sqrt{-\frac{d\sigma}{d\rho}}.$$

In the elastic range, the shift rate

(98.07) $$g_0 = \sqrt{\frac{E}{\rho_0}}$$

is constant while the sound speed $c = (\rho_0/\rho)g_0$ is not. The graph of the characteristic shift rate $g(\epsilon)$ is given below. In accordance

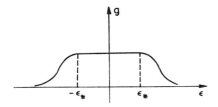

FIG. 87. Graph of the relationship between the shift rate g and the strain ϵ for an elastic-plastic material.

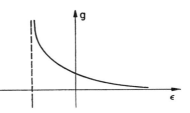

FIG. 88. Graph of the relationship between the "shift rate" and the "strain" for a gas.

with assumption (97.08), $g(\epsilon)$ decreases during tension when ϵ becomes larger than ϵ_*, and also during compression when ϵ becomes smaller than $-\epsilon_*$.

99. Impact loading

The basic problem of wave propagation in a bar of elastic-plastic material is concerned with the motion resulting from *impact-loading*, i.e., from a velocity being suddenly imparted to one end of the bar and then maintained there. Pushing in or pulling out the end of the bar corresponds to pushing in or withdrawing the piston in a tube filled with gas. From a receding piston a centered rarefaction wave is propagated into the gas. We proceed to discuss the corresponding phenomenon when the end of the bar is pulled out.

Imparting a constant velocity u_0 to the end cross section is, as we shall see, equivalent to imparting to it a constant strain ϵ_0. If this strain ϵ_0 is less than the critical strain ϵ_*, the strain resulting in the bar through wave propagation also remains below the critical strain. The wave propagation is, therefore, governed by linear differential equations with constant coefficients. If such is the case, initial discontinuities are propagated as discontinuities with constant characteristic shift rate.

Whenever the initial strain ϵ_0 is greater than the critical strain, the differential equations of propagation are nonlinear, and nonlinearity implies: whether the initial discontinuity is propagated through a shock wave or smoothed out by a rarefaction wave depends on whether the characteristic shift rate $g(\epsilon)$ increases or decreases with increasing ϵ. According to the assumptions made here, see (97.08), $g(\epsilon)$ decreases when $\epsilon > \epsilon_*$ increases. Consequently the influence of greater values of ϵ is propagated with smaller speed. This fact entails that a suddenly imparted initial strain ϵ_0, when it is greater than the critical strain, is propagated through a rarefaction wave.

In order to determine the resulting motion it is convenient to write the equations of motion in characteristic form, according to the methods of Section 22,

(99.01) $$da = \pm g\, dt,$$

(99.02) $$du \mp g\, d\epsilon = 0.$$

Introducing the function

(99.03) $$\phi(\epsilon) = \int_0^\epsilon g(\epsilon)\, d\epsilon,$$

IMPACT LOADING   241

We can write equation (99.02) in the form

(99.04) $$d(u \mp \phi) = 0.$$

Suppose now that the bar lies along the positive x-axis, $x \geq 0$, and the impact produces a velocity $u_0 < 0$ at the end cross section

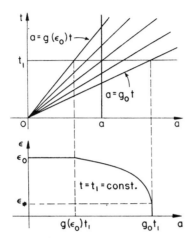

FIG. 89. Centered simple compression wave in an elastic-plastic material. The curve in the (a, ϵ)-plane shows the distribution of strain in the elastic-plastic material at time t_1 after a tensile impact.

$x = 0$. A centered simple wave will move in a forward direction. Across it $u + \phi(\epsilon)$ is constant; since $u = 0, \phi(\epsilon) = 0$ for $t = 0, x > 0$, we have, across the wave,

(99.05) $$u + \phi(\epsilon) = 0.$$

In particular, the strain ϵ_0 produced by the impact at the end cross section is such that

(99.06) $$u_0 = -\phi(\epsilon_0).$$

The quantity $\phi(\epsilon)$ is called the *impact velocity* because $-\phi(\epsilon)$ is the velocity u that must be imparted to the end of the bar in order to produce the strain ϵ there.

The influence of the impact travels with the shift rate g_0; hence we have

(99.07) $$\epsilon = 0, \quad u = 0 \quad \text{for} \quad 0 \leq t \leq \frac{a}{g_0}.$$

If $\epsilon_0 \leq \epsilon_*$, the strain (of a particle with original position a) jumps from $\epsilon = 0$ to $\epsilon = \epsilon_0$ at the time $t = a/g_0$, while the velocity jumps from $u = 0$ to $u = -\phi(\epsilon_0)$. Afterwards the state remains constant;

(99.08) $$\epsilon = \epsilon_0, \quad u = -\phi(\epsilon_0), \quad \text{for} \quad t \geq \frac{a}{g_0}.$$

If, however, $\epsilon_0 > \epsilon_*$, the strain jumps from $\epsilon = 0$ to $\epsilon = \epsilon_*$ at the time $t = a/g_0$, while the velocity jumps from $u = 0$ to $u = -\phi(\epsilon_*)$. Afterwards there is a simple centered rarefaction wave which can be described by the parametric representation

(99.09) $$u = -\phi(\epsilon), \quad \frac{a}{t} = g(\epsilon), \quad \text{for} \quad \epsilon_* < \epsilon \leq \epsilon_0.$$

Accordingly, to every time in the interval $(a/g(\epsilon_*) < t < a/g(\epsilon_0)$ values ϵ and u are uniquely assigned, since $g(\epsilon)$ decreases with increasing ϵ according to assumption (97.08). After the impact strain $\epsilon_0 = -\phi(u_0)$ has been reached, the state remains constant:

(99.10) $$\epsilon = \epsilon_0, \quad u = -\phi(\epsilon_0), \quad \text{for} \quad t > \frac{a}{g(\epsilon_0)}.$$

The motion would, of course, also be described by formulating the occurrence at a fixed time $t = t_1$. Figure 89 corresponds to such a description.

The discontinuity at the head of the wave is of particular interest. It does not deserve the name "shock" since it is "sonic", moving with the characteristic shift rate. It may be considered as a degenerate section of a simple wave due to the fact that all the characteristics $a = g(\epsilon)t$ for $0 \leq \epsilon \leq \epsilon_*$ have the same slope and thus coincide.

We now discuss the case of an initial impact producing a *positive* velocity u_0 and hence a compressive strain $\epsilon_0 < 0$. This case corresponds to that of a piston suddenly pushed into a gas-filled tube. In the gas, as we know, a shock wave results. *In the elastic-plastic material, however, the compression is propagated through a simple* wave with a discontinuous head just as a tensile impact would be propagated. The description of the wave can be obtained from that of the expansion wave by substituting $-u$ for u in formulas (99.05–.10). The simple wave derived from (99.09) is such that to every time in the interval $a/g(-\epsilon_*) < t \leq a/g(\epsilon_0)$ values of ϵ and u are uniquely assigned; this follows again from the fact that $g(\epsilon)$ decreases as ϵ decreases from $-\epsilon_*$ to ϵ_0 according to assumption (97.08).

Quite generally, an initial discontinuity, as stated before, is propagated through a simple wave if, and only if, the characteristic shift rate g for the state ahead of the discontinuity is greater than the shift rate for the state behind the discontinuity. For the material considered we saw that this is the case for both a tensile and a compressive impact since $g(\epsilon)$ decreases as $|\epsilon|$ increases from ϵ_* on. In a gas, however, $g(\epsilon)$ increases with decreasing ϵ; therefore a compressive impact in a gas is propagated through a shock wave.

It may be mentioned that there are materials for which assumption (97.07) is not satisfied and $dg/d\epsilon$ changes sign for sufficiently large strains. Then, if the impact is strong enough, the transition is propagated by a simple wave followed by a shock, the state in front of the shock being so determined that the characteristic shift rate of this state coincides with the shift rate of the shock wave

100. Stopping shocks

There is another peculiar situation in which shocks are propagated through an elastic-plastic material. So far, it has been assumed that the velocity imparted to one end is maintained there indefinitely by applying the appropriate stress, $\sigma = \sigma(\epsilon_0)$. It is, of course, important to investigate what happens when this stress is suddenly released. The influence of this new discontinuity can certainly not be propagated through a simple wave since the characteristic shift rate g is smaller before stopping than after stopping. It is therefore to be expected that the influence of this stopping is propagated through a shock wave. This shock is of a particularly simple type due to the phenomenon of *hysteresis:* When a plastic material is released from a strained position it will not, on its return, obey the same stress-strain relation as when the strain was produced. The general experience is that on returning, the stress depends linearly on the strain, and that $d\sigma/d\epsilon = E$ as in the elastic state, see Figure 90. Therefore, when the stress has returned to zero, a "permanent" strain different from zero remains. The same then is true for the transition

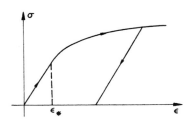

FIG. 90. Stress-strain graph illustrating the phenomenon of hysteresis in a plastic material.

across a stopping shock. Let $[\sigma]$ and $[\epsilon]$ be the differences of the values of σ and ϵ, respectively, in front of and behind the shock. Then, according to the property $d\sigma/d\epsilon = E$ formulated above,

(100.01) $$[\sigma] = E[\epsilon].$$

The shock transition relations for the Lagrangian representation were derived in Section 62. The first two of them can be written

(100.02) $$[u] = -\dot{a}[\epsilon], \quad [\sigma] = -\rho_0\dot{a}[u],$$

where $\dot{a}$ is the shift rate of the shock, $\rho_0\dot{a}$ is the mass flux per unit area crossing the shock from front to back. Eliminating $[u]$ we find

(100.03) $$[\sigma] = \rho_0\dot{a}^2[\epsilon].$$

Hence $\dot{a}^2 = E/\rho_0$ or $\dot{a} = g_0$. Thus it is seen that the shift rate $\dot{a}$ of the shock coincides with the characteristic shift rate g_0 belonging to the elastic state.

The third shock relation, expressing conservation of energy, can now be used to determine energy changes; but the shock is already determined by the first two conditions alone. It is in this respect that the present "stopping shock" is simpler in character than the shocks occurring in gas dynamics.

The decisive feature of gas dynamical shocks is that they produce **permanent** changes in the conditions of the gas by increasing the **entropy**. One is tempted to consider the change of entropy as **analogous** to the permanent strain resulting after a stopping process.

This analogy, however, does not carry very far. Permanent changes in elastic-plastic material appear to be linked with the nonlinear phase of the process; in contrast to gases, they would also occur if the stress were reduced in a gradual manner. Therefore, permanent deformations can not be ascribed to the shock transition as such.

101. *Interactions and reflections*

The stopping shock eventually catches up with the simple wave running ahead, and a more complex process of interactions ensues. Due to the simple nature of the shock it is possible to analyze this process of interaction in all detail. This has been done, but we shall refrain from reporting on the results here, mentioning only that the final permanent change of state of the material can be determined completely.

Wave motion in elastic-plastic material has also been analyzed in another direction. The motion in a bar of finite length can be described by a succession of reflections. It is appropriate to introduce as new independent variables the velocity u and the impact velocity $\phi = \phi(\epsilon)$. Then equations (98.02) and (98.04) are transformed into the linear equations

(101.01) $$a_\phi = gt_u, \qquad a_u = gt_\phi,$$

where g may be considered a function of ϕ. When the other end of the bar, $x = l$, is fixed, the velocity there is $u = 0$; hence the region in the (u, ϕ)-plane is the fixed strip

$$u_0 \leq u \leq 0, \qquad 0 \leq \phi.$$

It is to be borne in mind that in the (u, ϕ)-plane the image of a constant state is a point and the image of a simple wave is a line. The image of a region of interaction between incoming and reflected waves is a triangle. The motion corresponding to this triangle can then be determined by an approximate method using characteristic lines. It is seen that on successive reflections the strain increases. Accordingly, the characteristics which coincided in the first simple wave and formed the elastic discontinuity front will, when continued through reflection, come into the nonlinear range and thus spread. Therefore the reflected waves have a continuous front. The characteristic line resulting through reflection from the one with $\epsilon = \epsilon_*$ is shown as a dotted line in Figure 91.

One more remark might be made in conclusion. The nature of the stress-strain relation is in reality not so well established as that for gases and varies considerably for different materials. Various

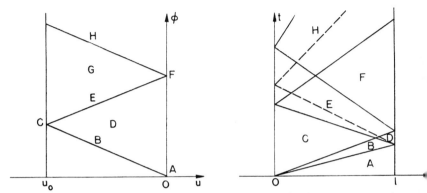

Fig. 91. Graphic representation of the motion of a bar as a succession of reflected waves in a (u, φ)-plane and an (a, t)-plane.

approximate assumptions can be made. In particular, the relation $\sigma = \sigma(\epsilon)$ can with sufficient approximation be so chosen that an explicit integration of the differential equations becomes possible. The assumption

$$g = \frac{b^2 g_0}{(b + \epsilon)^2},$$

for instance, has proved to be very suitable for this purpose. In particular, the problem of reflection can be treated rather explicitly under this assumption and the final state approached as time increases indefinitely can be determined.

CHAPTER IV

Isentropic Irrotational Steady Plane Flow

102. Analytical background

Next in simplicity to one-dimensional flow is *two-dimensional* or *plane flow* which is moreover steady, irrotational, and isentropic; see Sections 10, 13, and 16. Its theory parallels that of Chapter III.

Let us recall the analytical background developed in Section 16. Under our assumptions the flow is characterized by the two components u, v of the velocity $\vec{q}$ as functions of the Cartesian coordinates x, y in the plane; similarly, ρ, p, c are functions of x, y alone, not depending on z or on t and connected with $q^2 = u^2 + v^2$ by Bernoulli's equation

$$(102.01) \qquad q^2 + 2i = \hat{q}^2 = \text{constant},$$

see (16.02), which for polytropic gases can be written in the form

$$(102.02) \qquad \mu^2(u^2+v^2) + (1-\mu^2)c^2 = c_*^2 = \mu^2\hat{q}^2,$$

see (14.07–.08). As we saw in Section 16, see (16.06–.07), the differential equations of motion are

$$(102.03) \qquad v_x - u_y = 0,$$

$$(102.04) \qquad (\rho u)_x + (\rho v)_y = 0.$$

The differential of ρ as a function of u and v, derived from (102.01), is given by

$$(102.05) \qquad \tau d\rho = -c^{-2}q\,dq = -c^{-2}(u\,du + v\,dv),$$

see (16.04). Equation (102.04) can by (102.05) be written in the form

$$(102.06) \qquad (c^2 - u^2)u_x - uv(u_y + v_x) + (c^2 - v^2)v_y = 0,$$

see (16.08), in which c^2 is a given function of $q^2 = u^2 + v^2$ by (102.01) or (102.02).

By introducing a *potential function* $\varphi(x, y)$, such that

(102.07) $$\varphi_x = u, \; \varphi_y = v,$$

see (16.10–.11), equation (102.06) becomes an equation of second order,

(102.08) $$(c^2 - \varphi_x^2)\varphi_{xx} - 2\varphi_x\varphi_y\varphi_{xy} + (c^2 - \varphi_y^2)\varphi_{yy} = 0.$$

By introducing a stream function $\psi(x, y)$, such that

(102.09) $$\psi_x = -\rho v, \; \psi_y = \rho u,$$

see (16.12), equation (102.03) becomes a differential equation of second order for ψ.

A general mathematical theory would attempt to analyze flow phenomena by formulating and solving appropriate *boundary value problems* for the differential equation (102.08). We shall discuss this general point of view at the end of the chapter, in Part F. Tangible results, however, have so far been achieved primarily by studying special flow patterns. Most of this chapter is therefore concerned with methods of finding special solutions; these solutions are useful since they can be adapted to relatively simple boundary conditions.

A. Hodograph Method

103. Hodograph transformation

Before turning to our main subject, we insert a brief account of the approach by the hodograph transformation, already indicated in Chapter II, Section 21. The extensive work which has been done on the basis of this method, initiated by Chaplygin [74, 75], is primarily concerned with subsonic flow and can not be described here in any detail.

The differential equations (102.03), (102.06) are linear and homogeneous in the derivatives of u and v with coefficients depending on u and v only. Therefore these equations are reducible to a pair of linear differential equations by introducing x and y as functions of the velocity components u and v, provided that the Jacobian

(103.01) $$j = u_x v_y - u_y v_x$$

does not vanish, see Section 21. By means of the relations
$$u_x = jy_v\,,\ u_y = -jx_v\,,\ v_x = -jy_u\,,\ v_y = jx_u,$$
the two equations (102.03) and (102.06) are transformed into the two linear differential equations

(103.02)
$$x_v - y_u = 0,$$
$$(c^2 - u^2)y_v + uv(x_v + y_u) + (c^2 - v^2)x_u = 0.$$

The first equation implies that a function $\Phi = \Phi(u, v)$ exists such that the relations

(103.03) $$\Phi_u = x,\quad \Phi_v = y$$

hold, and the second equation then takes the form

(103.04) $$(c^2 - u^2)\Phi_{vv} + 2uv\Phi_{uv} + (c^2 - v^2)\Phi_{uu} = 0.$$

The function $\Phi(u, v)$, by the way, is the Legendre transform, see for example [32, p. 26],

(103.05) $$\Phi = ux + vy - \varphi,$$

of the potential function $\varphi(x, y)$ whose derivatives are $u = \varphi_x$, $v = \varphi_y$ and which obviously is represented by

$$\varphi = \int \varphi_x\,dx + \varphi_y\,dy = x\varphi_x + y\varphi_y - \Phi.$$

It is worth while mentioning that a Legendre transform $\Psi(\rho u, \rho v)$ can also be introduced for the stream function $\psi(x, y)$ by

(103.06) $$\Psi(\rho u, \rho v) = \rho u y - \rho v x - \psi(x, y).$$

To the relations $\psi_x = -\rho v$, $\psi_y = \rho u$ correspond the reciprocal ones

(103.07) $$\Psi_{\rho u} = y,\quad \Psi_{\rho v} = -x.$$

Equation (103.04) is a *linear differential equation* for $\Phi(u, v)$; therefore once several solutions of it are found, a manifold of solutions can be obtained by superposition. Every solution $\Phi(u, v)$ defined in a certain region of the (u, v)-plane leads to a flow given by u and v as functions of x and y provided that the Jacobian

(103.08) $$J = \Phi_{uu}\Phi_{vv} - \Phi_{uv}^2 = x_u y_v - x_v y_u$$

does not vanish and hence x and y can be introduced as new variables by (103.03). Equation (103.04) is elliptic for subsonic flow, hyperbolic for supersonic flow.

For subsonic flow the Jacobian J never vanishes unless $\Phi_{uu} = \Phi_{uv} = \Phi_{vv} = 0$. For, by (103.04) we have

$$(c^2 - u^2)\Phi_{uv}^2 + 2uv\Phi_{uv}\Phi_{uu} + (c^2 - v^2)\Phi_{uu}^2 = -(c^2 - u^2)(\Phi_{uu}\Phi_{vv} - \Phi_{uv}^2)$$

and, because $u^2 + v^2 = q^2 < c^2$, the left member here is a positive definite quadratic form in Φ_{uv} and Φ_{uu}. A similar argument shows that the Jacobian j never vanishes for a non-constant subsonic flow. Thus the problem of subsonic flow is essentially equivalent to that of solving the linear equation (103.04). The advantage of linearity gained in this way is paid for by a complication in the boundary conditions; the boundaries in the (u, v)-plane, corresponding to given walls in the (x, y)-plane, depend on the solution of the problem.

On the other hand the problem of supersonic flow is not just equivalent to solving equation (103.04) for the Jacobians j and J may change sign, as we saw in Section 30 and shall discuss again in Section 105. As was explained in Section 30, the hodograph of the flow is not simple if j changes sign; in this case, the image of the flow in the (u, v)-plane covers certain parts of the flow two or three times. If J changes sign, the functions $x(u, v)$, $y(u, v)$ do not represent a flow throughout the (x, y)-plane since the image in the (x, y)-plane possesses a fold covering certain parts two or three times. The edge of such a fold is called a "limiting line." We shall discuss these singularities in more detail in Section 105.

In spite of the possibility of such singularities, one can obtain a number of special significant flow patterns, supersonic and subsonic, by the hodograph method.

For this purpose it is convenient to use both the potential function φ and the stream function ψ, in addition to Φ and Ψ. Considered as functions of u and v they satisfy linear differential equations which are obtained from

$$\varphi_u u_x + \varphi_v v_x = u, \qquad \varphi_u u_y + \varphi_v v_y = v,$$

$$\psi_u u_x + \psi_v v_x = -\rho v, \qquad \psi_u u_y + \psi_v v_y = \rho u,$$

together with (102.03) and (102.06) by eliminating u_x, u_y, v_x, v_y.

The result is the linear system

(103.09)
$$\rho(v\varphi_u - u\varphi_v) + (u\psi_u + v\psi_v) = 0,$$
$$\rho c^2(u\varphi_u + v\varphi_v) + (q^2 - c^2)(v\psi_u - u\psi_v) = 0.$$

If polar coordinates q, θ are introduced in the (u, v)-plane through

(103.10) $\qquad u = q\cos\theta,\ v = q\sin\theta,$

equations (103.09) become

(103.11)
$$\rho\varphi_\theta = q\psi_q,$$
$$\rho q \varphi_q = \left(\frac{q^2}{c^2} - 1\right)\psi_\theta,$$

from which equations for φ and ψ can be found by eliminating ψ or φ respectively. It may be noted that the Legendre transforms Φ, Ψ as functions of q and θ satisfy a similar system:

(103.12)
$$\frac{1}{\rho} q\Psi_q = \Phi_\theta\left(1 - \frac{q^2}{c^2}\right),$$
$$\frac{1}{\rho} \Psi_\theta = -q\Phi_q,$$

which is easily derived from (103.03), (103.07), (103.10), and the relation

(103.13) $\qquad d(\rho q) = \left(1 - \dfrac{q^2}{c^2}\right)\rho\, dq,$

which follows immediately from (102.05). The connection between the quantities φ and Φ expressed by (103.05) assumes in polar coordinates, by (103.03), the form

(103.14) $\qquad \varphi = q\Phi_q - \Phi,\quad \varphi_q = q\Phi_{qq}.$

Equation (103.04) becomes

(103.15) $\qquad c^2\Phi_{qq} - (q^2 - c^2)(q^{-1}\Phi_q + q^{-2}\Phi_{\theta\theta}) = 0;$

this relation can also be obtained by eliminating Ψ from (103.12). The connection between the quantities ψ and Ψ, as expressed by (103.06), assumes by (103.07) and (103.13), the form

(103.16) $\qquad \psi = \rho q \Psi_{\rho q} - \Psi = \left(1 - \dfrac{q^2}{c^2}\right)^{-1} q\Psi_q - \Psi.$

252   PLANE FLOW   CHAP. IVA

It is frequently more convenient to employ the relations

(103.17)
$$q\psi_q = \rho(q\Phi_q - \Phi)_\theta,$$
$$\psi_\theta = \rho(q\Phi_q + \Phi_{\theta\theta}),$$

which follow from (103.11), (103.14), and (103.15) and determine the stream function ψ once a solution Φ has been found.

For later purposes we note that the Jacobian J given by (103.08) assumes in polar coordinates the form

(103.18) $$q^{-2}[\varphi_{qq}(q\varphi_q + \varphi_{\theta\theta}) - (\varphi_{q\theta} - q^{-1}\varphi_\theta)^2],$$

as is easily verified.

104. Special flows obtained by the hodograph method

We now proceed to discuss some *special flows*.

1. A simple solution of equation (103.15) is

$$\Phi = k\theta = k \arctan v/u.$$

By (103.03) we have

$$x = -kvq^{-2}, \quad y = kuq^{-2};$$

hence

$$q = kr^{-1}, \quad u = kyr^{-2}, \quad v = -kxr^{-2}.$$

These relations represent *a circulatory flow* with the angular velocity kr^{-2}. The corresponding stream function is found from (103.17)

$$\psi = -k \int_{q_0}^{q} \rho q^{-1} \, dq.$$

2. Another simple solution is given by

$$\psi = k\theta;$$

for a function $\varphi = \varphi(q)$ satisfying (103.11) can then be determined. Assuming that Φ depends only on q, $\Phi = \Phi(q)$, we derive from (103.17)

$$\Phi = k \int_{q_0}^{q} \rho^{-1} q^{-1} \, dq.$$

From (103.03) we have

$$x = k\rho^{-1}q^{-2}u, \quad y = k\rho^{-1}q^{-2}v$$

and
$$r = k\rho^{-1}q^{-1}.$$

Taking $k > 0$ we can obtain q as a function of $k^{-1}r$,
$$q = Q(k^{-1}r),$$
and
$$u = xr^{-1}Q(k^{-1}r), \qquad v = yr^{-1}Q(k^{-1}r).$$

These formulas represent a *purely radial flow*. It is important[*] that the inversion of the relation $r = k\rho^{-1}q^{-1}$ is uniquely possible only if $q > c_*$ or $q < c_*$; for the derivative $(\rho^{-1}q^{-1})_q = \rho^{-1}(c^{-2} - q^{-2})$, see (103.13), vanishes for $q = c$ and thus for $q = c_*$. Here c_* is the critical speed introduced in Section 15. The flow under consideration is therefore *either purely subsonic or purely supersonic*. Evidently, the quantity $\rho^{-1}q^{-1}$, considered as a function of q, has a minimum for $q = c_*$; hence $r \geq r_* = k\rho_*^{-1}q_*^{-1}$ holds for both flows. In other words, *both the subsonic and the supersonic flows are confined to the exterior of a certain circle $r = r_*$ at which the speed becomes sonic* while the acceleration $qdq/dr = \rho c^2 q^3(q^2 - c^2)^{-1}$ becomes infinite there. This circle $r = r_*$ is then a *limiting line* of the flow, see Sections 30 and 105.

3. Superimposing the functions Φ for the purely radial and the purely circulatory flow we obtain the function
$$\Phi = k_0\theta + k\int_{q_0}^{q} \rho^{-1}q^{-1}\,dq$$
with the stream function
$$\psi = -k_0\int_{q_0}^{q} \rho q^{-1}\,dq + k\theta.$$

Further
$$x = (-k_0 v + k\rho^{-1}u)q^{-2}, \qquad y = (k_0 u + k\rho^{-1}v)q^{-2},$$
and
$$r = \sqrt{k_0^2 + k^2\rho^{-2}}\,q^{-1}$$

The Jacobian J, see (103.08), is by (103.18), (103.13–.14),
$$J = k^2(\rho^{-1}q^{-1})_q\rho^{-1}q^{-2} - k_0^2 q^{-4} = -k_0^2 q^{-4} + k^2\rho^{-2}q^{-2}(c^{-2} - q^{-2}).$$

[*] See the discussion of nozzle flow later, Section 144.

Since this expression varies from a finite negative value to infinity as q varies from the sound speed c_* to the limit speed $\hat{q}$, it is clear that $J = 0$ for a certain value $q_l > c_*$ of q. It follows then that the solution represents *two branches of a flow, one being purely supersonic, the other being partly subsonic partly supersonic*. Both take place outside a certain circle $r = r_l$ which serves *as a limiting line*, see Figure 1.

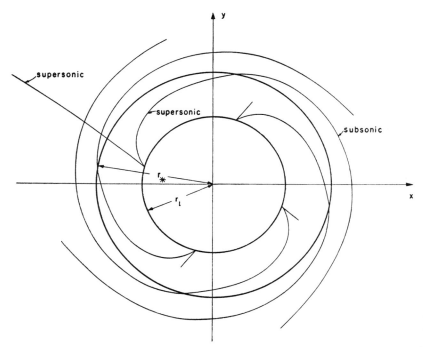

Fig. 1. Streamlines of a spiral flow showing a transition from subsonic to supersonic speed at the circle $r = r_*$ and a limiting circle $r = r_l$.

The streamlines $\psi = $ constant of both flows are evidently *spirals* which meet the limiting circle at the angle $\arcsin c/q$. This fact could be verified from the expression for the stream function by a direct calculation.

4. The spiral flow just discussed illustrates the possibility that gas may flow from a region in which the flow is supersonic to one in which it is subsonic or vice versa. A simple example of a flow being first subsonic, then supersonic, and finally subsonic again, was given by Ringleb [79]. Such flows are obtained by seeking solutions Φ, Ψ or φ, ψ which are products of functions of q with functions of θ. The simplest case is characterized by the functions

$$\psi = kq^{-1}\sin\theta, \qquad \varphi = k\rho^{-1}q^{-1}\cos\theta,$$

which satisfy equations (103.11). From (103.17) one finds

$$\Phi = kq\left\{\int_{q_0}^{q} \rho^{-1}q^{-3}\,dq\right\}\cos\theta,$$

from which by (103.03)

$$x = k\int_{q_0}^{q} \rho^{-1}q^{-3}\,dq + k\rho^{-1}q^{-2}\cos^2\theta,$$

$$y = k\rho^{-1}q^{-2}\cos\theta\sin\theta.$$

For small values of q we have

$$x \sim \frac{k}{2}\rho_0^{-1}q^{-2}\cos 2\theta, \qquad y \sim \frac{k}{2}\rho_0^{-1}q^{-2}\sin 2\theta, \qquad r \sim \frac{k}{2}\rho_0^{-1}q^{-2},$$

and hence

$$x \sim r\cos 2\theta, \qquad y \sim r\sin 2\theta, \qquad \psi \sim \sqrt{k\rho_0(r-x)}.$$

Consequently, at great distances, the streamlines are approximately parabolas, $r - x = $ constant. Furthermore the positive x-axis appears as a streamline since $y = 0$, $x = 0$, $\psi = 0$ for $\theta = 0$ and $\theta = \pi$. Thus it would appear that the gas flows around a section of the positive x-axis. This is, however, not the case since the flow reaches a limiting line before it can turn around the edge, see Figure 2.

The Jacobian J is found from (103.18),

$$J = k^2\rho^{-2}(c^{-2}\cos^2\theta - q^{-2}).$$

Therefore the limiting line is characterized by the condition $|\cos\theta| = q^{-1}c$. This line consists of four branches. Two branches enter the x-axis perpendicularly, the other two extend to infinity with

$x \to \infty$. These branches meet at a cusp characterized by $q^2 = 2c^2 - qc\dfrac{dc}{dq}$. For a polytropic gas we find that the cusp is at $q = \sqrt{2}c_*$, $c = \sqrt{(3-\gamma)/2}\,c_*$; hence we have $|\cos\theta| = cq^{-1} = \tfrac{1}{2}\sqrt{3-\gamma}$ there. A simple discussion then shows that there are streamlines passing near the cusp but not intersecting the limit line along which the flow is first subsonic, then supersonic, and finally subsonic again, see Figure 2.

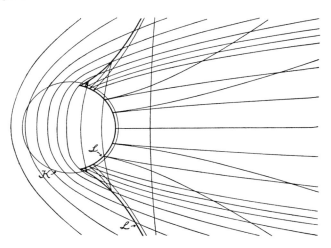

Fig. 2. Showing streamlines of Ringleb's flow with the limiting line $\mathcal{L}$ and the circle $\mathcal{K}$ on which the change from subsonic to supersonic flow takes place.

Many other special flows can be obtained from special solutions of equation (103.15). For details we refer to the literature in the bibliography.

105. The role of limiting lines and transition lines

The method of hodograph transformation to obtain solutions of the flow equations is applicable only if neither the Jacobian

$$j = u_x v_y - u_y v_x$$

of the desired flow $u(x, y)$, $v(x, y)$ nor the Jacobian

$$J = x_u y_v - x_v y_u$$

of the solution $x(u, v)$, $y(u, v)$ of the linear hodograph equations vanishes.

In Section 30 we have described the behavior of the hodograph mapping if either $J = 0$ or $j = 0$ along a "critical" curve. For steady two-dimensional flow now under consideration the description can be amplified by discussing the behavior of the streamlines and potential lines at a critical curve.

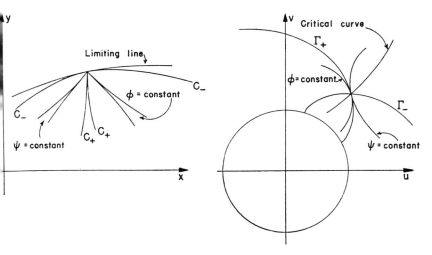

Fig. 3. Limiting line in the (x,y)-plane and critical curve in the (u,v)-plane.

Suppose the Jacobian J vanishes for a solution $x(u, v), y(u, v)$ of the linear hodograph equations along a critical curve in the (u, v)-plane. Then the mapping of the (u, v)-plane into the (x, y)-plane forms a fold in the (x, y)-plane. The edge of the fold, the *limiting line*, was shown to be the envelope of one set of C-characteristics while the characteristics of the other kind have a cusp on the edge. Streamlines and potential lines bisect the angles between the two C-characteristics, as will be shown in Section 106. Therefore they are not tangential to the edge and are, consequently, images of curves crossing the critical line in the (u, v)-plane in the exceptional direction. Hence, according to the statements made in Section 30, *streamlines and potential lines in the (x, y)-plane also have cusps at the limiting line.* Since the exceptional direction in the (u, v)-plane is characteristic it follows that *a critical curve in the (u, v)-plane can be characterized by the condition that the images of streamlines and potential lines pass through it in a characteristic direction,* viz. the exceptional direction.

This property may on occasion enable one to spot a critical curve: if one observes that the images of a potential line and a streamline in the (u, v)-plane are tangential to a characteristic or to each other, the occurrence of a critical curve in the (u, v)-plane, or a limiting line in the (x, y)-plane, is assured. Since the streamlines and potential lines have cusps at the limiting line, their curvature, in general, is infinite there.

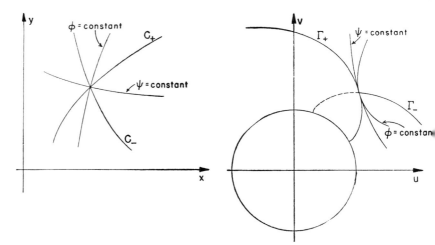

Fig. 4. Transition Mach line C_+ in the (x,y)-plane and edge characteristic Γ_+ in the (u,v)-plane.

A limiting line can never occur in an actual flow since the flow would have to perform a physically impossible reversal of its direction at a limiting line. Actually shock fronts will form before limiting lines are reached.

We may consider now the other case, in which along a "transition curve" in the (x, y)-plane the Jacobian j vanishes for a solution (u, v) of the flow equations. This transition curve, as was shown in Section 30, is a C-characteristic. The corresponding Γ-characteristic is the edge of the fold of the mapping of the (x, y)-plane into the (u, v)-plane. The other C-characteristics cross the transition line in the exceptional direction. Therefore, the streamlines and the potential lines cross the transition line in non-exceptional directions. Consequently, *the images of streamlines and potential lines in the (u, v)-plane are tangential to the Γ-characteristic which forms the edge of the fold.* Transition lines do occur in actual flow; for example in nozzle flow, as will be shown in Chapter V, Section 146.

B. Characteristics and Simple Waves

106. Characteristics. Mach lines and Mach angle

From now on we shall be concerned primarily with supersonic flow assuming $q^2 > c_*^2 > c^2 > 0$. Then a large part of the theory is based on the characteristic transformation of the differential equations. We recall from Section 23, the following facts:

The C-characteristics of equations (102.03) and (102.06) satisfy the differential equation, see (23.07),

(106.01) $\quad (c^2 - u^2)dy^2 + 2uv\,dy\,dx + (c^2 - v^2)dx^2 = 0,$

or

$$c^2(dx^2 + dy^2) = (u\,dy - v\,dx)^2,$$

which yields two roots ζ_+, ζ_- for the quotient dy/dx. The Γ-characteristics in the (u, v)-plane are the solutions of the equation, see (23.08),

(106.02) $\quad (c^2 - u^2)du^2 - 2uv\,du\,dv + (c^2 - v^2)dv^2 = 0,$

or

$$c^2(du^2 + dv^2) = (u\,du + v\,dv)^2.$$

Equation (106.01) for the C-characteristics depends on the particular flow, while the Γ-characteristics are two fixed families of curves in the (u, v)-plane; we shall describe them in detail in Sections 107 and 108.

Once the two roots ζ_+, ζ_- are determined from equation (106.01), this equation can be split into

(106.03) $\qquad C_+ : \ y_\alpha = \zeta_+ x_\alpha , \qquad C_- : \ y_\beta = \zeta_- x_\beta .$

As shown in Section 23, equation (106.02) then splits into

(106.04) $\qquad \Gamma_+ : \ u_\alpha = -\zeta_- v_\alpha , \qquad \Gamma_- : \ u_\beta = -\zeta_+ v_\beta .$

From these four characteristic equations the relations

(106.05) $\qquad u_\alpha x_\beta + v_\alpha y_\beta = 0, \quad u_\beta x_\alpha + v_\beta y_\alpha = 0$

immediately follow; they express the important fact:

If (u, v) and (x, y) are represented in the same coordinate plane, *the directions of the C-characteristics of one kind are perpendicular to the Γ-characteristics of the other kind.* Or, more precisely, *the directions of C_+ and Γ_- and of C_- and Γ_+ through corresponding points (x, y) and (u, v) are perpendicular.*

Geometrically, we interpret the characteristic equations (106.03) and (106.04) by introducing the angle A between the direction of the flow velocity (u, v) and that of the C-characteristic (dx, dy) at a point (x, y) and the angle A′ between the flow direction and that of the corresponding Γ-characteristic (du, dv) at the corresponding point (u, v). Then, with $q^2 = u^2 + v^2$, equation (106.01) can be written in the form

(106.06) $\qquad\qquad\qquad c^2 = q^2 \sin^2 A,$

and (106.02) as

(106.07) $\qquad\qquad\qquad c^2 = q^2 \cos^2 A'.$

Consequently,

(106.08) $\qquad\qquad\qquad A' = 90° - A.$

Relation (106.06) holds for both characteristics through a point. Consequently, both characteristics make the same angle, the *Mach angle* A, with the streamlines. Relations (106.06–.08) also imply: *the component of the flow velocity normal to the direction of a characteristic C is equal to the sound speed*, and similarly, *the component of the flow velocity in the direction of a Γ-characteristic is equal to the sound speed.*

Suppose we denote by C_+ that C-characteristic which forms with the flow direction the angle A in the positive sense, then the corresponding Γ-characteristic $Γ_+$ forms with the flow direction the angle $A' = 90° - A$ since its direction is perpendicular to that of C_-. Of course, C_- then forms the angle $-A$ with the flow direction and $Γ_-$ the angle $-A' = A - 90°$, see Figure 5.

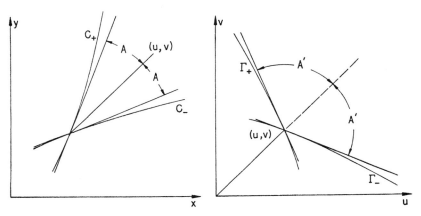

FIG. 5. Angles between the flow direction and the C- and $Γ$-characteristics.

The characteristic equations can be most conveniently expressed by introducing the angle $θ$ between the flow direction and the positive x-axis; then

(106.09) $$u = q \cos θ, \quad v = q \sin θ.$$

The roots $ζ_+$ and $ζ_-$, being the slopes of the characteristic directions, are then given by the relations

(106.10) $$ζ_+ = \tan(θ + A), \quad ζ_- = \tan(θ - A),$$

in which the choice of the sign in front of A corresponds to the convention just stipulated for the characteristics. With these notations the characteristic equations (106.03) and (106.04) assume the form

(106.11) $$y_α \cos(θ+A) = x_α \sin(θ+A),$$
$$y_β \cos(θ-A) = x_β \sin(θ-A),$$

(106.12) $$v_α \sin(θ-A) = -u_α \cos(θ-A),$$
$$v_β \sin(θ+A) = -u_β \cos(θ+A).$$

It should be noted that, in these equations, both $\theta = \arctan v/u$ and the *Mach angle* A are known as functions of u and v by (106.06) and Bernoulli's law (102.01). For a polytropic gas we have from (102.02) the relation

$$(106.13) \qquad \sin^2 A = \frac{c^2}{q^2} = \frac{\mu^2}{1-\mu^2}\left(\frac{\hat{q}^2}{q^2} - 1\right);$$

here $\hat{q} = c_*/\mu$ is the limit speed, see Chapter I, Section 15.

The quantity

$$(106.14) \qquad M = q/c = 1/\sin A$$

is called the *Mach number* of the flow. It is greater than one for supersonic flow. For sonic flow for which $q = c$ or $M = 1$, we have $A = 90°$; the Mach directions then coincide and are perpendicular to the flow direction. When $M \to \infty$ we have $A \to 0$, hence the Mach directions approach the flow direction. In particular this is the case when *cavitation* is approached, $c \to 0$ and $q \to \hat{q}$.

The characteristics C are often called *Mach lines*.

107. Characteristics in the hodograph plane as epicycloids

If the gas is polytropic, the differential equation (106.02) for the Γ-characteristics can be solved explicitly. Geometrically expressed: *For polytropic gases the Γ-characteristics in the (u, v)-plane are epicycloids generated by the points on the circumference of a circle of diameter $c_*\left(\frac{1}{\mu} - 1\right) = \hat{q} - c_*$ which rolls on the "sonic circle" $u^2 + v^2 = c_*^2$.*

A simple proof of this fact follows from the preceding geometrical interpretations: First, using Bernoulli's equation in the form (102.02) we write equation (106.06) in the form

$$(107.01) \qquad q^2\{\mu^2 + (1 - \mu^2) \sin^2 A\} = c_*^2.$$

Secondly, we note that a Γ-characteristic is a curve which makes the angle $A' = 90° - A$ with the direction of the vector $\vec{q} = (u, v)$. Now we consider Figure 6, in which a circle of radius $r = \frac{1}{2}\left(\frac{1}{\mu} - 1\right)c_*$ touches, at the point T, the "sonic" circle about O with radius c_*.

OQ represents the vector $\vec{q}$. If the outer circle rolls on the sonic circle, then the point T is the instantaneous center of rotation; hence the trajectory of Q is perpendicular to the segment TQ. To

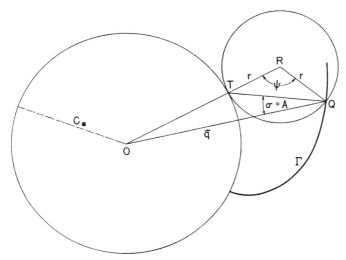

FIG. 6. Γ-characteristics as epicycloids.

identify Γ with the trajectory of Q, i.e. with an epicycloid as described, it is therefore sufficient to show that the angle OQT, temporarily called σ, is equal to the Mach angle A. Denoting the angle ORQ by Ψ, so that $\angle OTQ = 90° + \dfrac{\Psi}{2}$, we obtain from the triangle ORQ the relation

$$q^2 = (r + c_*)^2 + r^2 - 2r(r + c_*) \cos \Psi$$

or

$$\frac{q^2}{c_*^2} = \frac{1}{4}\left(\frac{1}{\mu} + 1\right)^2 + \frac{1}{4}\left(\frac{1}{\mu} - 1\right)^2 - \frac{1}{2}\left(\frac{1}{\mu^2} - 1\right)\cos \Psi$$

or

$$\frac{q^2}{c_*^2} = \frac{1}{\mu^2} - \left(\frac{1}{\mu^2} - 1\right)\cos^2 \Psi/2.$$

Through the relation

$$\cos \frac{\Psi}{2} = \frac{q}{c_*} \sin \sigma$$

found from the triangle OQT, we find in turn
$$q^2\{\mu^2 + (1 - \mu^2)\sin^2\sigma\} = c_*^2.$$
Comparing this result with (107.01) yields
$$\sigma = A.$$
Through each point of the annular ring $c_*^2 < u^2 + v^2 < \hat{q}^2 = \frac{1}{\mu^2} c_*^2$ in the (u, v)-plane there pass two epicycloids, so that this ring is covered with a net of two families of Γ-characteristics as described in the general theory of Chapter II, Section 22.

108. Characteristics in the (u, v)-plane continued

Independently of Section 107 we now determine the Γ-characteristics analytically as integral curves of the ordinary differential equation (106.02).

For this purpose it would seem natural to introduce polar coordinates q and θ instead of u and v by (103.10) and to try to integrate the differential equation

(108.01) $$\frac{dq}{d\theta} = \pm q \cot A' = \pm q \tan A,$$

which is equivalent to (106.12). It is simpler, however, to set up a system of differential equations for the components c and g of the vector (u, v) in, and perpendicular to, the direction of the Γ-characteristic considered. The angle which the direction of a Γ-characteristic makes with the positive u-axis is evidently

(108.02)
$$\omega = \theta + A' = \theta - A + 90° \text{ for } \Gamma_+,$$
$$\omega = \theta - A' = \theta + A - 90° \text{ for } \Gamma_-.$$

Accordingly,

(108.03) $$du \sin\omega - dv \cos\omega = 0$$

along Γ. From (106.07) and (108.02) we have

(108.04) $$c = u \cos\omega + v \sin\omega;$$

CHARACTERISTICS IN (u, v)-PLANE CONTINUED

introducing the component g by

(108.05) $$g = v \cos \omega - u \sin \omega,$$

we have

(108.06) $$\begin{aligned} u &= c \cos \omega - g \sin \omega, \\ v &= c \sin \omega + g \cos \omega. \end{aligned}$$

We now consider ω as a parameter along Γ and determine c and g as functions of ω.

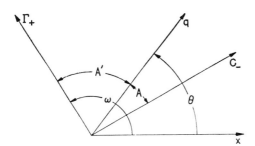

FIG. 7. Relationship between the angles θ, ω, A, and A'.

By differentiating equation (108.06) with respect to ω we obtain the relations

(108.07) $$\begin{aligned} \frac{du}{d\omega} &= c_\omega \cos \omega - g_\omega \sin \omega - (c \sin \omega + g \cos \omega), \\ \frac{dv}{d\omega} &= c_\omega \sin \omega + g_\omega \cos \omega + (c \cos \omega - g \sin \omega), \end{aligned}$$

which, when inserted into (108.03), yield

(108.08) $$g_\omega = -c.$$

From Bernoulli's relation (102.02) for a polytropic gas we have

(108.09) $$c_*^2 - c^2 = \mu^2(q^2 - c^2) = \mu^2 g^2,$$

from which, by differentiating with respect to ω and using (108.08),

(108.10) $$c_\omega = \mu^2 g.$$

The solution of the two equations (108.08) and (108.10) is evidently

(108.11)
$$g = -\mu^{-1} c_* \sin \mu(\omega - \omega_*),$$
$$c = c_* \cos \mu(\omega - \omega_*),$$

with an arbitrary constant ω_*. Inserting this result in (108.06), we find the parametric representation of the velocity components

(108.12)
$$\frac{u}{c_*} = \cos \mu(\omega - \omega_*) \cos \omega + \mu^{-1} \sin \mu(\omega - \omega_*) \sin \omega,$$
$$\frac{v}{c_*} = \cos \mu(\omega - \omega_*) \sin \omega - \mu^{-1} \sin \mu(\omega - \omega_*) \cos \omega.$$

The angle $\omega - \omega_*$ ranges from $-\pi/2\mu$ to $\pi/2\mu$ since $c \geq 0$. From (108.02) and (108.05) we have

(108.13)
$$g = -q \sin A' = -q \cos A \text{ for } \Gamma_+,$$
$$= q \sin A' = q \cos A \text{ for } \Gamma_-.$$

Consequently $g < 0$ or $g > 0$ for the Γ_+- or Γ_--branch respectively; hence $\omega > \omega_*$ corresponds to the Γ_+-branch and $\omega < \omega_*$ to the Γ_--branch. From our construction it is obvious that all the curves represented by equations (108.12) are obtained by rotating one of them about the origin O.

The characteristics Γ are again easily identified as epicycloids between the circles $u^2 + v^2 = c_*^2$ and $u^2 + v^2 = \frac{1}{\mu^2} c_*^2 = \hat{q}^2$. Figure 8 shows the various geometric quantities involved.

For later application we note the relation

(108.14)
$$\tan A = \mu \mid \cot \mu(\omega - \omega_*) \mid$$

which by (106.06) follows from (108.11) and the relation $|g| = \sqrt{q^2 - c^2}$, implied by (108.06).

109. Simple waves

The theory of simple waves is fundamental in building up the solutions of flow problems out of elementary flow patterns. Mathematically a simple wave was defined, in Section 29, as a flow in a region in the (x, y)-plane whose image in the (u, v)-plane is an arc of

SIMPLE WAVES

one Γ-characteristic. It was shown that the simple wave region is swept by a one-parametric family of straight C-characteristics, Mach lines, along each of which u, v and consequently c, p, ρ, τ remain constant. A significant property is, see Section 29: *A non-constant state of flow adjacent to a constant state is always a simple wave.* Two-dimensional steady simple waves were discovered by Prandtl and the theory was elaborated by Meyer [100].

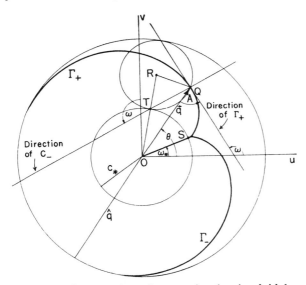

FIG. 8. Construction of one pair of epicycloidal characteristics.

The non-straight Mach lines in a simple wave may be called *cross Mach lines*. The image of each of them in the (u, v)-plane is the same Γ-characteristic. (On occasion we shall call a simple wave a Γ_+-wave or a Γ_--wave depending on whether the simple wave maps into a Γ_+- or into a Γ_--characteristic.) Each straight Mach line is mapped into one point of this Γ-characteristic and its direction is perpendicular to the direction of this Γ-characteristic at this corresponding point. The angle ω which the latter direction makes with the positive x-axis is therefore at the same time the angle which the straight Mach line makes with the positive y-axis, see Figure 8.

The set of straight characteristics can be prescribed in the form

(109.01) $\quad x = a(\sigma) - r \sin \omega(\sigma), \qquad y = b(\sigma) + r \cos \omega(\sigma),$

with arbitrary functions $a(\sigma)$, $b(\sigma)$, $\omega(\sigma)$ of a parameter σ. Here r is an abscissa along the straight line $\sigma =$ constant and ω, according to Section 108, is the angle between the straight characteristic and the positive y-axis. The values of u and v on each of these lines are given through the relations which represent the Γ-characteristic corresponding to the simple wave, in particular, for polytropic gases,

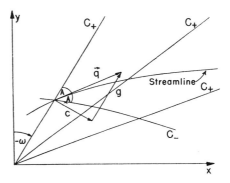

Fig. 9. Three straight Mach lines C_+, one streamline, and one cross Mach line C_- in a centered simple wave corresponding to a section of a Γ_--characteristic.

through relations (108.12). Then the value of i and hence those of c, p and ρ on these lines are found from Bernoulli's equation (102.01). That a flow so described really satisfies the differential equations (102.03) and (102.06) follows from the general theory of Chapter II, Section 29, but could, of course, be verified directly.

If all straight characteristics pass through one point, called the center (i.e. if $a(\sigma)$ and $b(\sigma)$ are constant), the wave is called a *centered simple wave*, see Figure 9.

When a simple wave is traversed by letting the parameter σ vary monotonically, the image point in the (u, v)-plane need not run monotonically through a section of the corresponding Γ-characteristic. Furthermore, the image point may touch the sonic circle $q = c_*$ and pass over from a Γ_+-branch to a Γ_--branch or vice versa. In such a transition the roles of C_+ and C_- as straight Mach lines interchange.

In the following discussion we consider simple waves for which the corresponding section of the Γ-characteristic is traversed monotonically, or in which the angle ω of the straight Mach lines with the y-axis changes monotonically with σ. We may then identify the

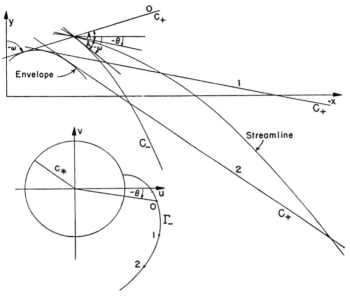

FIG. 10. Three straight C_+-Mach lines $(0,1,2)$ in a simple wave corresponding to a section $(0\text{-}1\text{-}2)$ of a Γ_--characteristic.

parameter σ with the angle ω. If the flow speed q of a gas particle increases as it crosses the simple wave and hence the pressure p, and also c and ρ, decrease, the wave is called an *expansion* or *rarefaction wave;* if the flow speed decreases, the wave is called a *compression* or *condensation wave*.

We consider sections of simple waves bounded by two straight Mach lines, a streamline or an envelope of the Mach lines, and extending indefinitely in one direction. Such sections occur as elements in more complicated flow patterns.

In Figure 10 we show expansion waves corresponding to a section of a Γ_--characteristic; the wave shown in Figure 11 turns from a compression wave into an expansion wave and corresponds to sections of both Γ_+ and Γ_- curves. The flow of Figure 12 is a com-

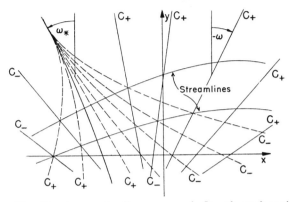

Fig. 11. Simple wave leading supersonic flow through sonic flow to supersonic flow.

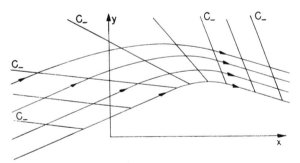

Fig. 12. Compression wave.

pression wave corresponding to a section of a Γ_+-characteristic. In all these cases the angle ω decreases in the direction of the flow. In Figure 13 we show a *complete simple wave* which corresponds to a whole arc Γ_-.

In a complete simple wave the flow is sonic at one end and the flow direction is perpendicular to the Mach line C_* there; at the other end, the speed is the limit speed $\hat{q}$, and pressure, density, and sound speed are reduced to zero; a zone adjacent to this end would therefore be a zone of *cavitation*. Since the Mach angle A is zero at this end, each Mach line C_+ or C_- approaches a streamline there.

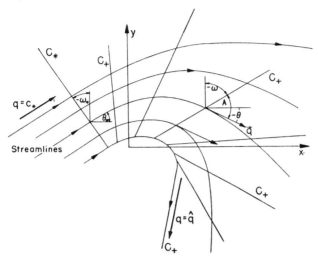

Fig. 13. Complete simple wave covered by C_+-Mach lines corresponding to the complete Γ-arc.

Cavitation is approached across a Γ_--wave in a polytropic gas when $\omega - \omega_*$ approaches the value $-\pi/2\mu$ since then the Mach angle A, by (108.14), approaches the value zero. The angle θ of flow direction, given by $\theta = \omega + A'$ for a Γ_--wave, see (108.02), varies from the value $\theta_* = \omega_*$ to the value $\hat{\theta} = \omega_* - \pi/2\mu + \pi/2$. The total change of the angle θ is therefore

(109.02) $$\hat{\theta} - \theta_* = -\left(\frac{1}{\mu} - 1\right)\frac{\pi}{2}.$$

Since $\mu < \frac{1}{2}$, we have $|\hat{\theta} - \theta_*| > \pi/2$. The values of $\left(\frac{1}{\mu} - 1\right)\frac{\pi}{2}$ for various values of γ are given in Table II, Section 116.

110. Explicit formulas for streamlines and cross Mach lines in a simple wave

It is easy to determine explicitly any *streamline* or *cross Mach line* in a simple wave if one streamline or one cross Mach line respectively is prescribed. We employ the representation (109.01) and assume that $x = a(\sigma)$, $y = b(\sigma)$ represents the given initial streamline or cross Mach line; the parameter r then measures the distance of a point from this initial curve along the straight Mach line through it.

Along a streamline we have by the definition of θ

(110.01) $$\frac{dy}{dx} = \tan \theta$$

and along a cross Mach line in a Γ_+- or Γ_--wave we have by (106.11)

(110.02) $$\frac{dy}{dx} = \tan (\theta \pm A).$$

Along the initial curve $x = a(\sigma)$, $y = b(\sigma)$ we prescribe the distribution of θ and A such that relation (110.01) or (110.02) holds.

It is convenient to use the angle ω defined by (108.02), and to consider A a given function of ω which for polytropic gases is given by (108.14). We assume that $\omega = \omega(\sigma)$ is prescribed on the initial curve.

The relations (110.01) and (110.02) may then be written in the form

(110.03) $$\cos(\omega + kA)x_\sigma + \sin(\omega + kA)y_\sigma = 0$$

with $k = \pm 1$ for a streamline, $k = \pm 2$ for a cross Mach line, the upper sign referring to a Γ_+-wave, the lower one to a Γ_--wave. We insert into relation (110.03) the expressions

(110.04) $$\begin{aligned} x_\sigma &= a_\sigma - r\omega_\sigma \cos \omega - r_\sigma \sin \omega, \\ y_\sigma &= b_\sigma - r\omega_\sigma \sin \omega + r_\sigma \cos \omega, \end{aligned}$$

which result from (109.01) by differentiation. Observing that the functions $x = a(\sigma)$, $y = b(\sigma)$, representing the initial curve, satisfy relations (110.03) we obtain

(110.05) $$r_\sigma = r \cot kA \, \omega_\sigma ,$$

and, through integration,

(110.06) $$r = r_0 \exp \int_{\omega_0}^{\omega(\sigma)} \cot kA \, d\omega$$

with arbitrary values of r_0 and ω_0. We note that A is a known function of ω, and $\omega(\sigma)$ is prescribed. For polytropic gases we have by (108.14)

(110.07)
$$\pm \cot A = \mu^{-1} \tan \mu(\omega - \omega_*),$$
$$\pm \cot 2A = \pm\tfrac{1}{2}(\cot A - \tan A)$$
$$= \tfrac{1}{2}[\mu^{-1} \tan \mu(\omega - \omega_*) - \mu \cot \mu(\omega - \omega_*)].$$

Hence (110.06) leads to

(110.08)
$$r = R \cos^{-\mu^{-2}} \mu(\omega - \omega_*)$$

as the representation of the *streamlines* and to

(110.09)
$$r = R \cos^{-\frac{1}{2}\mu^{-2}} \mu(\omega - \omega_*) \sin^{-\frac{1}{2}} \mu(\omega - \omega_*)$$

as the representation of the *cross Mach lines*, with appropriate constants R.

An interesting observation can be made from relations (110.08) and (110.09), namely that, as cavitation is approached, $\omega - \omega_* \to -\dfrac{\pi}{2\mu}$, both the distance between streamlines and the distance between cross Mach lines increase indefinitely.

For streamlines in centered simple waves in which $a = b = 0$ the relations (110.05) have a simple physical interpretation. Along a streamline we identify the parameter σ with the time t. Then $r_\sigma = \dot{r}$ is the radial velocity and $\omega_\sigma = \dot{\omega}$ the angular velocity. Comparing equations (110.04) with (108.06) we are led to identify

$$\dot{r} = g, \quad r\dot{\omega} = -c,$$

in agreement with the meaning of g as the component of the flow velocity in the direction of the straight Mach line and of c as the perpendicular component.

111. *Flow around a bend or corner. Construction of simple waves*

Supersonic flow around a bend or sharp corner, one of the most important elementary flows, is effected by a simple wave. We suppose, see Figure 14, that the flow arrives with a constant velocity q_0 along a wall which is straight up to a point A, then bends along a smooth bend K from A to B and continues straight beyond the point B. We further assume that the oncoming flow is of constant state in a region adjacent to the straight part of the wall before A. Then the question is: How does the flow turn the corner? Or how does the flow continue along the bend K and along the straight wall beyond B?

If the oncoming flow is subsonic, the problem involves potential flow, governed by an elliptic differential equation whose solution at any point depends on the boundary conditions even at remote parts of the boundary.

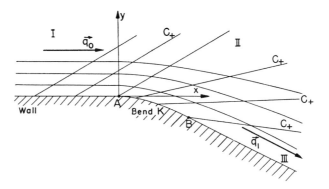

FIG. 14. Flow around a bend effected by a simple expansion wave.

We confine ourselves, however, to the case of supersonic flow. Then the solution is much simpler. It can be obtained by piecing together three domains of the flow that have essentially different analytic character. These are the zone (I) of constant state of the oncoming flow, a simple wave (II) which follows and through which the flow effects its turn, and finally a zone (III) also of constant state which may be either a zone of flow parallel to the straight wall beyond B (if the simple wave effects the complete turn prescribed by the bend) or a zone of cavitation (if the flow has expanded to zero density before the full turn around the bend has been achieved).

Let us construct this solution in detail. First, the zone (I) of constant state is terminated by a Mach line of characteristic C_+^A or

C^A_- which forms the Mach angle A_0, defined by

(111.01) $$\sin A_0 = \frac{c_0}{q_0},$$

with the direction of incoming flow, i.e., with the straight wall. This angle A_0 is known since the state (0) or (I), and hence the corresponding sound speed c_0, is known. The variation of the flow starts at one of two possible Mach lines, one inclined downstream and one inclined upstream. For the moment we discuss the first possibility in which the transition Mach line is C^A_+. The straight Mach lines are then C_+-lines and the adjacent simple wave corresponds to a section of a Γ_--wave if, as in Figure 15, the angle θ of flow direction decreases. The speed q then evidently increases, see the figure, and thus the wave is a *rarefaction or expansion wave*. To construct the simple wave we need only realize that the wall including the bend is a streamline and that hence at each point P of the bend the direction of the bend is that of the flow. The known flow velocity (u_0, v_0) on the transition Mach line C^A_+ is represented by a point in the (u, v)-plane through which the Γ_--characteristic representing the simple wave is determined. Along this Γ_--characteristic u, v, q and hence also c, A, p, and ρ are known functions of θ. For polytropic gases one can determine the angle ω as a function of θ by inverting the formulas (108.02) in which A is a given function of ω by (108.14) once ω_* is known. More about this will be said in Section 116.

Here we describe a geometric procedure for a polytropic gas. We pass the complete epicycloidal arc Γ_- through the point $(u_0, v_0) = (q_0, 0)$ in the (u, v)-plane corresponding to the given initial state. On it the point A_1 at the distance q_0 from O corresponds to the beginning A of the bend. The angle at O subtended by the epicycloidal arc SA_1 is the angle $\theta_* = \omega_*$. To any point P on the bend K we obtain the corresponding position P_1 on Γ_- by drawing OP_1 parallel to the tangent on K at P. The straight Mach line C_+ through P is then determined as the line perpendicular to the direction that Γ_- has at P_1, i.e., parallel to the line TP_1, T being the point of contact between the rolling and the fixed circle. Along C_+ the velocity $\vec{q}$ is parallel to the line OP_1 and the speed q is given by the length of OP_1.

If each point P of the bend has an image P_1 on Γ then the arc

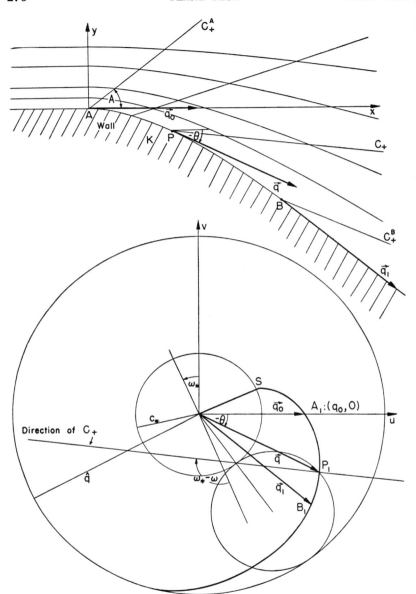

Fig. 15. Construction of simple expansion wave in flow around a bend.

A_1B_1 of Γ_- represents the incomplete simple wave from which the flow emerges parallel to the straight wall beyond B with the speed equal to the length of the segment OB_1.

If, however, the bend is too strong, i.e., if the end B_1 of the arc Γ, at which Γ touches the limit circle $q = \hat{q}$, corresponds to a point of the bend between A and B, then the simple wave is completed by the Mach line C_+ through B, which is then tangent to the bend at B; beyond this Mach line there will be cavitation, see Section 109,

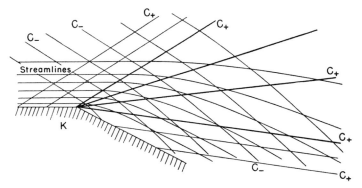

Fig. 16. Incomplete centered simple wave (air, $\gamma = 1.4$).

and the flow in the wave zone (II) will acquire asymptotically the direction of this terminal Mach line.

By using the non-dimensional quantities $\dfrac{u}{c_*}$, $\dfrac{v}{c_*}$, $\dfrac{q}{c_*}$, μ, we can carry out this continuation with one single epicycloidal arc Γ, depending on μ only. It is obvious how to proceed graphically; for example, by using an arc Γ drawn on transparent paper.

All of the preceding discussion has referred to a bend K with continuously turning tangent. The entire treatment, however, holds just as well for the idealized case where the gradual bend is replaced by a *sharp corner* K, see Figure 16. Then the flow arrives along the wall before K and suddenly turns at K into a new direction. The turn, discontinuous at the corner K, is smoothed out into a continuous turn inside the flow region; it is effected by a *centered simple wave*, swept by a set of characteristics C_+ all of which come from the center K. The previous discussion remains unchanged except that the angle indicating the direction of the flow along the bend K loses its meaning. A complete centered simple wave which is entered by the constant flow in the x-direction through a sonic Mach line in the y-direction, $\theta_* = \omega_* = 0$, is shown in Figure 17 with streamlines, straight characteristics C_+ meeting in the center O, and curved characteristics C_-.

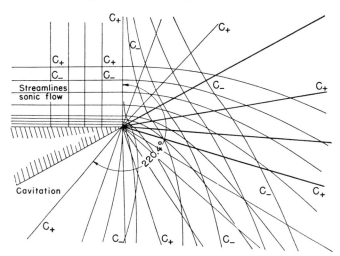

Fig. 17. Complete centered simple wave (air, $\gamma = 1.4$).

This complete simple wave is bounded by two regions of constant state, one of sonic flow, the other of cavitation. Figure 16 shows an incomplete centered simple wave bounded by two constant states. The incomplete wave corresponds to a sector, outlined in Figure 17, cut out of the complete wave, and the states on either side of the incomplete wave are constant. In the regions of constant state the straight Mach lines C_+ are parallel and the cross Mach lines C_- continue straight and parallel out of the incomplete simple wave; both cut the streamlines at a constant Mach angle.

112. Compression waves. Flow in a concave bend and along a bump

The simple wave considered in the preceding section was an *expansion wave* around a convex bend. However, *compression*

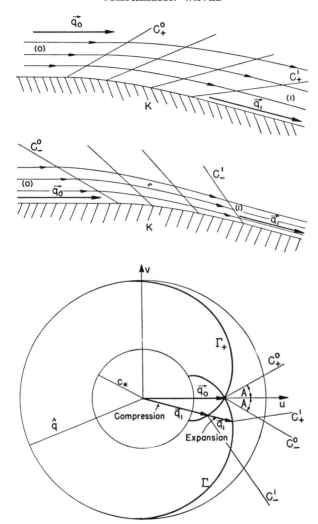

Fig. 18. Simple expansion (top) and compression wave (center) which can turn the oncoming flow with velocity $\vec{q}_0$ around a convex bend K.

waves in a flow around a convex bend or corner are equally possible, as is immediately seen, for example, by considering the flow which is the reverse of an expansion wave, see Figure 18. In the preceding sections we selected those solutions of the differential equations which yield expansion waves along the bend K by choosing the branch of the epicycloid from the point A_1 in the (u, v)-plane which leads to larger values of $q^2 = u^2 + v^2$, and hence to smaller values of p and ρ. However, for a given bend or corner we might just as

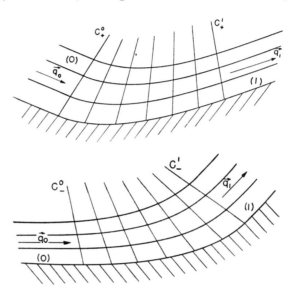

Fig. 19. Simple compression (top) and expansion wave (bottom) which can turn the oncoming flow with velocity $\vec{q}_0$ in a concave bend.

well have chosen the Γ_+-branch of the epicycloid through the point u_0, v_0 in the hodograph plane, which corresponds to decreasing speed q when the flow angle θ decreases and thus to increasing pressure and density. All our arguments and formulas remain essentially the same for the choice of a characteristic arc Γ_+ representing a compression.

For compression waves the characteristics C_- along which u, v, ρ, c, p remain constant are the Mach lines inclined, not downstream but upstream, against the streamlines, as indicated in Figure 18 (center).

What actually happens in an individual case, whether an expansion or a compression occurs in flow around a corner or a bend of a wall, depends on conditions on other parts of the boundary, as will be explained in the following section.

So far we have assumed that the flow turns around a convex bend. In the same way we can handle the case of *flow turning in a concave bend*. In the "normal" case in which the straight Mach lines issuing from the wall are inclined downstream we have a compression wave; in the "exceptional" case in which the straight Mach lines issuing from the wall are inclined upstream we have an expansion wave. Such an exceptional expansion wave can be considered as the reverse of a

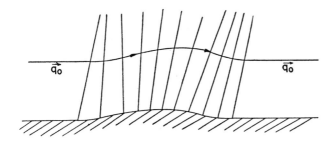

FIG. 20. Simple wave flow along a bump in a straight wall.

certain "normal" flow. We note that, in both cases the straight Mach lines eventually form an envelope so that the flow can be described by a simple wave only in a neighborhood of the wall which becomes smaller as the curvature of the bend increases. Therefore, in the limiting case of a sharp corner this neighborhood shrinks to the vertex itself; then no continuous flow in any neighborhood of the corner is possible. In the general case, the occurrence of an envelope, or more generally, the intersection of our straight characteristics indicates that at least at some distance from the corner a discontinuity must develop. We shall treat flows in concave bends or corners in the next part with the aid of the shock theory.

Of considerable interest is the *flow along a wavy wall* or simply the flow along a wall which is straight except for a "bump." In the normal case, as we shall show in the next article, the straight Mach lines issuing from the beginning of the bump are inclined downstream.

Suppose the bump is an indentation into the flow region, see Figure 20. Then the resulting simple wave is first a compression wave, then an expansion wave, and finally a compression wave again. The final flow velocity agrees with the initial one.

113. Supersonic flow in a two-dimensional duct

The reason why the flow around or in a bend is under normal circumstances a simple wave whose straight Mach lines viewed from the bend are inclined downstream becomes clear when such a flow is regarded as the limiting case of a flow in a duct, one wall of which

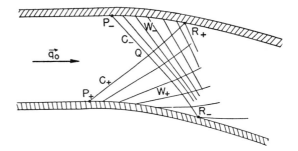

Fig. 21. Supersonic flow in a duct showing simple expansion and compression waves adjacent to the parallel flow with straight Mach lines inclined towards the flow.

is removed to infinity. We consider a two-dimensional duct bounded by two walls beginning with a straight parallel entrance section in which the flow is assumed to arrive with constant supersonic velocity $\vec{q}_0$. When the flow passes the points P_+ and P_- at the two walls at which the bend begins, it starts to deviate from a constant parallel flow. To determine where each gas particle changes its velocity, one draws those Mach lines issuing from the points P_+ and P_- which are inclined downstream, i.e. the C_+-line from the point P_+ at the wall to the right and the C_--line from the point P_- at the wall to the left. Each particle moves with the velocity $\vec{q}_0$ until it meets one of these two "transition" Mach lines; this follows from the theory of the uniqueness

SUPERSONIC FLOW IN A TWO-DIMENSIONAL DUCT 283

of the solutions of hyperbolic differential equations, see Section 24. Beyond the transition Mach line, the flow is a simple wave according to the fundamental theorem of Chapter II, see Section 24. It is thus well determined which type of a simple wave occurs.

Suppose now one of the walls is removed to infinity; then the remaining transition Mach line starting at the remaining wall is inclined downstream. Thus under normal circumstances this is what occurs. We shall see below under which special circumstances the straight Mach lines of the simple wave are inclined upstream.

In Figure 21 we have shown a flow in a duct for which the two transition Mach lines intersect inside the duct at a point Q. Therefore two simple waves W_+ and W_- arise. The continuations of the Mach lines C_+ and C_- through the waves W_- and W_+ respectively are cross Mach lines of these waves and intersect the walls (if at all) at the points R_+ and R_- respectively. These continued Mach lines then delimit the simple wave regions. Beyond them an interaction process begins which no longer results in a simple wave. About such interaction processes we shall say more in the next section.

A somewhat different flow pattern arises if one of the Mach lines, C_+ say, issuing from the point P_+ at the beginning of a bend, meets the opposite wall at a point R_+ at which this wall is still straight. Then this Mach line is the only transition Mach line and only one simple wave zone occurs, see Figure 22. This simple wave region is bounded by the cross Mach line, the "reflected" transition Mach line, which issues from the point R_+ at which the transition Mach line meets the opposite wall. The interaction flow beyond that "reflected" Mach line may be called in a convenient but not quite proper way "superposition of incoming and reflected simple wave."

There is a special case in which the "reflected wave" drops out, in which the flow even beyond the "reflected" transition Mach line remains a simple wave. This happens if, at the point R_+, the intersection of the transition Mach line and the opposite wall, the opposite wall bends in such a way that it coincides with a streamline of the simple wave W_+ which is then continued past the reflected transition Mach line. In that case the continued flow is just this continued simple wave, see Figure 23. We note that this simple wave flow when observed from the opposite wall is one whose Mach lines are inclined against the flow direction. Thus we see that such a simple wave flow is possible, but only under extremely special circumstances.

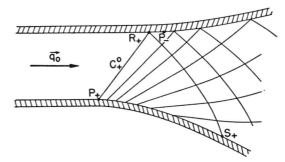

Fig. 22. Supersonic flow in a duct with one simple expansion wave and its "reflected" wave.

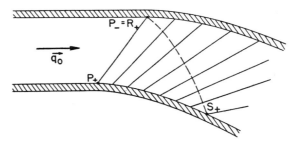

Fig. 23. Supersonic flow in a duct with one simple expansion wave whose "reflected" wave drops out.

From the properties of simple waves we can easily see how to construct a duct through which a uniform parallel flow can be turned into another flow which is also uniform and parallel but is compressed and has a different direction. The direction of the plane wall at a point R is simply changed into the desired new direction and the opposite wall is built in such a way that it forms the streamline of the backward inclined centered simple wave with center R, see Figure 24. (This construction plays a decisive role in the nozzle designed by Oswatitsch [148].)

SUPERSONIC FLOW IN A TWO-DIMENSIONAL DUCT

An exceptionally interesting problem is to shape the end section of an otherwise fixed two-dimensional duct in such a manner that the flow assumed to enter the duct with an appropriate Mach number *emerges from it with constant parallel velocity*. The solution of this problem can be found with the aid of simple waves. Suppose the duct is to be adjusted beyond two points A_+ and A_- at the upper

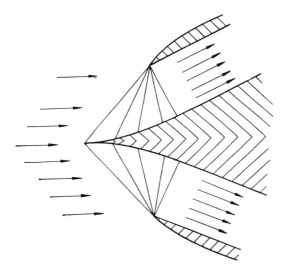

FIG. 24. Ducts through which supersonic parallel flow is changed through a centered simple wave into sonic flow with a different direction.

and lower wall respectively. The entrance Mach number and the shape of the walls up to these two points determine the flow in a region bounded by a C_--Mach line through A_+ and a C_+-Mach line through A_-, up to their intersection at a point O as indicated in Figure 25. We construct the continuation of the duct beyond A_+ and A_- in such a way that eventually the flow velocity everywhere equals the velocity $\bar{q}$ at the point O. More precisely, the two Mach lines C_+^0 and C_-^0 through O shall be the lines across which the transition to constant parallel flow takes place. Consequently, these Mach lines are to be straight and the flow adjacent to each of them should be a simple wave. Of these simple waves two cross Mach lines C_-

are known together with the distribution of velocity, sound speed, density, and pressure along them. As was shown earlier, in Section 110, a simple wave is determined by these data. In particular, the streamlines through the points A_+ and A_- are determined. They

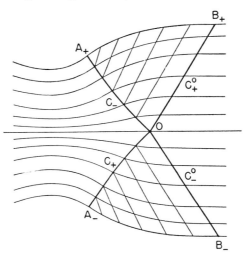

Fig. 25. Construction of the exit section of a duct through which the flow is straightened out.

are to be employed as the walls of the exit section of the duct up to the points B_+ or B_- at which they intersect the two straight Mach lines C_+^0 and C_-^0 issuing from O. From there on the duct is continued straight in the direction of the final flow velocity.

The direction of the flow discussed may just as well be reversed and therefore the construction to straighten out a two-dimensional flow can also be employed to construct an *entrance section of a two-dimensional duct through which a constant parallel flow may enter and be changed gradually without discontinuities*, see also Section 150.

114. Interaction of simple waves. Reflection on a rigid wall

When two simple waves, (I), (II) such as issue from opposite sides of a duct, interact, we must anticipate a situation as indicated

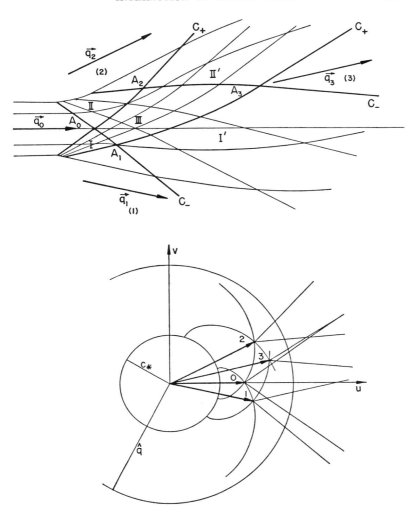

Fig. 26. Interaction of two centered simple waves showing the region of penetration III.

in Figure 26, analogous to the situation when two non-steady rarefaction waves in one dimension interact, see Section 82, Chapter III. There will be a zone (III) of penetration bounded by a characteristic quadrangle. Suppose the two sides of the duct possess two short bends and then continue straight; then two simple waves (I′), (II′) emerge again from the zone of penetration. If the two interacting waves (I) and (II) are known, then the emerging waves (I′) and (II′) can be found easily without solving the differential equations in the penetration zone; that is, we can determine the waves (I′) (II′) in the sense that the corresponding characteristic arcs Γ, or u and v as functions of the angle of direction θ of the straight Mach lines, can be found from Figure 26 (or from corresponding analytic relations).

Suppose that in the domain (0) we have a constant supersonic velocity u_0, v_0, say $u_0 = q_0 > c_*$, $v_0 = 0$. Then in the hodograph plane the two waves (I), (II) are represented by two arcs of epicycloids 0–1 and 0–2 respectively. The waves (I′), (II′) are again represented by two epicycloidal arcs, 1–3 and 2–3, and the latter define as their intersection the point 3, representing the ultimate state of the fluid after the particles pass both waves.

The outcome of a *reflection on a rigid wall* simply corresponds to the interaction of two symmetric waves with the wall as the line of symmetry.

The preceding reasoning does not give detailed information about the width of the transmitted waves, the distribution of the straight characteristics C in them, or the state in the zone of penetration (III). To obtain such detailed information we have to determine the flow in zone (III), and for this purpose we must revert to the general differential equations of flow (102.03) and (102.06). Along the two known characteristic arcs A_0A_1 and A_0A_2, see Figure 26, which bound the end of the incoming simple waves, the values of u and v are known corresponding to the two given arcs Γ which represent (I) and (II). With these initial data we must solve the equations (102.03) and (102.06). In other words, our task is the solution of a characteristic initial value problem as described in Section 24, Chapter II. Solving it determines the two families of characteristics C covering (III), and in particular the characteristic arcs A_1A_3 and A_2A_3. These arcs are cross Mach lines for the simple waves (I′), (II′), which can therefore be immediately constructed, see Section 110.

For the method of solving the characteristic initial value problem see the remarks in Section 24, and the literature quoted there.

Numerical or graphical integration is not difficult on the basis of this theory and has been carried out in many cases, see for example, [103]. The method is similar to that explained in Chapter II, Section 24, and Chapter III, Section 83.

115. Jets

As an example of the interaction of simple waves we indicate a description of phenomena created in a jet by gas streaming in parallel supersonic flow out of an orifice into the atmosphere*, see Figure 27.

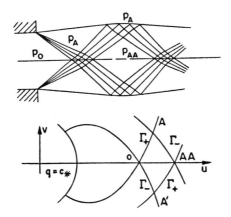

FIG. 27. Wave pattern in a jet resulting when a parallel flow enters a region of lower pressure.

On the basis of observation we suppose that the jet of escaping gas is separated from the quiet air at atmospheric pressure by a boundary consisting of a discontinuity surface (actually by a vortex layer which becomes thicker along the jet and may ultimately consume it). Furthermore, we consider two-dimensional, steady, and isentropic flow, and we assume that the pressure p_0 in the oncoming parallel gas flow is greater than the atmospheric pressure p_A. Then a simplified description of the phenomenon (as long as the jet is not yet destroyed by the vortex layer) is as follows, see also Section 148.

At the corners of the orifice the compressed gas expands in two

---

* In Chapter V we shall study jet flows in more detail.

symmetrical centered simple waves to atmospheric pressure. These two simple waves interact and emerge again as simple waves from their zone of penetration. From the discontinuity surface which forms the boundary of the jet the two simple waves are reflected again as simple waves which penetrate each other and continue as simple waves. These latter waves are compression waves inasmuch as the gas flowing through them increases in density. These waves therefore lead to shocks unless, as Prandtl assumed, they are each centered on the opposite side of the boundary. This assumption is approximately justifiable for small pressure differences $p_0 - p_A$. As in Figure 27, the pattern is assumed to repeat itself, and to continue periodically until the vortex layer on the boundary causes the disintegration of the phenomenon.

The velocity of the emerging gas determines a corresponding point 0 in the (u, v)-plane. Since the pressure $p = p_0$ of the gas at the beginning of the jet is known, the pressure is a known function of the speed q across the two simple waves issuing from the rim. Hence points A and A' on the Γ_+- or Γ_--characteristics through the point 0 are determined to which just the atmospheric pressure $p = p_A$ corresponds. Thus the strength of the two simple waves is determined. The two simple waves resulting after penetration correspond to the two arcs of Γ_-- or Γ_+-characteristics from A or A' up to their intersection AA. Thus the strength of the two emerging waves is also determined.

If the pressure difference $p_0 - p_A$ exceeds a certain value the two arcs of Γ_- or Γ_+ starting at A or A' do not intersect but meet the circle $q = \hat{q}$ instead. In that case the process of interaction goes on indefinitely and the region of penetration extends to infinity.

116. Transition formulas for simple waves in a polytropic gas

For many purposes it is desirable to express the changes of the quantities across a simple wave as functions of the change of the angle of flow direction θ. For this purpose it is sufficient to express the angle ω in terms of the angle θ, see Section 109. Suppose on one of the straight Mach lines the angle $\theta = \theta_0$, the flow speed $q = q_0$, and the sound speed $c = c_0$ are given. Then the Mach angle $A = A_0$ is given from (106.06) by

(116.01) $$\sin A_0 = c_0/q_0$$

and the critical speed c_*, from (102.02) by

(116.02) $$c_*^2 = \mu^2 q_0^2 + (1 - \mu^2) c_0^2.$$

The angles ω_0 and ω_* are then determined from

(116.03)
$$\omega_0 = \theta_0 - A_0 + 90° \text{ for } \Gamma_+,$$
$$\omega_0 = \theta_0 + A_0 - 90° \text{ for } \Gamma_-,$$

see (108.02) and, see (108.14),

(116.04) $$|\omega_0 - \omega_*| = \mu^{-1} \operatorname{arc\,cot} \mu^{-1} \tan A_0.$$

The angle ω as function of θ is finally determined by solving the equation

(116.05) $$\theta = \omega \pm A \mp 90° = \omega \pm \arctan \mu \cot \mu(\omega - \omega_*),$$

the upper sign referring to a Γ_+-wave, the lower one to a Γ_--wave.

It is useful for practical purposes to give the resulting relations in a table for a complete simple wave, a table which depends only on γ or μ and which can be used in an obvious way, with proper interpolations if necessary. Such a tabulation is reproduced in Table I. To apply it in a specific case one determines the portion of the complete wave to be used from the ratio q_0/c_* or the Mach number $M_0 = q_0/c_0$ of the incoming flow with $\theta = \theta_0$. The table then gives the corresponding angles $\theta_0 - \omega_*$; further quantities characterizing the appropriate simple wave can be read off.

At the end of Section 109 we saw that the ends of the complete arcs Γ, at which they touch the limit circle, are given by $\omega - \omega_* = \pm \frac{\pi}{2\mu}$. Thus from equation (109.02) or from our diagrams, we see that the maximum angle through which the flow can be bent in a Γ_+ or a Γ_- wave is

(116.06) $$|\hat{\theta} - \theta_*| = \left(\frac{1}{\mu} - 1\right)\frac{\pi}{2},$$

which is attained only in the ideal case of a complete wave, in which the flow starts at sonic speed c_* and ends at the limit speed $\hat{q}$. Table II, on page 293, gives numerical values for this maximum angle for various values of γ. If the initial speed q_0 is greater than the critical speed c_*, obviously the whole angle through which the flow can possibly turn before cavitation occurs is smaller.

For large values of $\frac{q_0}{c_0}$ or small Mach angles A one finds from (116.04–.05) for the maximum angle $|\theta_0 - \theta_*|$ through which the flow can turn, the approximate value $\left(\frac{1}{\mu^2} - 1\right)\frac{c_0}{q_0} = \frac{2}{\gamma - 1}\frac{c_0}{q_0}$.

For many cases in which the change across a simple wave is not very great, it is sufficient to employ *expansions* with respect to powers of $\theta - \theta_0$. We may assume that the incoming flow is in the positive x-direction, i.e. $\theta_0 = 0$, and that the simple wave is a Γ_--wave with straight C_+-characteristics. Setting

(116.07) $\qquad t_0 = \tan A_0 = 1/\sqrt{M_0^2 - 1},$

the expansions are

$$q/q_0 = 1 - t_0\theta - \frac{1}{2(1-\mu^2)}(\mu^2 + 2\mu^2 t_0^2 + t_0^4)\theta^2 + \cdots$$

$$u/q_0 = 1 - t_0\theta - \frac{1}{2(1-\mu^2)}(1 + 2\mu^2 t_0^2 + t_0^4)\theta^2 + \cdots$$

$$v/q_0 = \theta - t_0\theta^2 + \cdots$$

$$c/c_0 = 1 + \frac{\mu^2}{1-\mu^2}(t_0 + t_0^{-1})\theta$$

(116.08) $\qquad\qquad + \frac{\mu^2}{2(1-\mu^2)^2}(1 + t_0^2)[t_0^2 - (1 - 2\mu^2)]\theta^2 + \cdots$

$$p/p_0 = 1 + \frac{1+\mu^2}{1-\mu^2}(t_0 + t_0^{-1})\theta$$

$$+ \frac{1+\mu^2}{2(1-\mu^2)^2}(1 + t_0^2)[t_0^2 + 2\mu^2 + t_0^{-2}]\theta^2 + \cdots$$

$$\rho/\rho_0 = 1 + (t_0 + t_0^{-1})\theta + \frac{1}{2(1-\mu^2)}(1 + t_0^2)[t_0^2$$

$$+ (1 - 2\mu^2)t_0^{-2}]\theta^2 + \cdots$$

$$A = A_0 + \frac{1}{1-\mu^2}(\mu^2 + t_0^2)\theta + \cdots$$

$$90° + \omega = 90° + \omega_0 + \frac{1}{1-\mu^2}(1 + t_0^2)\theta + \cdots,$$

the latter angle being the angle of inclination $A + \theta$ of the Mach line with the positive x-axis.

TABLE I
Tabulation of quantities characterizing a complete simple wave
(For air $\gamma = 1.405$)

p/p_*	c/c_*	q/c_*	$M = q/c$	$|\omega - \omega_*|$	$|\theta - \omega_*|$
.00	0.000	2.437	∞	219.32°	129.32°
.01	.515	2.151	4.178	143.79	67.64
.02	.569	2.083	3.661	134.80	60.65
.03	.603	2.035	3.373	123.90	56.14
.04	.629	1.996	3.175	124.37	52.73
.05	.649	1.964	3.024	120.64°	49.95°
.07	.682	1.909	2.801	114.60	45.52
.09	.707	1.863	2.636	109.73	41.82
.11	.728	1.823	2.506	105.57	39.09
.13	.745	1.788	2.399	101.91	36.55
.15	.761	1.755	2.307	98.62°	34.30°
.20	.793	1.683	2.123	91.47	29.58
.25	.819	1.621	1.979	85.35	25.70
.30	.841	1.565	1.861	79.90	22.40
.35	.860	1.513	1.760	74.89	19.50
.40	.876	1.465	1.672	70.19°	16.92°
.45	.891	1.420	1.593	65.71	14.60
.50	.905	1.376	1.521	61.38	12.48
.55	.917	1.335	1.455	57.13	10.55
.60	.929	1.295	1.394	52.92	8.78
.70	.950	1.218	1.282	44.39°	5.66°
.80	.968	1.144	1.181	35.22	3.09
.90	.985	1.071	1.088	24.27	1.10
.95	.993	1.036	1.043	16.96	.39
1.00	1.000	1.000	1.000	0.00	.00

TABLE II
Angle $|\hat{\theta} - \theta_*|$ through which a complete simple wave turns a flow for various values of γ

γ	1.00	1.20	1.25	1.30	1.40	1.67	2.00	3.00	7.00
$\|\hat{\theta} - \theta_*\| = \left(\dfrac{1}{\mu} - 1\right)\dfrac{\pi}{2}$	∞	221.8°	180.0°	159.2°	130.4°	90.0°	65.9°	37.3°	13.9°

C. Oblique Shock Fronts

117. Qualitative description

As in the case of one-dimensional flow, the assumption of continuous flow in two or three dimensions is often incompatible with the boundary conditions; discontinuities may arise as in one-dimensional flow. Again the fortunate situation prevails that many phenomena of actual flow are adequately represented by the relatively simple model of *shock fronts*, i.e. surfaces across which density, pressure, and velocity undergo jump discontinuities in agreement with the conservation laws.

To visualize how discontinuities arise automatically even from continuous boundary conditions, we consider the flow in a concave bend or corner and carry the analysis somewhat farther than in Section 112.

Suppose a constant two-dimensional supersonic flow arriving along a straight wall is forced to turn in a concave bend K.

In principle, our previous construction of a simple wave remains valid *near* the wall. There is a Mach line through the point A in the (x, y)-plane, across which the constant flow passes into a simple wave. But in contrast to the case of flow outside a bend, the subsequent straight Mach lines of the simple wave now turn so that an envelope originates inside the flow. In general this envelope has a cusp. The mathematically ambiguous state in the region between the two branches of the envelope (u and v would not be uniquely defined) is physically impossible. As observations indicate, it is avoided by a shock discontinuity, i.e., a surface of discontinuity for the quantities u, v, ρ, p, T, S. Since the flow is two-dimensional, this surface is characterized by its intersection with the (x, y)-plane, the *shock curve*. This shock curve $\mathcal{S}$ starts with strength zero at the cusp of the envelope, and runs between the two branches of this envelope.

As we shall see presently, the shock conditions imply that different particles crossing the shock front $\mathcal{S}$ from a zone of constant entropy, in general, suffer different entropy changes, and the flow behind the shock front ceases to be isentropic. Behind a shock front consideration of non-uniform entropy is unavoidable then. However, in many important cases (the only ones that lend themselves to relatively simple analysis) the variation in entropy change is either absent or negligible, so that our simple differential equations of isentropic flow, (102.03) and (102.06), remain practically valid on both sides of the shock line $\mathfrak{S}$.

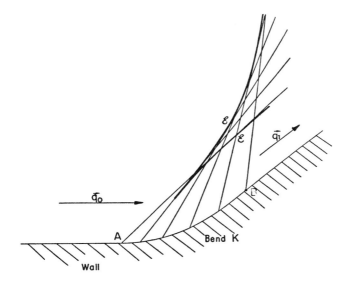

FIG. 28. Envelope $\mathcal{E}$ of a straight characteristic issuing from a bend K.

This situation prevails when the shock line $\mathfrak{S}$ is straight and the state is constant on either side of $\mathfrak{S}$. Typical of this situation and basically important in itself is the limiting case of a bend K concentrated in a sharp corner K; the flow, traveling with constant supersonic speed parallel to one leg of the angle, arrives at the corner K and turns discontinuously into the direction of the other leg, again at constant velocity. The sudden transition of direction and speed of the flow is effected across a straight shock line $\mathfrak{S}$, provided the

296   PLANE FLOW   CHAP. IVC

angle through which the wall turns is below a certain bound, see Section 122. In this case the *oblique shock front* connects two zones of constant state, (I) and (II), and there is no complication from

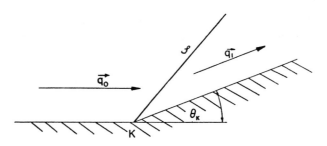

Fig. 29. Straight shock line resulting from a flow in a sharp corner K.

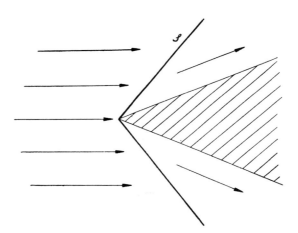

Fig. 30. Supersonic flow against a wedge involving two shock fronts.

variations of entropy. The situation is qualitatively indicated in Figure 29.

Our remarks apply also to two-dimensional flow against a wedge or the head of a two-dimensional airfoil. Such a flow can be obtained by piecing together two flow patterns, each corresponding to the flow in a corner formed by the streamline reaching the tip of the wedge, and one of the edges. Three-dimensional supersonic flow against a conical projectile will be treated later in Chapter VI, Part B.

Before carrying out the quantitative analysis we establish the general shock relations.

118. Relations for oblique shock fronts. Contact discontinuities

For shock fronts which exist in two dimensional flow and for those in three dimensional flow as well, the discontinuity conditions could be derived from the principles of conservation of mass, momentum, and energy in exactly the same way as for one-dimensional flow. For this purpose, one would consider a smooth portion of a discontinuity surface, express the conservation laws in integral form for a flat cylindrical volume enclosing the discontinuity surface and moving with it, and then allow the cylindrical volume to shrink to the flat discontinuity surface.

Instead of carrying out this procedure in detail, we shall make use of our previous results and simply reduce the conditions for oblique shocks to those for straight shocks derived in Chapter III.

We start with some general remarks: First, a sufficiently small smooth portion of the discontinuity surface S may be considered as plane with any degree of approximation. Secondly, within a sufficiently small time interval the normal component U of the shock front velocity may be considered as constant. Third, in a small neighborhood of the nearly plane portion of S, and at the instant under consideration, each of the states (0) and (1) on the two sides of S is approximately constant. Now, the laws of conservation contain no derivatives of p, ρ, $\bar{q}$, U; therefore it is plausible, and not difficult to prove, that in deriving the shock conditions by expressing the conservation laws for a small volume, the result remains the same if we assume S plane, U constant, and likewise the two states (0) and (1) constant.

After this preparation the shock conditions are easily reduced to those of one-dimensional flow by referring the phenomenon to a coordinate system which moves with a suitable constant velocity relative to the original one and by remembering that, according to Galileo's principle of relativity, the system of shock relations remains invariant under translatory motions of the coordinate system.

Thus, we may obtain the shock relations by using the shock front $\mathcal{S}$ as a coordinate surface or, what is equivalent, by regarding the *shock front* as *stationary*, no matter whether the flow under consideration is steady or not. We still have one more degree of freedom, inasmuch as we can move the origin of the coordinate system with constant velocity along the line $\mathcal{S}$. Hence, without restriction of generality, we may assume that the velocity component of the oncoming flow parallel to $\mathcal{S}$ is zero, and thus visualize the flow as a flow of constant speed, meeting a stationary surface of discontinuity $\mathcal{S}$ at a right angle. If the speed q_0 is not zero, i.e., if mass is transported through $\mathcal{S}$, then the law of conservation of momentum requires that the tangential component of the momentum flux is continuous across this front. For, any jump in this component would have to be balanced by tangential forces acting on the front; but there are no such tangential forces, since all forces on the fluid are assumed to be pressure forces perpendicular to the surface against which they act. Continuity of the tangential component of the momentum flux implies continuity of the tangential component of the velocity since the flux of mass across the surface is continuous. Consequently, if the velocity of the oncoming flow is perpendicular to the front, the velocity in the state (1) behind the front is likewise perpendicular to $\mathcal{S}$. In other words, *observed from a suitable coordinate system, an oblique shock front is always equivalent to a stationary one-dimensional shock front.*

If, however, $q_0 = 0$, as seen from this coordinate system, then in the state (1), by the law of conservation of momentum, the normal component N of the velocity $\vec{q}$ likewise vanishes, while the tangential component may be arbitrary. In this case we have a *contact discontinuity*, described generally for an arbitrary surface as follows. A contact discontinuity $\mathcal{D}$ is a surface through which there is no mass flux (so that the flow is tangential as seen from both sides), across which, however, density, temperature, and entropy are discontinuous. Yet, as we shall see, the pressure is the same on both sides. Such a contact surface may be considered a "vortex sheet," along which two different layers of the substance (or even of different substances) slide along each other.

For genuine shock fronts through which there is a mass flux we distinguish, as in one-dimensional flow, between the *front side* and the *back side* by saying that *the fluid passes through the shock front from the front side to the back side*.

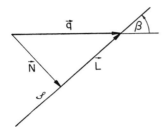

FIG. 31. Normal and tangential components of the flow velocity ahead of a shock front.

To formulate the shock relations we denote by N the normal, and by L the tangential component of the flow vector $\vec{q}$, and by U the normal component of the velocity of the shock front, see Figure 31. Then the shock relations are:

Conservation of mass

(118.01) $$\rho_0(N_0 - U) = \rho_1(N_1 - U) = m,$$

Conservation of momentum

(118.02) $$\rho_0 N_0(N_0 - U) + p_0 = \rho_1 N_1(N_1 - U) + p_1,$$

Continuity of tangential component

(118.03) $$L_0 = L_1.$$

That is, *the difference $\vec{q}_0 - \vec{q}_1$ of the velocity vectors is perpendicular to the shock line.*

Conservation of energy

(118.04) $$\tfrac{1}{2}\rho_0(N_0 - U)q_0^2 + \rho_0(N_0 - U)e_0 + N_0 p_0$$
$$= \tfrac{1}{2}\rho_1(N_1 - U)q_1^2 + \rho_1(N_1 - U)e_1 + N_1 p_1$$

with

$$q^2 = L^2 + N^2.$$

In addition we require that upon crossing a shock front the entropy of a particle increases, $S_1 - S_0 > 0$.

For a *stationary shock front* with $U = 0$ the relations simplify to:

(118.05) $$N_0 \rho_0 = N_1 \rho_1 = m$$

(118.06) $$\rho_0 N_0^2 + p_0 = \rho_1 N_1^2 + p_1 = P$$

(118.07) $$L_0 = L_1$$

(118.08) $$\tfrac{1}{2} q_0^2 + i_0 = \tfrac{1}{2} q_1^2 + i_1 = \tfrac{1}{2} \hat{q}^2;$$

$q^2 = N^2 + L^2$ is the square of the flow speed, and $\hat{q}$ is the *limit speed* of the flow. The last relation expresses the important fact that the Bernoulli constant $\tfrac{1}{2} q^2 + i$ remains unchanged across a stationary shock front. Thus, *Bernoulli's law for steady flow holds even if the flow crosses shock fronts.*

Of course, the form (118.05–.06) of the conditions remains valid if N is replaced by $N - U$, and thus a moving shock front introduced. (The same relations (118.05–.06) and (118.08) hold for contact discontinuities when $m = 0$; then $p_0 = p_1$, and $N_0 = N_1 = 0$ follow, while L_1 and L_0, as well as ρ_0 and ρ_1 may be different.) In the following we shall be concerned with *stationary shocks*, unless otherwise noted.

A slightly different and more symmetric way of writing the first two conditions for stationary shocks, expressing the conservation of the mechanical quantities, mass and momentum, is obtained as follows. We insert (118.05) into (118.06) and make use of (118.07); then we find

$$(p_1 - p_0) = \rho_0 N_0 (N_0 - N_1) = \rho_0 \{ N_0 (N_0 - N_1) + L_0 (L_0 - L_1) \}$$

or

(118.09) $$\tau_0 (p_1 - p_0) = \vec{q}_0 \cdot (\vec{q}_0 - \vec{q}_1),$$

and similarly

(118.10) $$\tau_1 (p_0 - p_1) = \vec{q}_1 \cdot (\vec{q}_1 - \vec{q}_0);$$

from which

(118.11) $\qquad (p_1 - p_0)(\tau_0 - \tau_1) = (\vec{q}_0 - \vec{q}_1)^2,$

and

(118.12) $\qquad (p_1 - p_0)(\tau_0 + \tau_1) = \vec{q}_0^2 - \vec{q}_1^2$

follow. Equations (118.07), (118.09–10) together are equivalent to the equations (118.05–.07).

As in the case of one-dimensional shocks it is important to note that the conditions (118.05–.07) and (118.09–.10) are valid on the basis of mechanics alone. The thermodynamical nature of the medium enters only through condition (118.08). As stated in Section 61, Chapter III, there are many instances in which the solution of flow problems is eased by the possibility of determining the flow without using the thermodynamical shock condition.

For stationary phenomena the shock conditions show: *to a fixed state (0) ahead of a shock front, there corresponds a one-parametric family of states (1) behind the shock which satisfy the shock conditions.* Omitting the subscript (1) we may then consider the state behind the shock as described by functions p and $\vec{q}$ of τ or of $\tau - \tau_0$ or of any other measure of the shock strength.

The change of entropy across a shock is a quantity of only third order in the shock strength. This statement is implied by the analogous statement for one-dimensional shocks, since an oblique shock is equivalent to a one-dimensional shock if observed from a suitable uniformly moving coordinate system, and since entropy, pressure, and density do not depend on the frame of reference.

If the shock is weak, the two mechanical conditions (118.11–.12) lead in the limit, when $\tau - \tau_0 \to 0$, to the result: *An infinitely weak shock is a sonic disturbance;* or, *for infinitely weak shocks the shock direction becomes characteristic.*

The proof is simple: Omitting the subscript, we obtain for a family of shocks by differentiating relations (118.05–.06),

$$N\,d\rho + \rho\,dN = 0, \quad m\,dN + dp = 0;$$

hence

$$-mN\,d\rho + \rho\,dp = 0;$$

or, since $m = N\rho$,
$$-N^2 \, d\rho + dp = 0.$$
Since $dS/d\rho = 0$ for $\tau = \tau_0$, we have
$$N^2 = \frac{\partial p}{\partial \rho} = c^2 \quad \text{for} \quad \tau = \tau_0.$$

Thus the normal component of the flow velocity, with respect to the limit position of the shock line as $\tau \to \tau_0$, is just the sound speed. According to relation (106.06) this fact assures that the limit direction of the shock line $\mathcal{S}$ is characteristic.

119. Shock relations in polytropic gases. Prandtl's formula

As in the case of one-dimensional flow the thermodynamical condition for stationary shocks in a polytropic gas (with which we shall be primarily concerned) is, from (118.08) and (9.06),

(119.01) $\quad \mu^2 q_0^2 + (1 - \mu^2) c_0^2 = \mu^2 q_1^2 + (1 - \mu^2) c_1^2 = c_*^2 \, ;$

here c^2 is the sound speed, $c_* = \mu \hat{q}$ the critical speed, and $\mu^2 = \dfrac{\gamma - 1}{\gamma + 1}$. As in the case of normal shocks, we may replace relation (119.01) by the relation

(119.02) $\quad \dfrac{p_1}{p_0} = \dfrac{\mu^2 - \dfrac{\rho_1}{\rho_0}}{\mu^2 \dfrac{\rho_1}{\rho_0} - 1} = \dfrac{\mu^2 \tau_1 - \tau_0}{\mu^2 \tau_0 - \tau_1},$

which connects pressures and densities only and is invariant under translations, see (67.01).

Of great importance for the study of shocks in polytropic gases is the generalization of Prandtl's relation $q_0 q_1 = c_*^2$, previously obtained for normal stationary shocks, see Section 66. To apply this relation in the case of an oblique shock, we write

(119.03) $\qquad\qquad \vec{q} = \vec{N} + \vec{L},$

where $\vec{L}$ is the vectorial component of the flow parallel, and $\vec{N}$ the vectorial component normal to the stationary shock line $\mathcal{S}$, so that

$\vec{N}.\vec{L} = 0$. By substituting in Bernoulli's equation (119.01),
$$\mu^2 q^2 + (1 - \mu^2)c^2 = c_*^2,$$
we find
$$\mu^2 N^2 + \mu^2 L^2 + (1 - \mu^2)c^2 = c_*^2,$$
or
(119.04) $$\mu^2 N^2 + (1 - \mu^2)c^2 = c_*^2 - \mu^2 L^2 = \tilde{c}_*^2,$$
where $\tilde{c}_*$ is the critical speed in a new coordinate system in which

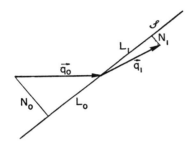

FIG. 32. Showing how the flow is turned toward the shock line.

the shock front is normal. Hence by Prandtl's relation for this case, see Section 66, we have

(119.05) $$N_0 N_1 = \tilde{c}_*^2$$

or

(119.06) $$N_0 N_1 = c_*^2 - \mu^2 L^2,$$

a relation which may be used instead of the preceding forms of the thermodynamical shock condition.

Some general and significant information follows from the relations developed so far. Equation (119.06) shows that

(119.07) $$q_0 q_1 = \sqrt{N_0^2 + L^2} \sqrt{N_1^2 + L^2} > N_0 N_1 + L^2$$
$$= c_*^2 + (1 - \mu^2)L^2 > c_*^2,$$

because of the relation $\mu^2 < 1$. Hence, if the state (0) is at the front side of the shock and therefore $N_0 > N_1$, then $N_0 > c_*$; but we cannot, as for normal shocks, conclude that $N_1 < c_*$.

120. General properties of shock transitions

The following facts are easily derived from the transition formulas for polytropic gases. They hold, however, quite generally (except for the reference to the critical speed c_*).

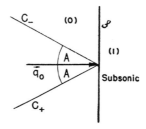

FIG. 33. Mach lines C exist only in front of a normal shock line.

The speed of a flow ahead of a shock front (observed from the shock front) is supersonic; the speed on the back side may be subsonic or supersonic. The latter possibility becomes obvious if we recall that the tangential component L is quite arbitrary and may be chosen larger than c_* , whereupon q_0 and q_1 both become larger than c_*.

An even more immediate remark is: *Through an oblique shock in a gas the flow direction is always turned towards the shock line* $\mathfrak{S}$. For, the normal component N becomes smaller when the flow crosses $\mathfrak{S}$, while the tangential component L remains unchanged.

A further important remark concerns the *relative position of Mach lines and flow vectors.*

For a normal shock line $\mathfrak{S}$ there are two Mach lines, C_+ and C_-, in the region of supersonic flow (0) ahead of $\mathfrak{S}$. In the zone (1) behind $\mathfrak{S}$, however, the flow is subsonic; consequently there are no Mach lines in the zone behind a normal shock front, see Figure 33.

GENERAL PROPERTIES OF SHOCK TRANSITIONS

For oblique shocks the latter statement remains true as long as $q_1 < c_*$, i.e., as long as the flow in region (1) is subsonic. If $q_1 = c_*$, a Mach line appears perpendicular to the flow velocity in (1). If $q_1 > c_*$, there are two different Mach lines in region (1). It is important to realize that their position is as indicated in Figure 34. (Note that the sum of the angles which C_+ and C_- make with $\vec{q}_1$ equals 180°).

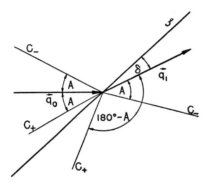

Fig. 34. Positions of the Mach lines C in front and in back of an oblique shock line.

In other words, *behind the shock front S the two Mach directions* (if they exist) forming the same angle A_1, the Mach angle, with the flow direction *lie in the obtuse angle between S and the flow vector $\vec{q}_1$*. The statement is obviously equivalent to $\delta < A_1$, where δ is the acute angle between the flow in (1) and the line S. For the proof we recall $\sin A_1 = \dfrac{c_1}{q_1}$, $\sin \delta = \dfrac{N_1}{q_1}$; hence we must prove that c_1 is greater than N_1. This, however, follows immediately when we reduce the oblique shock to a normal shock by referring the motion to a coordinate system moving with the velocity $\vec{L}$. Then the normal components N_0, N_1 of the flow velocities remain unchanged, as do the sound speeds c_0, c_1, which depend only on pressures and densities. In the new coordinate system the speed of the flow behind the shock front is N_1, and since the shock front is normal, the speed behind it is subsonic, i.e., $N_1 < c_1$.

121. Shock polars for polytropic gases

The quantitative analysis of oblique shocks can be carried out by geometrical constructions in two dimensions or by corresponding analytical methods.

We assume the shock line $\mathfrak{S}$ in the (x, y)-plane of the flow to be straight and the states (0) and (1) on both its sides to be constant, assumptions always valid "in the small." Furthermore we assume the shock to be stationary and the medium to be a polytropic gas.

As stated before, to a fixed state (0) there corresponds a one-parametric family of possible states (1) which can be reached through a shock. Since such a state can be characterized by two variables, the possible states (1) are given by a relation between these variables geometrically represented by a curve. Such curves, called *shock polars*, can be introduced in a variety of ways according to which variables we choose for defining the state (1).

Let us characterize the shock direction by the angle β between the shock line $\mathfrak{S}$ and the direction of the oncoming flow. The angle between the vector $\vec{q}_0$ of the oncoming and $\vec{q}_1$ of the outgoing flow, i.e. the angle through which the flow is turned by the shock, will be denoted by θ. Placing the coordinate system so that the oncoming flow is parallel to the x-axis and using our previous notation, N and L, for the components of $\vec{q}$ normal and tangential to the shock line $\mathfrak{S}$, respectively, we have

$$u_0 = q_0, \quad v_0 = 0,$$
(121.01) $\quad u_1 = L_1 \cos \beta + N_1 \sin \beta, \quad v_1 = L_1 \sin \beta - N_1 \cos \beta,$
$$L_0 = L_1 = q_0 \cos \beta, \quad N_0 = q_0 \sin \beta.$$

Prandtl's basic relation (119.06) yields $N_1 = \dfrac{c_*^2 - \mu^2 L_0^2}{N_0}$. Using (119.01), (106.06), and this form of Prandtl's formula, we obtain the relations,

(121.02)
$$u = (1 - \mu^2)q_0 \cos^2 \beta + \frac{c_*^2}{q_0}$$
$$= q_0 - (1 - \mu^2)[\sin^2 \beta - \sin^2 A_0]q_0,$$
$$v = (q_0 - u) \cot \beta,$$

in which the subscript characterizing the state (1) behind the shock front has been suppressed. These equations show that, for a given critical speed c_* and given speed q_0 of the oncoming flow, the angle β between the shock line S and the direction of the vector $\vec{q}_0$ deter-

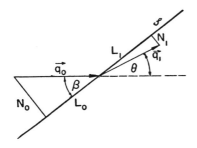

FIG. 35. Angles of flow and shock directions.

mines the outgoing flow velocity $\vec{q}$. If β varies, the point (u, v) in the hodograph plane describes a shock polar, associated with the value q_0 and to the critical speed c_*, and which is represented by equations (121.02) with β as a parameter.

Eliminating the angle β from equations (121.02) we find

$$(121.03) \qquad v = (q_0 - u) \sqrt{\frac{u - \tilde{u}}{U - u}}$$

for a shock through which the y-component v of the flow becomes positive or, without this restriction,

$$(121.04) \qquad v^2 = (q_0 - u)^2 \frac{u - \tilde{u}}{U - u}.$$

Here,

$$(121.05) \qquad \tilde{u} = \frac{c_*^2}{q_0} = [\mu^2 + (1 - \mu^2) \sin^2 A_0] q_0$$

is the velocity of the flow that would result from a normal shock, and

$$(121.06) \qquad U = (1 - \mu^2) q_0 + \tilde{u} = [1 + (1 - \mu^2) \sin^2 A_0] q_0.$$

(Note that, by Bernoulli's equation (102.02), c_* is known when the state (0) is given.)

The curve in the (u, v)-plane given by equation (121.04), the *shock polar* introduced by Busemann [3], is the well-known "Folium of Descartes," with a double or isolated point at $u = q_0$, $v = 0$, and with an asymptote $u = U$.

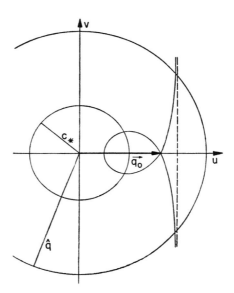

Fig. 36. Shock polar for $q_0 > c_*$.

Of the shock polar only the part with $u^2 + v^2 \leq \hat{q}^2 = c_*^2/\mu^2$ corresponds to shock transitions. It is the geometrical locus of all points $(u\ v)$ characterizing a state (1) which can be connected with the fixed state (0) by a stationary shock. The branch $u < q_0$ represents states behind shock fronts when the given state (0) refers to its front side. The branch $u > q_0$ represents states ahead of the shock front when the state (0) behind the front is given.

The one-parametric family of possible shock transitions from the state (0) can also be illustrated by a shock polar in the (θ, p)-plane, which shows the pressure on the other side of the shock and the angle θ through which the flow is turned (again the state (0) is assumed to be fixed, e.g., q_0, ρ_0, p_0 or q_0, p_0, c_* prescribed). In

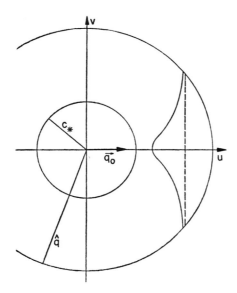

Fig. 37. Shock polar for $q_0 < c_*$.

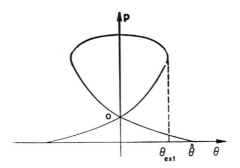

Fig. 38. Shock polar in the (θ, p)-plane, showing angles θ_{ext} and $\hat{\theta}$.

parametric form with u as parameter, we find from (118.09) and (121.03) the equations

(121.07) $$p - p_0 = \rho_0 q_0 (q_0 - u),$$

(121.08) $$\tan \theta = \frac{q_0 - u}{u} \sqrt{\frac{u - \tilde{u}}{U - u}};$$

These equations define the image of the Descartes' loop in the (θ, p)-plane with u as parameter; this (θ, p)-shock polar is shown in Figure 38. The loop gives possible states behind a shock front when the state (0) is in front; the lower branches give states in front when the state (0) behind is given.

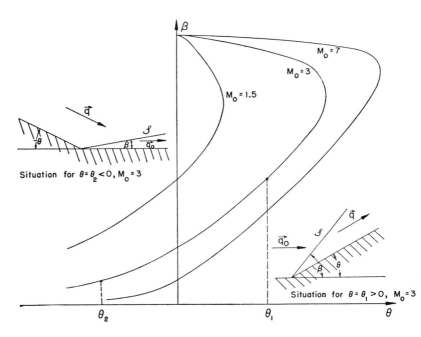

Fig. 39. Relation between shock angle β and flow angle θ if the flow velocity on one side of the shock front is given. The cases $\theta > 0$ and $\theta < 0$ correspond to the situations in which the velocity $u = q_0$, $v = 0$ is given ahead of the front and behind the front respectively.

Still another shock polar representing the relationship between the angles θ and β is useful. This relation is given by (121.08) and (121.02.1). The polar is shown in Figure 39. The branch $\beta > A_0$, $\theta > 0$ refers to oblique shock fronts S for which the state (0) ahead of them is prescribed, while the branch $\beta < A_0$, $\theta < 0$ refers to shock fronts for which the state (0) behind them is prescribed. Of course other shock polars, such as that given by the relations between p and β, could be considered.

122. Discussion of oblique shocks by means of shock polars

Shock polars exhibit the quantitative situation in shock transitions. Let us make use of the shock polar in the (u, v)-plane, associated with c_* and q_0; it can be used to construct oblique shocks as follows.

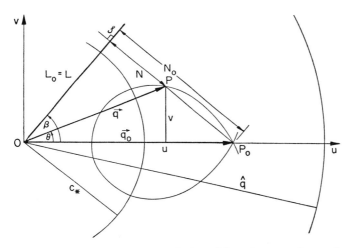

FIG. 40. Determination of a shock line with a given turning angle θ from the shock polar in the (u,v)-plane.

Figure 40 shows such a shock polar for finding the state behind the shock front when $q_0 > c_*$; it has a double point at $P_0 = (q_0, 0)$, the end point of the vector $\vec{q}_0$. For any point P on the shock polar we find the velocity behind the shock front as the vector OP. The angle θ through which the shock turns the incoming flow is the angle between OP and the u-axis. The direction of the shock line S, making the angle β with the incoming flow, is perpendicular to the line connecting the double point P_0 with the point P, in accordance with the fact stated before that the vector $\vec{q} - \vec{q}_0$ is perpendicular to the shock line. The components $L_0 = L$, N_0, N of the velocity vectors $\vec{q}_0$ and $\vec{q}$ with respect to the shock line can be read off from the diagram as indicated there.

The shock polar in the (u,v)-plane yields immediately only relations between velocities. The pressure behind the shock front is then determined by (121.07), the density by (119.02).

Our diagrams lead to the following observations which are easily confirmed by calculation: the angle β between the shock line and the vector $\vec{q}_0$ is greater than the angle θ between $\vec{q}$ and $\vec{q}_0$, $\beta > \theta$.

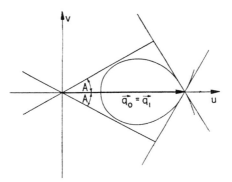

Fig. 41. Showing that the shock lines of weak shocks are near to the Mach lines.

The flow past the shock front may be supersonic, $q > c_*$, *or subsonic*, $q < c_*$, the first case arising for relatively weak, the second case for relatively strong shocks. A normal shock, in particular, is represented by the intersection of the shock polar with the u-axis, see (121.05), $q = c_*^2/q_0 = \tilde{u}$.

A point P with the coordinates (u, v) near the double point $(q_0, 0)$ on the loop of the polar represents a weak shock with little change in velocity and pressure. As P tends to the double point the shock becomes sonic, the vector $\vec{q} - \vec{q}_0$ becomes tangential, and hence the shock line S tends to a Mach line. This fact, corresponding to the statement proved in Section 118, can be inferred immediately from equations (121.02). Consequently, the two tangents to the loop at the double point are the normals to the Mach lines at the double point, since they are orthogonal to the limiting shock line.

The polar diagram shows immediately how to determine a shock line from a given state (0) in front of it if either the shock direction β or the turning angle θ is given. It also shows that *there is an extreme angle* $\theta = \theta_{\text{ext}}$, *such that for* $\theta > \theta_{\text{ext}}$ *no shock transition exists,*

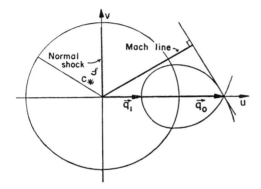

FIG. 42. Sonic disturbance and strong normal shock for $\theta = 0$.

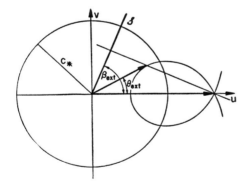

FIG. 43. Extreme angles θ_{ext} and β_{ext}.

see Figure 43, *while for* $\theta < \theta_{\text{ext}}$ *there are two possible shocks, a weak and a strong one, with small or large change in velocity, respectively.* If θ becomes small, we have in the limit either an infinitely weak sonic disturbance or a strong normal shock, see Figure 42. For $\theta = \theta_{\text{ext}}$ the two possible shocks coincide. The shock angle β reduces to the Mach angle A for a sonic shock and becomes 90° for a normal shock. In between, it passes through a value β_{ext} corresponding to the extreme value θ_{ext} of the angle θ.

Incidentally, we have

$$(122.01) \qquad \sin \hat{\theta}_{ext} = \frac{1}{\gamma}$$

for the limiting case $q_0 = \hat{q}$. For air, γ is approximately 1.4 and we find $\hat{\theta}_{ext} = 45.5°$. The corresponding angle $\hat{\beta}_{ext}$ is

$$(122.02) \qquad \hat{\beta}_{ext} = \frac{\pi}{4} + \frac{1}{2}\hat{\theta}_{ext},$$

which for air is about 67.75°, see Figure 46.

In the other limiting case, $q_0 = c_*$, θ_{ext} approaches 90°:

$$(122.02) \qquad \theta_{ext} = 90°, \quad \beta = 90°, \quad \text{for } q_0 = c_*,$$

as seen from (121.07).

A few more words should be said concerning the limiting cases which arise when q_0 approaches either c_* or $\hat{q}$. In the first case, see Figure 45, the loop shrinks to a point, and the two forward branches form a cusp at this point.

The limiting case $q_0 = \hat{q} = c_*/\mu$ corresponds to a situation in which there is either cavitation ahead of the shock front, $\rho_0 = 0$, $p_0 = 0$, or in which the pressure behind the shock front is infinite, $p = \infty$. In either case the shock strength is infinite. By (121.04) the shock polar reduces to the circle $v^2 = (\hat{q} - u)(u - \mu^2\hat{q})$, see Figure 46.

It should be noted that for weak shock transitions and sufficiently small θ the state on the back side is supersonic, while it is subsonic for the strong shock transition. As a simple calculation shows, the backward state is subsonic for the extreme angle θ_{ext}.

The same shock polar can also be used to determine the state ahead of a shock front with the given state (0) on the back side, see Figures 36 and 37. In this case, however, the unknown state is determined from the two "forward branches" with $q > q_0$ which approach asymptotically the line $u = U$. Within the limit circle $u^2 + v^2 = \hat{q}^2$ these two branches represent possible shock transitions. Again there appears a limit angle for θ, namely $\theta = \hat{\theta}$, the angle corresponding to the intersection of the forward branch with the limit circle.

OBLIQUE SHOCKS BY MEANS OF SHOCK POLARS 315

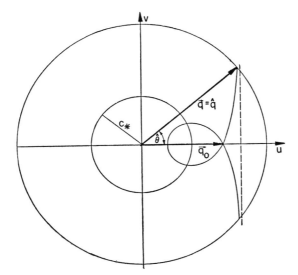

Fig. 44. Limit angle $\hat{\theta}$.

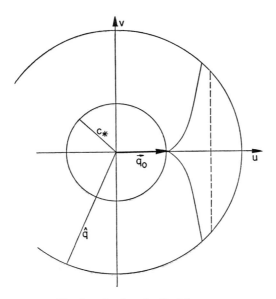

Fig. 45. Shock polar for the limiting case $q_0 = c_*$.

316 PLANE FLOW CHAP. IVC

In the preceding discussion we assumed the state (0) to be supersonic. Figure 37 shows the shock polar for a subsonic state (0), i.e., for $q_0 < c_*$. In this case the polar consists of one single branch without a loop; determination of the shock line is the same as before and, of course, always yields shock fronts for which the given state (0) is on the back side. Evidently, if the shock polar for the given point P_0 passes through the point P_1, then the shock polar for the

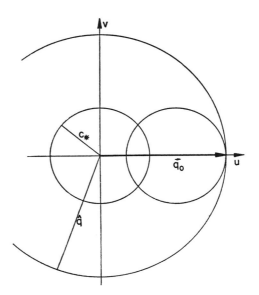

FIG. 46. Shock polar for the limiting case
$q_0 = \hat{q}$.

given point P_1 passes through the point P_0. It is also clear that shock polars for points P_0 with $q_0 > c_*$ cover all possible situations.

For practical purposes it is convenient to prepare a set of such polars with different values of $q_0 > c_*$ on transparent paper.

Alternative methods for the discussion of oblique shocks are given by the various other shock polars. Particularly suited for many purposes is the (p, θ)-polar in Figure 38. By intersecting it with vertical lines $\theta = $ constant one visualizes clearly the two possible transitions, the strong and the weak one, producing the same turning angle θ; the extreme situations are also obvious from this diagram.

A similar statement is true of the (β, θ)-polar in Figure 39.

123. *Flows in corners or past wedges*

The discussion of oblique shocks given in the preceding articles enables us to determine quantitatively the flow in a corner or past a wedge, which was considered in Section 117. If the angle θ_K of the corner, see Figure 28, is less than the extreme angle θ_{ext} associated with the state (p_0, ρ_0, q_0) of the incoming flow, then *two oblique shock*

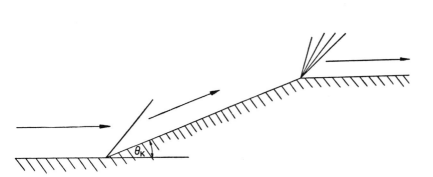

FIG. 47. The flow in a corner as a limiting case of a flow in a duct.

fronts are possible through which the flow is turned through the angle θ_K, a *weak* and a *strong* one. The question arises which of the two actually occurs. It has frequently been stated that the strong one is unstable and that, therefore, only the weak one could occur. A convincing proof of this instability has apparently never been given. Quite aside from the question of stability, the problem of determining which of the possible shocks occurs cannot be formulated and answered without taking the boundary conditions at infinity into account. The flow may be considered as the limiting case of the flow in a duct as the duct becomes infinitely wide and the inclined section infinitely long, see Figure 47. As will be explained in Section 143, the flow configuration depends on the conditions imposed on the downstream end of the duct. If the pressure prescribed there is below an appropriate limit, the weak shock occurs in the corner. If, however, the pressure at the downstream end is sufficiently high, a strong shock may be needed for adjustment. Under appropriate circumstances this strong shock may begin just in the corner and thus, of the two possibilities mentioned, the one giving a strong shock may actually occur.

Similar considerations answer the following question: what happens if the angle θ_K of the corner is greater than the extreme angle θ_{ext}? In this case, independently of the pressure at the downstream end, a shock front across the duct stands ahead of the corner. The flow behind this shock front is subsonic, and subsonic flow in a corner is possible; the instantaneous change of direction at the corner is achieved at vanishing velocity. The corner is thus a point of stagnation. The distance at which the shock stands ahead of the corner depends essentially on the length of the straight inclined section of the wall. If this length is increased, and the distance across the duct is increased at the same rate, the shock front moves to infinity in the upstream direction, as can be seen by considerations of similarity.

All statements made here are conjectures so far. While there is little doubt that they are in general correct, they should be supported, if possible, by detailed theoretical investigation.

D. Interactions. Shock Reflection

124. Interactions between shocks. Shock reflection

In this section we shall deal with the interaction between several shock fronts, a subject which in connection with shock reflection has aroused such great interest during recent years that it may be accorded a somewhat disproportionately detailed discussion here. We shall, however, omit the related topic of *reflection and refraction of shocks on an interface*, a problem quite amenable to the methods developed here, see Taub [103].

We assume that the shock fronts under consideration separate zones of constant states or at most of simple waves (an assumption justified at least in sufficiently small regions). Then no difficulties arise from the differential equations in the respective zones, and the problem of constructing solutions is an algebraic one involving merely the transition relations between the zones to be pieced together.

An important success from using such a procedure is the construction of flow patterns which represent *reflection of oblique shocks on rigid walls*. Reflection on non-rigid interfaces can be treated in a similar way.

125. Regular reflection of a shock wave on a rigid wall

Problems of reflection occur in connection with physical situations of the following type.* Suppose a shock wave produced by an

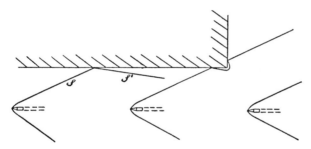

Fig. 48. Regular reflection of a shock wave produced by a projectile.

explosion or by a fast flying projectile hits an obstacle at such a distance that the shock front may be considered plane. When this shock front moves along a plane section of the surface of the obstacle a flow pattern may result which can be described in terms of an "incident" and a "reflected" shock front, see Figure 48. In that case one speaks of *regular reflection*. There is another type of reflection, which will be discussed in Section 128.

A phenomenon similar to that of shock reflection is the interaction of two shock waves produced, for example, by the explosion

*An entirely different physical situation leading to "reflection" of shocks will be discussed later in connection with flow in jets, see Section 148.

of two charges, see Figure 49. If the two shock waves are symmetric this process of interaction is equivalent to a reflection, the plane of symmetry playing the role of the wall.

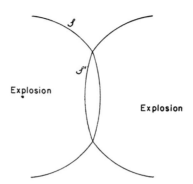

FIG. 49. Regular reflection of a shock wave produced by simultaneous explosions of equal charges.

If the incident and reflected shock fronts may be considered as planes moving with constant velocity along a plane wall, the flow appears to be steady from a frame of reference moving along the wall with an appropriate velocity; the two shock fronts then appear to be stationary.

For very weak stationary shocks the shock lines, as seen in Section 122, are approximately Mach lines, and therefore form the Mach angle with the wall, which is a streamline. Thus a stationary flow pattern containing a weak "incident" and "reflected" shock agrees with the law of reflection of geometrical optics, i.e., both shock lines form the same angle with the wall.

There is no reason why the situation should be similar when the incident or reflected shock (or both) has appreciable strength. As a matter of fact, observations show definite deviations of the flow pattern from that of weak or sonic waves; the angles between incident and reflected waves are in general not equal. The difference between the incident and reflected angle is indicated in Figures 50 and 51.

Reflection of shock waves can easily be analyzed theoretically on the basis of the results in Section 122. A systematic development of the theory was initiated by von Neumann and his collaborators, see [51, 114, 116]. The theoretical discussion consists in finding a mathe-

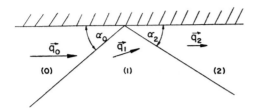

Fig. 50. Regular stationary reflection of shock fronts on a rigid wall.

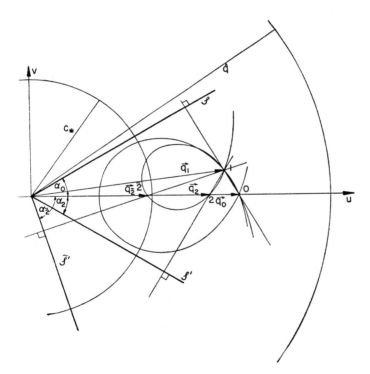

Fig. 51. Construction of weak ($\mathcal{S}'$) and strong ($\tilde{\mathcal{S}}'$) reflected shocks.

matically possible flow pattern compatible with the observed qualitative and quantitative features of the phenomena and yielding quantitatively correct predictions for the experimental facts, see [114, 124].

For the *mathematical formulation* of the problem an important preliminary step is the following. The phenomenon, as it presents itself to the observer, need not be stationary but is reduced to one of steady flow by subtracting from all velocities the velocity vector, parallel to the wall, of the point O at which reflection takes place; in other words, by referring the flow to a coordinate system moving with the point O. Then we are to find a steady flow in the (x, y)-plane for $y \leq 0$ such that the lower half-plane $y < 0$ is divided, as in Figure 50, into three regions (0), (1), (2), each of constant state, separated by two stationary shock lines S and S' and such that in the regions (0) and (2) the flow is parallel to the wall $y = 0$, i.e., $v_0 = v_2 = 0$. The configuration thus consists of two oblique shocks, the incident shock S and the reflected shock S', and it is clear from the previous sections that on passing the incident shock front the incident flow with the velocity $\vec{q}_0$ is turned toward the wall into a flow which is still supersonic with the velocity $\vec{q}_1$ and that this flow on crossing the reflected shock is turned into a flow with velocity $\vec{q}_2$ parallel to the wall, supersonic or subsonic as the case may be. In Figure 50 both shocks are facing toward the left. From our knowledge of the reflection of normal shock waves, see Section 70, we should expect a considerable rise in pressure behind the reflected shock front.

The mathematical objective is to find all such configurations and the corresponding interrelations between the velocities, pressures, angles, densities, etc. What physical quantitaties are given or observable may vary from case to case; as long as we know all the relevant relations between u, v, p, ρ, c, c_*, we may base the mathematical analysis on whatever quantities are mathematically most suitable as independent variables.

Having transformed the problem into one involving stationary shocks, we find it convenient to consider the critical speed c_* as one of the given quantities. Regular reflection can then be discussed with the aid of the shock polars in the (u, v)-plane. We start from state (0) and draw the corresponding backward loop of the shock polar. The state (1) is represented by a point on this loop. If, for example, the shock direction between (0) and (1) is prescribed, then this point is determined as the point 1 of intersection of the loop with the normal to the shock direction through 0. Through the point 1 we again draw the backward loop of the shock polar, symmetric about the direction of $\vec{q}_1$. If, as in Figure 51, this second loop intersects the line $v = 0$, then, as is easily seen, one intersection $\tilde{2}$ is subsonic while the other, 2, may be supersonic. Both (2) and ($\tilde{2}$) are possible states behind a "reflected" shock, inasmuch as they are compatible with the conservation laws, and the reflected shock line is easily found as the line perpendicular to $1 - 2$ or $1 - \tilde{2}$ respectively. Hence, there are two possibilities for a reflected shock: the regular strong reflection corresponding to ($\tilde{2}$) involving a high value of p_2, and the regular weak reflection represented by (2) and involving a smaller value of p_2. The weak reflected shock front makes a smaller angle α_2 with the wall than the strong reflected shock front. Ordinarily one should expect that it is the weak reflected shock that occurs in phenomena of reflection as described above. The weak reflection is nearer to the sonic case, approaching it when the strength of the incident shock decreases. The question of what determines the occurrence of the weak or the strong reflection will be discussed in Section 143. Instances of strong reflection are pointed out in Chapter V, Section 148.

If the point 1 on the first loop recedes from the point 0, i.e., if the incident shock becomes stronger and the angle of incidence greater, the second loop shrinks and the points 2 and $\tilde{2}$ come closer together. There is a last extreme case in which the two possible reflected states (2) and ($\tilde{2}$) coincide, and from there on no intersections 2, $\tilde{2}$ exist; thus, *regular reflection becomes impossible if the incident shock is made sufficiently strong*, keeping q_0 and c_* fixed.

The following is a modification of the preceding analysis. We start from state (1) and rotate the diagram about the origin O so that the vector $\vec{q}_1$ is horizontal. Then we draw the complete polar through the point 1. The state (0) now lies on the forward branch as in Figure 52. From the intersection of the vector $\vec{q}_0$ with the polar we again obtain two possible states (2) and ($\tilde{2}$), one state, or none at all, according as our line intersects at two points, touches, or misses the backward loop. The points 2 and $\tilde{2}$ correspond to states behind the reflected shock. The shock directions are again immediately determined as perpendicular to the lines 0–1, 2–1, and $\tilde{2}$–1.

It is clear that both constructions to determine regular reflection

324　PLANE FLOW　CHAP. IVD

configurations are equivalent. For, the shock polar through the point 0 passes through the point 1 since the shock polar through the point 1 passes through the point 0, as was mentioned in Section 122, see page 316.

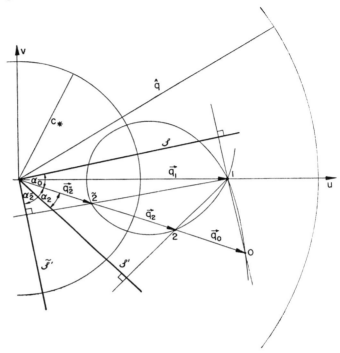

Fig. 52. Alternative construction of reflected shocks.

Both of these constructions yield all the relevant information about regular reflections; the quantities ρ, p, etc., are determined by the relations previously established; see also the next section.

Still a different way for discussing oblique shocks and regular reflection is offered by the (θ, p)-shock polar, described in Section 121. How regular reflection is represented in such a diagram is shown in Figure 53.

One important feature becomes evident from all these constructions. Regular reflection, no matter whether weak or strong, can occur only under restrictive conditions. From the diagram these conditions are obtained in a manner immediately adapted to the case of stationary shock fronts with given values of c_* and q_0. If however we are interested not in steady flow, but in shock waves

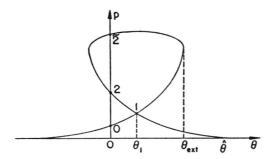

FIG. 53. The two possibilities of regular reflection shown with the aid of a (θ, p)-shock polar.

impinging obliquely on a zone of rest (0), the extreme situations, in which regular reflection ceases to exist, must be characterized with respect to the state (0) of rest. The corresponding conditions can easily be inferred from the preceding results.

Without referring to these graphic or analytic representations of shock transitions, we can show intuitively that *regular reflection is impossible if a strong shock front $\mathfrak{S}$ impinges on a zone of rest at an angle almost perpendicular to the rigid wall.* Then, referred to a frame in which the phenomenon is stationary, the shock $\mathfrak{S}$ is assumed to be so nearly normal that the state (1) behind it is subsonic. Now, from a subsonic state (1) there cannot exist a transition into a state (2) by a shock $\mathfrak{S}'$ facing towards region (1). Thus in such a case a configuration of regular reflection as in Figure 50 is impossible.

We return now to the situation of a shock impinging on a fixed state of rest (0) next to a wall. We consider the angle α_0 between the shock front and the wall and the excess pressure ratio $\dfrac{p_1 - p_0}{p_0}$ as the two independent variables. Through c_0, p_0, and p_1, the normal component N_0 of the velocity ahead of the incident shock front is found from the shock conditions, and consequently $q_0 = N_0/\sin \alpha_0$ and $c_*^2 = \mu^2 q_0^2 + (1 - \mu^2) c_0^2$ are determined. Thus the reflected shock can be determined by reduction to a steady state. To describe the variety of possible regular reflections we consider, in a plane with the coordinates α_0 and p_0, the rectangle

$$0 \leq \alpha_0 \leq 90°, \qquad 0 \leq \frac{p_0}{p_1} \leq 1.$$

This rectangle is divided by a curve E: $\alpha_0 = \alpha_{\text{ext}}$ as a function of p_0/p_1, into two domains as in Figure 54. To each point in the

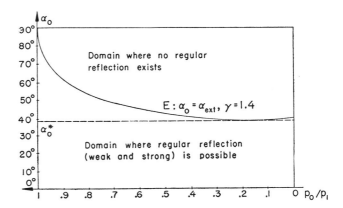

Fig. 54. Angles of incidence α_0 and the pressure ratios p_0/p_1 of the incident shock for which regular reflection is or is not possible.

domain $0 < \alpha_0 < \alpha_{\text{ext}}$ there correspond two possible regular reflections, a weak one and a strong cne, as characterized above. For the data corresponding to the points in the other domain no regular reflection exists.

A comprehensive picture of the manifold of regular reflection patterns is obtained by studying a sequence of incident shock waves of the same strength, i.e., having the same pressure ratio p_1/p_0, but with different angles of incidence α_0. If this angle α_0 tends to zero then for the weak reflections the three states (0), (1), (2) tend to those occurring in a head-on reflection of a shock wave which impinges normally on the wall. Thus for $\alpha_0 \to 0$ the reflected pressure ratio approaches the limit

(125.01) $$\frac{p_2}{p_1} = \frac{(2\mu^2 + 1)\frac{p_1}{p_0} - \mu^2}{\mu^2 \frac{p_1}{p_0} + 1}$$

given in (70.04). As α_0 increases we reach the "extreme situation" represented by a point on the separating curve E. At this position the weak and the strong reflected shocks coincide. As remarked above, for larger angles α_0 no regular reflection exists.

Figure 54 shows: in the limit $p_1 = p_0$, when the shock front becomes a Mach line, the extreme angle is 90°; for infinitely strong shocks, i.e., for $p_1/p_0 \to \infty$, the extreme angle approaches the value arc sin $\frac{1}{\gamma}$, which for air with $\gamma = 1.4$ is approximately 39.970°; for details see [116].

126. Regular shock reflection continued

Von Neumann and Seeger, see [51, 116], have presented a full discussion with many details of regular reflection for various values of γ; they use a different approach, starting at the outset with p_0/p_1 and α_0 as parameters. Of the results, the information concerning the reflected pressure ratio p_2/p_1 and angle α_2 between the reflected shock and the wall is of particular interest. We give a summary for the weak reflected shock. For small values of the angle of incidence α_0 we have $\alpha_2 < \alpha_0$. As α_0 rises from its value in the "head-on situation," i.e., from the value $\alpha_0 = 0$, the reflected pressure ratio p_2/p_1 first decreases below the head-on value given in (125.01) and then increases, attaining again the head-on value for a certain value $\alpha_0 = \alpha_0^*$. It is most remarkable that this value α_0^* is independent of the incident pressure ratio, and further that the angles α_0 and α_2 of the incident and the reflected shock front with the wall become equal for $\alpha_0 = \alpha_0^*$. For polytropic gases $\cos 2\alpha_0^* = (\gamma - 1)/2$ hence $\alpha_0^* = 39.23°$ for air with $\gamma = 1.4$. After α_0 has passed this value α_0^* the reflected pressure ratio rises and thus exceeds the head-on value. However, only for weak or moderately strong shocks $(p_1/p_0) < 7.02$ for air are α_0 and α_2 equal before the extreme situation is reached. Hence for strong shocks oblique reflection will never result in reflected pressures as high as those given by (125.01).

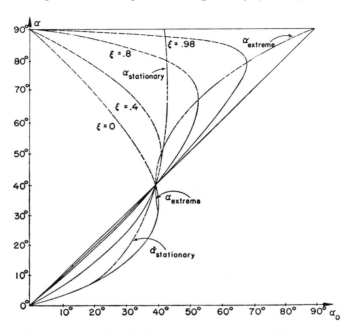

Fig. 55. Relationship between the angle of incidence α_0 and the angle of reflection α_2 with configurations of regular shock reflection for various ratios $\xi = p_0/p_1$ of the pressures ahead of and behind the incident shock front, for $\gamma = 1.4$.

These remarks about increase of pressure by reflection are obviously of practical importance; it should therefore be emphasized that for water and waterlike substances the situation is significantly different inasmuch as the rise of the reflected pressure ratio p_2/p_1 starts immediately from the head-on situation, $\alpha_0 = 0$, that is, α_2 is always greater than α_0. Therefore, oblique reflection in water always results in higher pressure than head-on reflection, if equal strength for the impinging shock waves is assumed.

The phenomena of regular reflection may be illustrated by graphs showing the relation between α_0 and α_2 for fixed strength $(p_1 - p_0)/p_0$ of the incident shock, see Figure 55 in which $\xi = p_0/p_1$.

127. Analytic treatment of regular reflection for polytropic gases

Without relying on our polar diagrams we may use the shock conditions for an independent algebraic approach to the problem of reflection and, as we shall see, to other problems concerning shock configurations. Assume that in the three domains (0), (1), (2), the thermodynamical quantities ρ and p are known, subject to condition (119.02) for adjacent domains. Then we have four quadratic equations, see (118.09–.10):

(127.01)
$$\frac{p_1 - p_0}{\rho_0} = \vec{q}_0 \cdot (\vec{q}_0 - \vec{q}_1), \qquad \frac{p_2 - p_1}{\rho_1} = \vec{q}_1 \cdot (\vec{q}_1 - \vec{q}_2),$$
$$\frac{p_0 - p_1}{\rho_1} = \vec{q}_1 \cdot (\vec{q}_1 - \vec{q}_0), \qquad \frac{p_1 - p_2}{\rho_2} = \vec{q}_2 \cdot (\vec{q}_2 - \vec{q}_1),$$

from which the three velocity vectors $\vec{q}_i$ can be determined (except for an arbitrary rotation of the coordinate system) provided that one additional condition is imposed. The advantage of this scheme is that relations (127.01) do not depend on the equation of state, and thus a clear separation is possible of such features of reflections as are perfectly general and such as are due to the specific character of the medium. The additional condition to be imposed is:

(127.02)
$$\vec{q}_0 \times \vec{q}_2 = 0$$

or $\vec{q}_0$ and $\vec{q}_2$ are parallel (in the direction of the wall). If the wall is given by $y = 0$, this means $v_0 = v_2 = 0$, and we are left with the problem of determining u_0, u_1, v_1, u_2 from the four quadratic equations (127.01). We refrain from presenting the analysis here in detail but give an account of some relevant formulas obtained.

Always assuming a polytropic gas we introduce, referring to Figure 50, the following quantities besides the sound speeds c_0, c_1, c_2, c_* and the speeds q_0, q_1 and $q_2 = |\vec{q}_2|$:

(127.03)
$$\left(\frac{c_*}{c_1}\right)^2 = x_*, \qquad \left(\frac{q_0}{c_1}\right)^2 = x_0, \qquad \left(\frac{q_2}{c_1}\right)^2 = x_2,$$
$$\frac{p_0}{p_1} = \xi, \qquad \frac{p_2}{p_1} = \zeta, \qquad \tan \alpha_0 = t_0, \qquad \tan \alpha_2 = t_2.$$

Then, after some algebraic manipulations, the shock relations yield the following selection of formulas symmetrical with respect to the states (0) and (2), which lend themselves to various numerical uses required by practical problems:

(127.04) $$\frac{(1 - \xi)t_0}{1 + \mu^2\xi + (\xi + \mu^2)t_0^2} = \frac{(\zeta - 1)t_2}{1 + \mu^2\zeta + (\zeta + \mu^2)t_2^2} = M,$$

(127.05) $$\frac{(1 - \xi)^2}{(\mu^2 + \xi)(1 + t_0^2)} = \frac{(\zeta - 1)^2}{(\mu^2 + \zeta)(1 + t_2^2)} = L,$$

(127.06) $$x_* = \mu^2 x_0 + (1 - \mu^2)\xi \frac{1 + \mu^2\xi}{\mu^2 + \xi}$$
$$= \mu^2 x_2 + (1 - \mu^2)\zeta \frac{1 + \mu^2\zeta}{\mu^2 + \zeta},$$

(127.07) $$\frac{(1 - \xi)^2}{\mu^2 + \xi}\left[1 - \frac{\mu^2 + \xi}{(1 + \mu^2)x_0}\right] = \frac{(\zeta - 1)^2}{\mu^2 + \zeta}\left[1 - \frac{\mu^2 + \zeta}{(1 + \mu^2)x_2}\right],$$

(127.08) $$M^2(1 - \mu^2)^2(t_0 - t_2) + M\{(1 - \mu^2)^2 - (t_0 - t_2)^2$$
$$- (\mu^2 + t_0 t_2)^2\} - (t_0 - t_2) = 0,$$

(127.09) $$x_*^2(1 + \mu^2)\{1 + 2\mu^2 - \xi\zeta - \mu^2(\xi + \zeta)\}$$
$$+ x_*\{2(\mu^4 - \mu^2 - 1) - \mu^4(2 - \mu^2)(\xi + \zeta) + 2(1 - \mu^6)\xi\zeta$$
$$+ \mu^4(\xi^2 + \zeta^2) + \mu^2\xi\zeta(\xi + \zeta)\} + (1 - \mu^2)(1 + \mu^2\xi)$$
$$(1 + \mu^2\zeta)(1 - \xi\zeta) = 0,$$

(127.10) $$L^2(1 + \mu^2)(1 - \mu^2)^3[1 + \xi\zeta + \mu^2(\xi + \zeta)]$$
$$+ L(1 - \mu^2)^2\{2 - (2 - 2\mu^2 - \mu^4)(\xi + \zeta)$$
$$+ 2(2 - \mu^2 - \mu^4)\xi\zeta - \mu^2(\xi^2 + \zeta^2) - \xi\zeta(\xi + \zeta)\}$$
$$+ (1 - \xi)^2(\zeta - 1)^2 = 0.$$

As an example of the use of these formulas we treat the problem of determining the regular reflection if the angle α_0 of incidence and the strength $\xi^{-1} - 1$ of the incident shock are given. Then one determines the quantity M from (127.04), t_2 from (127.08), ζ from (127.04), x_* from (127.09), x_0 and x_2 from (127.06), and finally (127.07) may be used as a check.

The properties discussed in the previous section are immediately read off from these formulas. For example, we see from equation (127.05) that the relation $\alpha_2 = \alpha_0$ implies a relation between ζ and ξ which is independent of the value of $\alpha_2 = \alpha_0$. Hence it is clear that the reflected shock strength $\zeta - 1$ for a given incident shock strength $\xi^{-1} - 1$ is the same for $\alpha_0 = 0$ with $\alpha_2 = 0$ as for $\alpha_0 = \alpha_0^*$ with $\alpha_2 = \alpha_0^*$.

The value of the extreme angle α_{ext} can be determined from the condition that equation (127.08) have just one real root t_2 for given t_0 and ξ. For $\xi = 1$, for example, this equation reduces to $t_2 = t_0$; hence $\alpha_{\text{ext}} = 90°$ for $\xi = 1$.

From the preceding relations and from the various graphs the fact appears repeatedly that regular reflection can occur only when the given parameters remain within certain limits.

Our analytic discussion was based on the assumption that the gas is polytropic. It should be emphasized that qualitative facts concerning regular reflection—and for that matter, concerning the more complex phenomena of "Mach reflection" to be discussed presently—remain unchanged for a much wider class of media. Whenever one-parametric families of possible shock transitions from a given state are qualitatively represented by shock polars of a shape similar to those considered above, conclusions of the same type can be drawn. This is the underlying reason why numerous reflection phenomena in nonlinear motion, not only in gas dynamics, show features as described here.

128. Configurations of several confluent shocks. Mach reflection

As we have seen, regular reflection is impossible in a great many cases, in particular, when the incident shock is very strong, for given $\dfrac{q_0}{c_*}$, or when a shock impinges on a quiet medium so that the angle α_0 between the incoming shock front and the wall is greater than the extreme angle α_{ext}.

What happens in these cases under experimental conditions similar to those which otherwise would have produced regular reflection? The answer is provided by innumerable phenomena of wave interaction examined long ago in experiments and papers by E. Mach. Yet this "Mach reflection," as it is now called, was all but overlooked and forgotten until attention was drawn to it some years ago, primarily by von Neumann.

To interpret physical phenomena of non-regular shock reflection or interaction it is better to start from a slightly more general point of view: The flow pattern of regular reflection on a wall $y = 0$ can immediately be extended by symmetry into the half-plane on the other side of the wall. Then we obtain a flow pattern with four shock lines issuing from the point at the wall. One could raise the general questions: What patterns of steady irrotational flow are possible with a given number of shock lines and possibly with additional centered simple waves and contact discontinuity lines issuing from a point Z? Furthermore: How can such patterns be used to interpret observed physical phenomena of "reflection" on a straight rigid wall?

Fortunately, observation indicates that at least qualitatively the mathematically simplest flow patterns are the ones that actually appear in many cases when regular reflection does not or cannot occur.

129. Configurations of three shocks through one point

It is of great importance to note the following fact: Three shocks separating three zones of different continuous states are impossible. To prove this statement, we may consider a small neighborhood of the point Z at which the (assumed) three shock lines meet; then for the assumed three states (0), (1), (2) we have, near Z, from the three shock relations (119.02),

$$(129.01) \qquad \frac{p_i}{p_k} = \frac{\lambda \rho_k - \rho_i}{\lambda \rho_i - \rho_k}$$

with $\lambda = \mu^2 = \dfrac{\gamma - 1}{\gamma + 1}$. After multiplication we find

$$(129.02) \qquad \begin{aligned} D(\lambda) &= (\lambda\rho_0 - \rho_1)(\lambda\rho_2 - \rho_0)(\lambda\rho_1 - \rho_2) \\ &\quad - (\lambda\rho_1 - \rho_0)(\lambda\rho_0 - \rho_2)(\lambda\rho_2 - \rho_1) = 0; \end{aligned}$$

$D(\lambda)$ is obviously a polynomial of second degree in λ and equation (129.02) is assumed to be satisfied for the value $\lambda = \mu^2$, which lies between 0 and 1. But we also see immediately that $D(0) = D(-1) = 0$. Hence the quadratic polynomial $D(\lambda)$ vanishes for three different values of λ; consequently, it vanishes identically, and we have, in particular, $D(1) = 0$, which yields

$$(\rho_0 - \rho_1)(\rho_2 - \rho_0)(\rho_1 - \rho_2) = 0.$$

Thus two of the adjacent states are identical, and our assumption of three confluent shocks is refuted.

Configurations involving three shock fronts must therefore involve an additional discontinuity. The simplest assumption, in agreement with many observations, is that there occurs in addition to the three shock fronts a single *contact discontinuity line*.

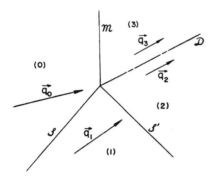

FIG. 56. Steady Mach configuration.

A "three-shock configuration" is then described as follows: one portion of the gas passes through two shock fronts, the "incident" and the "reflected" front, the other portion of the gas passes through one shock front, the "Mach front"; these two portions of the gas are, after crossing, separated by a line of contact discontinuity. In a steady three-shock configuration the three straight shock fronts and the contact line remain stationary, while the velocities in the four flow fields are constant, see Figure 56. By moving this configuration with an appropriate constant velocity one can obtain a progressing three-shock configuration moving into a gas at rest.

130. Mach reflections

A three-shock configuration* moving into gas at rest can now be so adapted as to describe the reflection of a shock front incident on

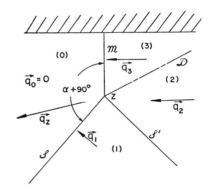

FIG. 57. Direct Mach reflection.

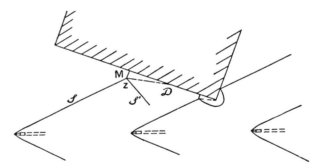

FIG. 58. Mach configuration of shock waves caused by reflection.

a wall. The reflected shock front branches off from the incident shock front at a point Z which does not move along the wall but moves away obliquely from the wall. The branch point Z is connected with the wall by a perpendicular "Mach shock front" through which the flow is normal. Finally, there is a contact discontinuity line $\mathcal{D}$ reaching from the branch point Z toward the wall.

* Also known as a λ- configuration.

Mach configurations can be observed just as frequently as regular reflections and shock interactions. Figures 58 and 59 indicate such

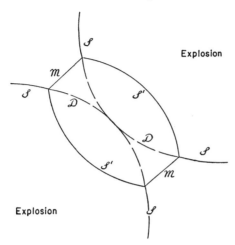

FIG. 59. Mach interaction resulting from simultaneous explosion of two charges.

cases, the latter configuration resulting from the simultaneous explosion of two charges.

131. Stationary, direct and inverted Mach configuration

A configuration of particular interest is the *stationary Mach reflection* in which the line $\mathcal{D}$ is parallel to the wall and the point Z moves parallel to the wall. Obviously, a stationary Mach configuration is equivalent to a configuration of regular reflection subject to the additional condition that the states in front of the incident and behind the reflected shock fronts are connected by the relations of transition through a single shock front, see Figure 60. Stationary Mach configuration reduced to steady flow is shown in Figure 61. In its duplicated form (reflected at the wall) it could occur when a steady jet of gas impinges on two parallel wedges, see Figure 62.

It is important not to be misled by the nomenclature "stationary." "Stationary Mach configuration" means simply a configuration in which the flow crosses the Mach shock front perpendicularly, and in which, therefore, the discontinuity line is normal to the Mach shock. This normal direction of the flow is then automatically in agreement with the existence of a wall along which the gas flows.

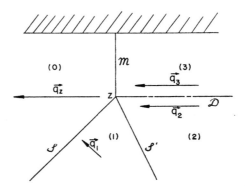

Fig. 60. "Stationary" Mach reflection.

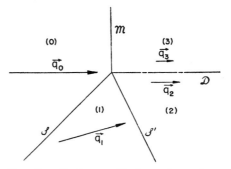

Fig. 61. Stationary Mach configuration reduced to steady flow.

Often Mach configurations are observed in steady flow involving non-constant states; then the shock lines are curved so that the stationary character of the flow and existence of a wall in no way imply that the Mach configuration at the branch point characterized as a local pattern is "stationary" in the sense defined above.

In general, a three-shock configuration is not stationary; the velocity $\vec{q}_0$ in the state (0) is in general not perpendicular to the Mach shock line. In the case shown by Figure 57 we call the configuration *direct;* in the case shown by Figure 63 we call it *inverted.* If, by subtracting $\bar{q}_0$ from all velocity vectors, these configurations are reduced to reflection configurations, moving into quiet gas in the

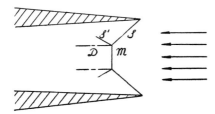

FIG. 62. Stationary Mach configuration in flow against parallel wedges.

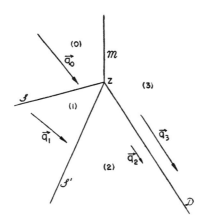

FIG. 63. Inverted Mach configuration.

state (0), we observe that in the direct Mach configuration the branch point moves away from the wall, see Figure 57, while in the inverted Mach configuration the branch point moves toward the wall, see Figure 64. An inverted Mach configuration would be quickly destroyed; it could never represent a proper reflection in cases such as are shown, for example, in Figures 58 and 59. We therefore exclude it from further consideration.

As was stated earlier, the contact line $\mathfrak{D}$ in a direct Mach reflection meets the wall at a point E. It is natural to ask how the gas flows near the point E at which the line $\mathfrak{D}$ intersects the wall. Observed from this point, the flow ahead of $\mathfrak{D}$ is at rest, while behind $\mathfrak{D}$ it has the direction of $\mathfrak{D}$. Along the wall the flow should have the direction of the wall. Adjustment is brought about by a simple wave with center E in case the flow in region (2) is supersonic when observed from E. If the flow in region (2), observed from E, is

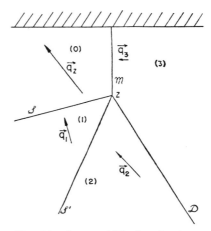

FIG. 64. Inverted Mach reflection.

subsonic, a non-constant flow in a corner results. In the following discussion we shall ignore this question; we shall consider only the local Mach configuration at Z assuming that the states in each of the regions meeting at the branch point Z are constant.

132. Results of a quantitative discussion

Before presenting methods for the mathematical construction of Mach configurations we summarize the results.

First, we focus our attention on three-shock configurations in steady flow. Introducing as parameters the reciprocal shock strength of the incident shock $\mathfrak{S}$, $p_0/p_1 = \xi$, and the quantity $x = q_0^2/c_*^2$ related to the Mach number $M_0 = q_0/c_0$ by $x^{-1} = \mu^2 + (1 - \mu^2)M_0^{-2}$, we consider in a (ξ, x)-plane the rectangle

$$1 \leq x \leq \mu^{-2}, \quad 0 \leq \xi \leq 1,$$

see Figure 65. In it we indicate two regions (C) and (P). In both (C) and (P) each point corresponds to a three-shock configuration. The region (C), shaded vertically (||||||), covers the "main branch

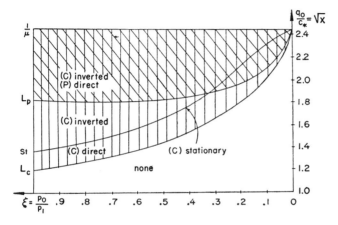

FIG. 65. Possible Mach reflections.

of three-shock configurations, first determined by S. Chandrasekhar [115], see also [116]. It is bounded by a line L_C corresponding to the limit case in which the strength of the reflected shock S' and the discontinuity of the line $\mathfrak{D}$ have shrunk to zero while the incident shock S is aligned with the Mach shock $\mathfrak{M}$. The region (C) contains further a curve St corresponding to stationary Mach reflections. The points between the curves L_C and St correspond to direct, those above the line St to inverted Mach configurations. The region (P), shaded obliquely (\\\\\\\), bounded by a curve L_P, represents an independent second branch of direct three-shock configurations, first determined by H. Polachek [116]. Note that *Mach reflection exists only for* $2/(1+\mu) \leq x < 1/\mu^2$ *or* $(2+2\mu)/(1+2\mu) \leq M_0^2 < \infty$.

A direct, though somewhat lengthy, algebraic computation leads to the following relations: The angle β_0 between the impinging shock and the direction of oncoming flow is found from

(132.01) $$\sin^2 \beta_0 = \frac{1 + \mu^2 \xi}{(1 + \mu^2)\xi} \cdot \frac{1 - \mu^2 x}{(1 - \mu^2)x}.$$

The curve St characterizing stationary reflection is given by the quadratic equation

(132.02) $$\left[\xi(x - 1) - \frac{1 - \xi}{\xi + \mu^2} (1 - \mu^2 x) \right] \cdot \left[\xi(x - 1) - \frac{1 - \xi}{1 + \mu^2} (1 - \mu^2 x) \right] = \xi(1 - \mu^2 x).$$

The equations of L_C and L_P are given by

(132.03) $$x = \frac{1 + \xi \pm [1 + 2\xi - (1 - 2\mu^2)\xi^2]R}{(\mu^2 + \xi)[1 \pm (1 + \xi)R]}$$

with

(132.04) $$R = \frac{\mu}{\sqrt{(\mu^2 + \xi)(1 + \mu^2 \xi)}}.$$

For these formulas, compare [114] and [116].

So far we have described the simplest possible stationary three-shock configurations. We now turn to the second question: How to characterize reflection processes proper, taking into account a rigid wall and allowing the flow to deviate from a steady state by a translatory motion. In other words we envisage now the original situation: An incident shock wave penetrating into a zone of rest and reflected at a rigid wall. The situation is then characterized by parameters depending solely on the state (0) of rest and on the incident shock wave. As such parameters we choose the ratio $p_0/p_1 = \xi$ of the pressure in front of the incident wave to the pressure behind the wave, and the angle α_0 between the incident shock front and the wall. In comparing theoretical and experimental results concerning three-shock configurations, the angle α_0 should be identified with the local angle between the incident shock and the normal to the Mach shock at the branch point; in reality the Mach shock is frequently curved and therefore its direction at the branch point is in general not exactly perpendicular to the wall.

The flow for which the state (0) is at rest is obtained from the flow with a stationary branch point by subtracting from all velocities the velocity in the region (0). It is not difficult to transform the results previously described in terms of ξ and $q_0/c_* = \sqrt{x}$ into results in terms of ξ and α_0.

We confine ourselves to direct Mach reflections belonging to the "main branch." In Figure 66, the region of points in the (ξ, α_0)-

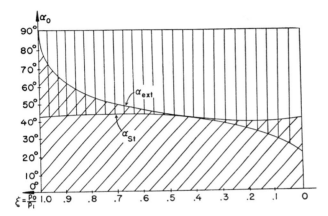

FIG. 66. In the region (/ / / / /) regular reflection is possible, while in the region (| | | | | |) Mach reflection is possible.

rectangle to which such configurations correspond is shaded vertically (||||||). This region is bounded by a curve whose points correspond to stationary Mach reflections. This curve is given by the equation, see [41],

$$(132.05) \quad \cot^4 \alpha_0 - \frac{\gamma\mu^2(\xi + \mu^2) + (1 - \xi)^2}{(\xi + \mu^2)(1 + \mu^2\xi)} \cot^2 \alpha_0 - \frac{\gamma(\xi + \mu^2)}{(1 + \mu^2\xi)^2} = 0.$$

For weak incident shocks, $p_0 = p_1$, for instance, one obtains

$$(132.06) \quad \cot^4 \alpha_0 - \left(\frac{1}{1 - \mu^2} - 1\right) \cot^2 \alpha_0 - \frac{1}{1 - \mu^2} = 0,$$

whence $\tan^2 \alpha_0 = 1 - \mu^2$, $\alpha_0 = 42°$, for $\gamma = 1.4$, $\mu^2 = \frac{1}{6}$.

By oblique shading (//////) we further show the region of values ξ and α_0 for which regular reflection is possible, see Figure 54. Figure 66 is drawn for $\gamma = 1.4$.

It is clear that the two regions overlap since the stationary Mach reflection can also be interpreted as regular reflection. For $\gamma \leq 3.59$, see [117] and [116, p. 11], the boundaries touch each other at a certain point, this point being at $p_1/p_0 = 2.61$ for $\gamma = 1.4$.

The theory does not determine which possibility, regular or Mach reflection, actually occurs in the common region. According to experimental evidence, the situation is as follows. Suppose we keep the strength of the incident shock wave fixed and vary the angle α_0 from $0°$ to $90°$; then the reflection is of the weak regular type approximately until stationary Mach reflection is reached. From then on Mach reflection appears. Where exactly this change of type takes place, and whether or not it occurs discontinuously, does not seem to be definitely settled. It should be mentioned that while Mach reflections with strong incident shocks agree, those with weak ones disagree quantitatively with predictions based on the preceding theory. This fact is evident from the great volume of experimental material which von Neumann, Seeger, Smith, and others (see [116] and [124]) have accumulated and compared with extensive calculations. The generalizations of the theory which will be indicated in Section 134, do not seem to be sufficient to account for the discrepancies. In Section 136 we shall resume this discussion.

133. Pressure relations

For Mach reflection as well as for regular reflection much of the interest is concerned with the resulting pressures. Let us consider the "reflected pressure ratio" p_2/p_1 of the pressure behind and in front of the reflected shock front S'. Then the reflected pressure ratio varies when the angle α_0 of incidence varies from $\alpha_0 = 0°$ to $\alpha_0 = 90°$ while the strength of the incident shock $\dfrac{p_1 - p_0}{p_0}$ is kept fixed.

As remarked in Section 125, see (125.01), for a head-on collision, $\alpha_0 = 0$, we have

$$(133.01) \qquad \frac{p_2}{p_1} = \frac{(2\mu^2 + 1)\dfrac{p_1}{p_0} - \mu^2}{\mu^2 \dfrac{p_1}{p_0} + 1}.$$

When α_0 increases, the ratio p_2/p_1 first decreases, but when α_0 approaches the extreme value, it may again increase beyond the head-on value, see the remarks on page 327. After a Mach configuration appears in the flow, this rise eventually terminates and the ratio p_2/p_1 declines again, approaching the value 1 as α_0 approaches 90°, since in this limit case the reflected shock becomes sonic.

134. Modifications and generalizations

The three-shock configuration described above is a simple mathematical pattern into which some phenomena fit surprisingly well. There exists, however, much experimental material that is clearly in

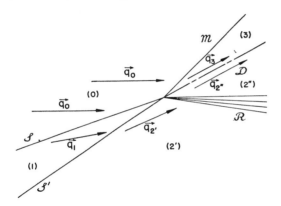

FIG. 67. Three-shock configuration modified by a rarefaction wave.

disagreement with these Mach configurations. One possible explanation for such deviations is that the actual flow pattern may correspond to a simple three-shock configuration only locally in an unobservably small region. There are, however, other possibilities to reconcile observed facts with mathematical flow patterns involving three shocks. As was stated in Section 129, the flow pattern described above is only one, the simplest, among infinitely many possibilities, all involving three confluent shocks. The reflection condition at a wall, i.e., flow parallel to the wall, can be satisfied by many flow patterns involving three shocks and a contact discontinuity through the branch point Z if one admits, in addition, simple centered expansion or compression waves through Z.

Complete success, as far as covering the available experimental evidence is concerned, has not yet been attained; the most direct generalization, namely, modification of Mach patterns by additional simple waves with the center in the branch point, may perhaps be sufficient to explain some of the more elusive phenomena.

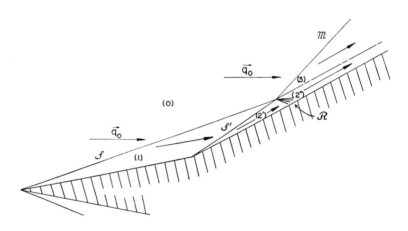

Fig. 68. Flow against an "arrowhead."

Such waves can be inserted in the region (2), see Figure 67, if the flow direction on crossing the reflected shock front from region (1) to region (2) is turned toward the branch point and the flow in region (2) (when observed from the branch point) is supersonic. The region (2) is then divided into two regions (2') and (2") and the flow on crossing from (2') through the simple wave into (2") is turned further toward the branch point. The conditions mentioned can be seen to be satisfied if the angle of incidence α_0 is sufficiently near to 90°.

Such configurations can be expected when the flow is deflected by an arrow-head, i.e., a wedge whose slope after some distance changes abruptly, see Figure 68. Actually observed "arrow-head" flows seem to confirm this interpretation (note that the simple wave would hardly show a trace on a shadow photograph).

A further, more radical, *modification of the simple Mach pattern* consists in assuming, instead of the contact discontinuity line, an angular zone $\mathcal{D}$ inside which the velocity is zero while the velocities in the adjacent zones are not parallel to each other but parallel respectively to the two edges of the zone $\mathcal{D}$; the pressures in the angular zone and in the two adjacent regions should be equal.

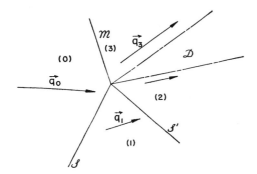

Fig. 69. Three-shock configuration with an angular zone instead of a contact discontinuity.

All these and other mathematically possible flow patterns with a singular center Z are at our disposal for interpreting experimental evidence. Which, if any, of these possibilities occurs under given circumstances is a question that cannot possibly be decided within the framework of a theory with such a high degree of indeterminacy. Here we have a typical instance of a theory incomplete and oversimplified in its basic assumptions; only by going more deeply into the physical basis of our theory, i.e. by accounting for heat conduction and viscosity, can we hope to clarify completely the phenomena at a three-shock singularity. It may well be that the boundary layer which develops along the constant discontinuity line modifies the flow pattern sufficiently to account for the observed deviation; an explanation on this basis was recently advanced by H. W. Liepmann [97]. The influence of the boundary layer along the discontinuity line is somewhat comparable to that of an angular zone $\mathcal{D}$ as mentioned above. As to the deeper significance of the indeterminacy of the theory see also the general remarks in Chapter IV, Part F.

135. *Mathematical analysis of three-shock configuration*

In spite of the indeterminacy of the problem in its present formulation it is important to substantiate the statements of the preceding section by a mathematical derivation. At least we shall indicate the proper procedure for the construction of three-shock configurations.

To construct Mach reflection patterns algebraically one could start from the relations between pressures and velocities in adjacent domains, expressed by the six equations, see (118.09–.10):

$$(135.01) \quad \begin{array}{ll} \tau_0(p_1 - p_0) = \vec{q}_0 \cdot (\vec{q}_0 - \vec{q}_1), & \tau_1(p_0 - p_1) = \vec{q}_1 \cdot (\vec{q}_1 - \vec{q}_0), \\ \tau_1(p_2 - p_1) = \vec{q}_1 \cdot (\vec{q}_1 - \vec{q}_2), & \tau_2(p_1 - p_2) = \vec{q}_2 \cdot (\vec{q}_2 - \vec{q}_1), \\ \tau_0(p_3 - p_0) = \vec{q}_0 \cdot (\vec{q}_0 - \vec{q}_3), & \tau_3(p_0 - p_3) = \vec{q}_3 \cdot (\vec{q}_3 - \vec{q}_0). \end{array}$$

These relations are supplemented by the condition

$$(135.02) \qquad \vec{q}_2 \times \vec{q}_3 = 0$$

or $\vec{q}_2$ parallel to $\vec{q}_3$. Any system of four vectors satisfying these seven equations will give a possible Mach configuration. Once the velocities are known, the shock lines are immediately determined. As we saw previously, in Section 127, we may consider the pressures and densities in these equations as given parameters subject to the relations (119.02). One direction, e.g., that of $\vec{q}_0$, might be arbitrarily prescribed; then we have to determine seven quantities from seven equations, (135.01–.02).

The stationary Mach reflection is simply characterized by one condition in addition to (135.01–.02), namely:

$$(135.03) \qquad \vec{q}_0 \times \vec{q}_3 = 0$$

or $\vec{q}_0$ parallel to $\vec{q}_3$. By eliminating the velocities a condition involving only densities and pressures can be obtained for stationary Mach effect. We shall not, however, follow this line of treatment here, but rather discuss the Mach effect by a more intuitive and comprehensive method based on a geometrical analysis using one or several of the shock polars introduced before.

It should be noted that such a procedure in contrast to a more algebraic one is open to generalization for non-polytropic media.

136. Analysis by graphical methods

Instead of the polar in the (u, v)-plane we use the shock polar in the (θ, p)-plane, see Section 121, which permits us immediately to take care of the condition that the vectors $\vec{q}_3$ and $\vec{q}_2$, and in the stationary case, the vectors $\vec{q}_0$ and $\vec{q}_3$, be parallel and that the pressures in the corresponding zones be equal. We saw that if a given state (0) is connected with another state by a stationary shock, the second state can be characterized by the angle θ through which the shock turns the flow and the pressure p in the new state. All possible states that can be connected with the state (0) by a shock may, therefore, be represented in a (θ, p)-plane by a shock polar having the shape shown in Figure 53 and represented by equations (121.07–.08), or explicitly by

(136.01)
$$\tan \theta = \frac{\frac{p}{p_0} - 1}{\gamma M_0^2 - \frac{p}{p_0} + 1} \cdot \sqrt{\frac{(1 + \mu^2)(M_0^2 - 1) - \left(\frac{p}{p_0} - 1\right)}{\frac{p}{p_0} + \mu^2}}.$$

From this (θ, p)-polar Mach configurations are obtained in such a way that numerical procedure can easily be supplied afterwards. To find Mach configurations from our shock polar Λ_0 through the point 0 as double point, consider a shock leading from a state (0) to a state (1) with higher pressure and a shock leading from (1) to (2), as well as a shock from (0) to (3), see Figure 57. In the (θ, p)-diagram, see Figure 70, the two states (2) and (3), on both sides of the vortex line $\mathfrak{D}$ are represented by the same point, since in these states the pressures and the flow directions are identical. This simplification is the reason for using the (θ, p)-diagram. We draw through the point 0 the shock polar loop Λ_0, and likewise another shock polar loop Λ_1 through the point 1 on Λ_0. Since the state (3) is connected with the state (0) through a shock the point 3 lies on Λ_0. On the other hand, the state (2) is reached by a shock from state (1); hence the point 2 in our diagram must lie on Λ_1, and since the points 2 and 3 in our representation are identical, we obtain 2 and 3 by intersecting the loops Λ_0 and Λ_1.

If $\theta > 0$ for the points 2,3 of intersection, see loop Λ_D in Figure 71, then we have a direct or ordinary Mach reflection, and stationary Mach reflection corresponds to the case in which the point of inter-

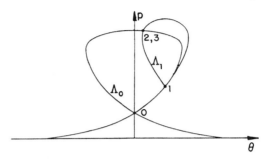

FIG. 70. (θ, p)-shock polars for direct Mach reflection. The point 2,3 represents the states (2) and (3) on either side of the discontinuity line $\mathfrak{D}$.

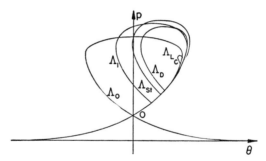

FIG. 71. (θ, p)-shock polars Λ_D for direct, Λ_I for inverse, and Λ_{St} for stationary Mach reflections. In each case, the intersections of these loops with the loop Λ_0 represent the states on either side of the discontinuity line $\mathfrak{D}$.

section occurs for $\theta = 0$, exactly at the top of the loop, see loop Λ_{St} in Figure 71.

Knowing the point 2,3 we obtain the shock lines $\mathfrak{S}$, $\mathfrak{S}'$ and $\mathfrak{M}$ immediately, and then proceed to determine the densities from the relation (119.02), and the velocity vectors from our previous relations (135.01). Thereby the difference between the states (2) and (3) becomes apparent automatically since we obtain the quantities for (2) by starting from (1) and those for (3) by starting from (0).

Altogether, the search for Mach configurations is replaced by a discussion of our shock polar loops and their intersections. For a detailed analysis of the possible intersections and in particular of various limiting cases see [117–120]; see also [116].

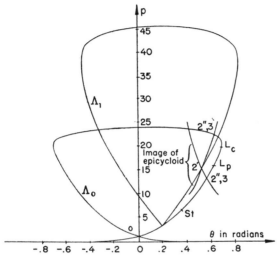

FIG. 72. Representation of a modified three-shock configuration in (θ, p)-plane.

The possibility of inserting a simple wave under the conditions indicated earlier can also be read off from the (θ, p)-diagrams. The curves of Figure 72 are self-explanatory. Through a point $2'$ on the loop Λ_1 one draws the image of a Γ-characteristic; its intersection with the loop Λ_0 gives the point $2''$, 3. This is possible only if the state $(2')$ is supersonic. A discussion of the relative position of shock lines and Mach lines shows that *a possible flow configuration results if* $\theta_0 < \theta_1 < \theta_{2'} < \theta_{2''}$; *in other words, if all the elementary waves through which one portion of the flow passes turn the flow in the same direction.*

In a similar way it is possible for one portion of the flow, instead of passing through only one incident and one reflected shock front, to

350   PLANE FLOW   CHAP. IVE

pass through a sequence of several shock fronts. A detailed discussion shows that *such configurations are possible only if all shocks of the sequence turn the flow in the same direction;* the other portion of the flow can then cross only one shock front.

After these discussions it is not difficult to analyze the other possibilities suggested earlier.

E. Approximate Treatments of Interactions.   Airfoil Flow

137. Problems involving weak shocks and simple waves

The results of the preceding part concerning the interaction of shocks with shocks were all derived by algebraic and sometimes cumbersome procedures. Such an elementary approach fails as soon as interaction processes involving simple waves are considered. Then the analytical difficulties inherent in the partial differential equations of the flow become unavoidable. Nevertheless, satisfactory approximate treatments of interactions are possible if the shocks and simple waves which interact are of small or moderate strength.

The first problem we shall treat concerns the interaction of a stationary shock front and a steady simple wave behind it. We shall see that in good approximation the shock does not influence the simple wave behind it and that its course can be determined by solving a linear differential equation, a treatment analogous to that for the interaction of a shock and a simple wave in one-dimensional flow in Section 74.

Two-dimensional flow past an airfoil leads to a weak simple wave flow if the airfoil is sufficiently slender. We shall discuss two variants of the perturbation method for determining this flow. The effect of shocks on the flow can once again be disregarded for the approximation considered. The course of such shock fronts can be approximately determined by the method mentioned before.

138. Comparison between weak shocks and simple waves

The basis for the treatment of interactions of weak shocks and simple waves is the fact that the changes across weak shocks and weak simple waves differ only in the third order of their strength.

We consider first a set of oblique backward-facing shock fronts all having the same front state given by τ_0, p_0, and the velocity $u_0 = q_0 > 0$, $v_0 = 0$, see Figure 73. We characterize the various

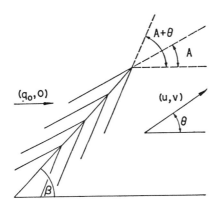

FIG. 73. Backward-facing shock front. Mach lines are shown on both sides.

shocks in the set by the value of the specific volume τ behind them; then all other quantities, pressure p, and velocity (u, v) behind the shock fronts are determined provided we stipulate $v > 0$, that is, that the shock front is inclined toward the flow. Then we expand p, u, and v in powers of $\tau - \tau_0$. To do so we first use Bernoulli's equation (118.08) which in the present case takes the form

(138.01) $\qquad \frac{1}{2}(u^2 + v^2) = i_0 - i + \frac{1}{2}q_0^2.$

Differentiating this relation twice in succession we obtain for $\tau = \tau_0$ the relations

(138.02) $\qquad q_0\, du = -\tau_0\, dp,$

(138.03) $\quad q_0\, d^2u + (du)^2 + (dv)^2 = -\tau_0\, d^2p - d\tau\, dp.$

From relation (118.11) in the form

(138.04) $\qquad (u - q_0)^2 + v^2 = -(p - p_0)(\tau - \tau_0)$

we obtain, by differentiating it two and three times (or by inserting into it expansions of u, v, p in powers of $\tau - \tau_0$),

(138.05) $$du^2 + dv^2 = -dp\,d\tau,$$

(138.06) $$du\,d^2u + dv\,d^2v = -\tfrac{1}{2}d^2p\,d\tau.$$

By (138.05) relation (138.03) reduces to

(138.07) $$q_0\,d^2u = -\tau_0\,d^2p.$$

The differentials here are of course to be taken for $\tau = \tau_0$. The relationship between p and τ is the same for oblique as for normal shocks; thus the values of dp and d^2p are known, as a matter of fact they are the same as if p and τ were related adiabatically, see Section 65

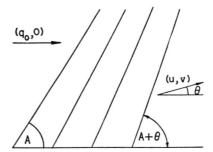

Fig. 74. Backward-facing simple wave indicated through a set of straight Mach lines.

From equations (138.02), (138.07), (138.05), (138.06) the values of du, d^2u, dv, d^2v are then determined since $dv > 0$ for $d\tau < 0$ according to the stipulation made above.

Next we consider a compressive simple wave facing to the left with a state in front given by τ_0, p_0, $u_0 = q_0 > 0$, $v_0 = 0$, see Figure 74. We compare the various sections of this simple wave between the initial straight Mach line and the straight Mach line carrying the constant specific volume τ with the various oblique shock fronts with the same state ahead of them and the same value τ behind them. Again we expand the values of p, u, and v on the Mach line carrying the value τ with respect to powers of $\tau - \tau_0$. Since the pressure and the density are adiabatically related across a simple wave the values of dp and d^2p are the same as for shocks with the same initial state. Since Bernoulli's law holds across a simple wave, relations (138.02) and (138.03) also hold. Relation (138.05) is equivalent to the characteristic equation for the cross Mach lines, see (106.02); hence it holds throughout the simple wave. By differentiating (138.05) relation (138.06) follows. It is then clear that the values of du, dv, d^2u, d^2v are the same for a simple wave transition as for a shock transition since this is the case for dp and d^2p. Thus we have proved that *the expansions of the quantities behind shock fronts and sections of a simple wave with the same initial state agree up to terms of second order*, that is, in first and second order. This statement derived for the expansion with respect to powers of $\tau - \tau_0$ evidently holds also for expansions with respect to powers of any quantity which may serve as the *strength* of the shock or the section of the simple wave.

The statement can also be expressed by saying that the shock polar and the epicycloid for the same critical speed c_* through the same point $(q_0, 0)$ in the (u, v)-plane have a contact of second order. If, therefore, the state behind a shock, represented by a point on the shock polar, is replaced by a state represented by a point on the epicycloid with the same entropy as the state ahead of the shock, then the shock relations are satisfied up to terms of second order in the shock strength.

Although the state behind a weak or moderately strong shock can thus be determined from the corresponding simple wave transition, the position of the shock front must be determined separately. We know that the shock front coalesces with a Mach line when the shock strength is zero. To find the position of the shock front for weak or moderately strong shocks we use an expansion of β, the angle which the shock front makes with the direction of the incoming flow. An expansion of β with respect to powers of the angle θ between the directions of the flow behind and ahead of the shock front can be derived from relations (121.02) and $\tan \theta = v/u$. Comparing the result with the last formula (116.08), which holds across simple waves and hence across shocks, we find the relation

(138.08) $$\beta = \tfrac{1}{2}A_0 + \tfrac{1}{2}(A + \theta) + \ldots$$

in which A may be considered a function of θ or vice versa. Here A and A_0 are the Mach angles behind and ahead of the shock front respectively. $A + \theta$ and A_0 are the angles between the direction of the incoming flow and those Mach lines behind and ahead of the shock front which are inclined in the same direction as the shock front. Thus, relation (138.08) shows that *in first order the shock front bisects the angles between the Mach lines behind and ahead of it.*

139. Decaying shock front

The fact that the shock transition agrees up to second order in the shock strength with a simple wave transition can be used for the approximate determination of a shock front which is influenced by a simple wave from behind while the state ahead of the shock front is constant. This treatment is quite similar to that used in

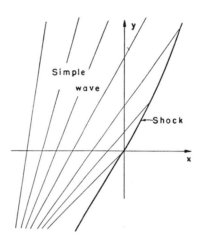

FIG. 75. Interaction between a shock front and a simple wave behind it.

determining the decay of a one-dimensional shock under the influence of a simple wave, see Section 75.

Suppose a constant supersonic stream of air with the velocity $u = q_0 > 0$, $v = 0$ is first deflected through a shock front whose slope dy/dx is positive. For $y < 0$ the deflected constant flow is assumed to be further deflected through a simple wave whose straight Mach lines have also a positive slope dy/dx. The simple wave is assumed to meet the shock front at the point $x = 0$, $y = 0$. The problem is to determine the flow behind the shock front for $y > 0$. Since the flow ahead of the shock front is supersonic the flow there remains unchanged, but the course and the strength of the shock is influenced by the flow behind it, see the discussion in Section 143 (Part F). If the strength of the shock is weak or at most moderate, we may approximately identify the shock transition with a simple wave transition. That means, to the degree of approximation considered, the flow behind the shock front is the same as that which would have occurred if the first deflection of the flow had been effected by a simple compression wave instead of a shock. Such a simple compression wave would form one simple wave with the given simple wave behind the shock front for $y < 0$ and therefore also for $y > 0$. In other words, to the degree of approximation considered, the shock does not affect the flow behind it. This flow is just the given simple wave continued for $y > 0$ up to the shock front. Thus the flow at each point would be determined if the position of the shock front were known. This position which depends on the simple wave flow behind the shock can be determined in the following way:

We assume that the simple wave, which is a Γ-wave, can be described by

(139.01) $\quad x = a(\sigma) - r \sin \omega(\sigma), \quad y = b(\sigma) + r \cos \omega(\sigma);$

Here a, b and the angle ω between the straight Mach line and the positive y-axis, see (109.01), are given functions of the parameter σ. The velocity $u = q \cos \theta$, $v = q \sin \theta$, and also p and τ are then given as functions of ω and the constants ω_*, c_* as explained in Section 109. The course of the shock front will be described by giving r, the abscissa along the straight Mach lines, as a function of σ. For this purpose we express β, the angle that the shock front makes with the positive x-axis, in terms of the angle θ of the flow direction behind the shock as given by (121.08) and (121.02.1) for polytropic gases. The angle θ again can be expressed through $\omega(\sigma)$ by relations (116.05) which hold for the simple wave. The function $\beta = \beta(\sigma)$ obtained in this way satisfies along the shock front the relation $\sin \beta \, dx - \cos \beta \, dy = 0$. Differentiating equations (139.01) we find

(139.02) $\quad \cos (\beta - \omega) \dfrac{dr}{d\sigma} + \sin (\beta - \omega) \dfrac{d\omega}{d\sigma} r = \sin \beta \dfrac{da}{d\sigma} - \cos \beta \dfrac{db}{d\sigma},$

a linear differential equation for $r = r(\sigma)$. Its solution determines the course of the shock front. For the details see a separate publication [55].

We only mention that relation (139.02) simplifies very much if we expand β in powers of $\omega - \omega_0$ and retain only terms up to the first or second order. From relation (138.08) and $\omega = A + \theta - 90°$, $\omega_0 = A_0 - 90°$ we obtain in first order

(139.03) $\qquad \beta = 90° + \omega_0 + \tfrac{1}{2}(\omega - \omega_0).$

Insertion in (139.02) gives, in first order, the equation

(139.04)
$$\tfrac{1}{2}(\omega - \omega_0)\frac{dr}{d\sigma} + \frac{d\omega}{d\sigma} r = \sin\frac{\omega + \omega_0}{2}\frac{db}{d\sigma} + \cos\frac{\omega + \omega_0}{2}\frac{da}{d\sigma} = f(\sigma)$$

which can easily be solved explicitly,

(139.05) $\qquad r = \dfrac{2}{(\omega(\sigma) - \omega_0)^2} \displaystyle\int_{\sigma_1}^{\sigma} (\omega(\zeta) - \omega_0) f(\zeta)\, d\zeta + r_1$

when $\sigma = \sigma_1$ and $r = r_1$ correspond to the point at which the shock first meets the simple wave. Inserting relation (139.05) in (139.01) yields a parametric representation of the shock curve.

140. Flow around a bump or an airfoil

An important application of the fact that the influence of a shock on a simple wave behind it is only of the third order in the shock strength is the following treatment of supersonic flow around airfoils or similar obstacles, see Figure 76.

We consider a supersonic flow arriving with the velocity q_0 along a straight wall, the x-axis, and meeting a bump which begins with a sharp angle θ_1 of inclination against the straight wall; we assume θ_1 is less than θ_{ext}, see Sections 122 and 123, so that the flow can turn through the angle θ_1 by means of a shock S. The continued shock front varies in strength and inclination (unless the bump is straight); hence the flow behind the bump is rotational and carries a non-constant entropy. In second order, however, this flow is just the simple wave flow that would have resulted if the flow had turned through the angle θ_1 by means of a simple compression wave. This simple wave flow corresponds in the (u, v)-plane to the epicycloid which begins at the point $(q_0, 0)$ and carries the same entropy as the flow ahead of the shock front.

These facts are sufficient to determine the simple wave along the bump and thereby approximately, for example, the pressure of the flow on the obstacle, without paying any attention to the shock line S itself. To determine the latter as the transition line between the

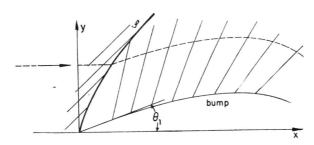

FIG. 76. Flow along a straight wall with a bump. The resulting shock and forward-facing Mach lines are shown.

zone of the incoming constant flow and the simple wave zone issuing from the obstacle we may proceed as in Section 139.

An incidental result of the procedure is that, in the approximation considered, the shock line S cannot penetrate the straight Mach line C_+ issuing from the highest point of the bulge. For, at this Mach line $\theta = 0$ and beyond it $\theta < 0$; the shock would therefore not effect any turn of the flow on this line, and beyond it the flow would turn from $\theta = 0$ in the direction of negative θ, which would contradict the fact that a shock always turns the flow toward the shock line.

141. *Flow around an airfoil treated by perturbation methods (linearization)*

We assume that the airfoil is an infinitely long cylinder; the cross section is the "profile." When the airfoil faces a stream of air perpendicular to the cylinder the steady flow is two-dimensional.

If the profile is of the type used for subsonic flight, with a blunt leading nose and a sharp trailing edge, a curved shock front forms ahead of the profile. Behind this front, the flow is subsonic (except possibly for a section at the upper contour); the flow pattern around the airfoil is the same as for subsonic flow and involves a point of stagnation somewhere at the nose and finite velocity at the sharp trailing edge, see Figure 77.

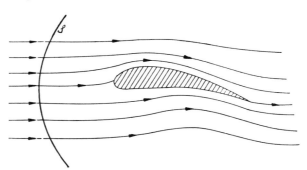

FIG. 77. Supersonic flow against an airfoil with a blunt nose.

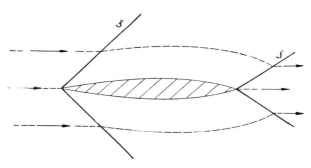

FIG. 78. Supersonic flow against an airfoil with a sharp leading edge. Shock fronts are attached at the edges.

In a supersonic stream one may use a profile consisting of two arcs forming sharp leading and trailing edges without fear that the flow will be detached at the leading edge. For, in supersonic flow, an instantaneous change of flow direction through a shock is possible. Thus at a sharp leading edge two shock fronts develop (Figure 78) provided that the angles made by the arcs at the tip with the incoming flow direction are below the extreme angle θ_{ext}. If either angle at the tip exceeds the extreme angle the flow pattern resembles that for a blunt nose; that is, it has a curved shock front ahead of the edge, see

Figure 79. If at the leading edge one of the arcs makes a negative angle with the flow direction, a centered rarefaction wave develops at the

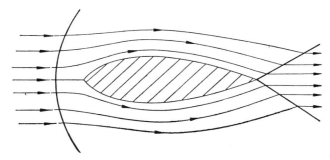

FIG. 79. Supersonic flow against an airfoil with a blunt leading edge. A shock front starts ahead of the airfoil.

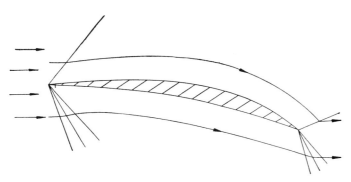

FIG. 80. Supersonic flow against an airfoil with a sharp leading edge. A shock front and a rarefaction wave are attached to the leading edge.

edge instead of a shock, see Figure 80. Also at the trailing edge two shock fronts or rarefaction waves develop according to what the angles permit. Under certain circumstances, however, a shock front begins at some point along the profile and causes flow detachment. (This is apparently produced by a viscous boundary layer even when the angles at the trailing edge permit shocks there.)

In this section we shall discuss approximation methods for determining the flow around profiles which are so flat that shocks at the leading and trailing edges are expected to occur. According to the discussion in Section 140 we obtain a good approximation for the flow around flat profiles if we assume that the flow at both sides of the profile consists of simple waves connected with the state ahead of the profile through the simple wave relations. The two arcs of the profile are then prescribed streamlines across these simple waves. We know from Section 111 that simple waves are always uniquely determined by such streamlines (provided again that the angles through which the streamlines are to turn are not so large that cavitation would result). The four shock fronts can be drawn in afterwards by employing the procedure described in Section 139.

This method for approximately determining the supersonic flow around airfoils is simple enough. However, a further approximation using the method of *perturbation* often gives sufficiently accurate results. Suppose the two arcs forming a profile are given by

(141.01) $$y = Y^+(x) \quad \text{and} \quad y = Y^-(x),$$

$$Y^-(x) \leq Y^+(x) \quad \text{for} \quad 0 \leq x \leq a,$$

(141.02) $$Y^+(a) = Y^-(a) = b.$$

Then we consider the variety of profiles given by

(141.03) $$y = \epsilon Y^+(x), \quad y = \epsilon Y^-(x), \quad 0 \leq x \leq a$$

depending on a parameter ϵ. Accordingly we have a set of problems depending on ϵ. We then assume that the solutions of these problems can be expanded in powers of ϵ and determine successively the coefficients of this expansion. Here we shall derive the terms of first order of the expansion of the flow. We may assume from the beginning that the flow is isentropic and that the entropy is the same as the entropy of the air ahead of the airfoil. We could set up an expansion of every quantity at a fixed point (x, y); this procedure will be indicated in Section 142. The calculations and results are, however, very much simpler if all quantities are treated as functions of two other parameters instead of x and y. The characteristic through a point which is inclined towards the profile intersects the profile at a point with the coordinates ξ and $\epsilon Y^\pm(\xi)$, see Figure 81. We then take x and ξ instead of y as the two parameters. This procedure is suitable for deriving terms of first and second order. To derive terms of higher order it is better to introduce as parameters the abscissas ξ and ξ' of the two points at which the two characteristics through the point (x, y) intersect the profile.

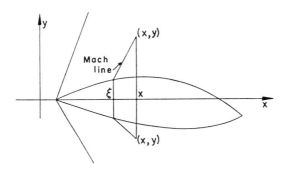

FIG. 81. Parameter ξ characterizing the position of a point (x,y) in the flow around a flat profile.

We know that up to the second order the flow is a simple wave and that, accordingly, all quantities are constant along each characteristic, $\xi =$ constant, and thus depend only on the parameter ξ. The angle θ of the flow direction is given by $\tan \theta = \epsilon Y_x^{\pm}(\xi)$. Since $\tan \theta = \theta$ up to second order, we have

(141.04) $\qquad \theta = \epsilon Y_x^{\pm}(\xi)$ up to second order.

To express the quantities u, v, c, ρ, p in terms of ξ we therefore need only express them in terms of θ. This can be done immediately by using the expansions, given in Section 116, of the quantities at the tail of a simple wave in powers of the angle through which the flow turns. The incoming flow is assumed to have constant sound speed c_0, Mach number $M_0 = q_0/c_0$, pressure p_0, density ρ_0, and velocity $(q_0, 0)$ with $q_0 > 0$. With the notation

(141.05) $\qquad t_0 = \dfrac{1}{\sqrt{M_0^2 - 1}},$

362   PLANE FLOW   CHAP. IVE

we obtain from the end of Section 116, in first order

(141.06) $\quad u = q_0[1 \mp t_0\theta + \cdots],$

$\quad\quad\quad\quad\quad v = q_0[\pm \theta - \cdots],$

(141.07) $\quad c = c_0\left[1 \pm \dfrac{\gamma-1}{2}(t_0^{-1} + t_0)\theta + \cdots\right],$

(141.08) $\quad \rho = \rho_0[1 \pm (t_0^{-1} + t_0)\theta + \cdots],$

(141.09) $\quad p = p_0 + \rho_0 q_0^2[\pm t_0\theta + \cdots];$

here $\rho_0 c_0^2 = \gamma p_0$ or $\rho_0 q_0^2 = \gamma M_0^2 p_0$, see (3.06), have been used. The upper and lower signs correspond to the upper and lower sides of the profile respectively.

In order to express y in terms of x and ξ we note that the straight Mach line through the point $(\xi, \epsilon Y^\pm(\xi))$ is given by

(141.10) $\quad y = \epsilon Y^\pm(\xi) + \tan(\pm A + \theta)(x - \xi).$

Here the slope $\tan(\pm A + \theta)$ of the Mach line is a known function of the angle θ of the flow direction on it. The expansion of $\tan(\pm A + \theta)$ in powers of $\theta = \epsilon Y_x^\pm(\xi)$ is by (108.02) and (108.14)

(141.11) $\quad \tan(\pm A + \theta) = \pm t_0 + \dfrac{\gamma+1}{2}(1 + t_0^2)^2\theta \pm \cdots,$

Hence the Mach lines are given to order zero by

(141.12) $\quad y = \pm t_0(x - \xi).$

This relation, expressing y in terms of x and ξ, can be used to represent the flow to first order in terms of x and y. By inserting the relation

(141.13) $\quad \xi = x \mp t_0^{-1} y$

in the first order terms of (141.06–.09) and expressing θ through the slope of the profile $Y_x(\xi)$, we obtain

(141.14) $\quad u = q_0[1 \mp t_0 \epsilon Y_x^\pm(x \mp t_0^{-1}y) + \cdots],$

$\quad\quad\quad\quad\quad v = q_0[\epsilon Y_x^\pm(x \mp t_0^{-1}y) + \cdots],$

(141.15) $\quad c = c_0\left[1 \pm \dfrac{\gamma-1}{2}(t_0^{-1} + t_0)\epsilon Y_x^\pm(x \mp t_0^{-1}y) + \cdots\right],$

(141.16) $\quad \rho = \rho_0[1 \pm (t_0^{-1} + t_0)\epsilon Y_x^\pm(x \mp t_0^{-1}y) + \cdots],$

(141.17) $$p = p_0 \pm \rho_0 q_0^2 t_0 \epsilon Y_x^{\pm}(x \mp t_0^{-1} y) + \cdots].$$

Since the strength of the shock approaches zero as $\epsilon \to 0$, the shock line is to the order zero just the Mach line

(141.18) $$y = \pm t_0 x$$

issuing from the tip.

To the next order, that is to the first, the Mach lines are given by the formula

(141.19) $$y = \epsilon Y^{\pm}(\xi) + \left[\pm t_0 + \frac{\gamma + 1}{2} (1 + t_0^2)^2 \epsilon Y_x^{\pm}(\xi) \right] (x - \xi),$$

as we see from (141.10–.11) and (141.04). From the discussion in Section 139 we know that to first order the shock line bisects the angle between the two Mach lines issuing from the tip $\xi = 0$. Consequently, we derive from equation (141.19) the following representation of the shock line to first order

(141.20) $$y = \left[\pm t_0 + \frac{\gamma + 1}{4} (1 + t_0^2)^2 \epsilon Y_x^{\pm}(0) \right] x.$$

We are especially interested in the pressure distribution along the profile. Setting $\xi = x$ we obtain from (141.17)

(141.21) $$p = p_0 + \rho_0 q_0^2 [\pm t_0 \epsilon Y_x^{\pm}(x) + \cdots.$$

For the total resulting force with the x-component F and the y-component G we obtain in obvious notation

(141.22) $$F = \epsilon \int_0^a [pY_x]^{\pm} dx$$
$$= \epsilon^2 \rho_0 q_0^2 t_0 \int_0^a \{(Y_x^+(x))^2 + (Y_x^-(x))^2\} dx + \cdots,$$

(141.23) $$G = -\int_0^a [p]^{\pm} dx = -2\epsilon \rho_0 q_0^2 t_0 b + \cdots;$$

here b is given by (141.02). We see that the drag F is of second order in the thickness parameter ϵ. The lift G is of first order in ϵ unless $b = 0$, i.e., unless the chord connecting the leading and the trailing tips, $(0, 0)$ and (a, b), has the direction of the oncoming flow. These formulas for lift and drag of first order were first derived by Ackeret [126].

Proceeding in the same way in which we have derived the terms of first order, one can determine terms of higher order for all quantities in question. Terms of second order were given by Busemann [127–129]. It turns out that the third order contribution to the drag vanishes for a symmetrical profile. Terms of third and fourth order for lift and drag have been derived by Donov [130]. In deriving such terms the change of entropy across the shock and the deviation of the shock relations from the simple wave relations are to be taken into account. Since the shock strength is constant in first order, the change of entropy is constant in third order. Therefore the flow past the shock is still isentropic and hence irrotational in third order, see Section 13. The rotational character of the flow appears only in fourth order. It is further remarkable that the lift in third order and the drag in fourth order depend on the slope of the profile at the tip, and in fourth (or fifth) order on the curvature of the tip. In addition these quantities depend, of course, on the shape of the profile as a whole.

142. *Alternative perturbation method for airfoils*

We stated earlier that there is another method for the expansion of the flow in powers of the thickness parameter, using the expansion of all quantities at a fixed point (x, y). This method is by far more complicated than the one outlined above. Nevertheless, we shall present it as far as first order terms are concerned because it leads to the *linear wave* equation and can be generalized for the case of three-dimensional flow, in which the previous method has apparently no analogue, see Chapter VI, Part A.

Since we are concerned only with the terms of first order we may assume from the beginning that the flow is isentropic and that the entropy is the same as that of the undisturbed air ahead of the airfoil. Then the flow problem can be formulated in terms of a stream function $\psi(x, y)$, see Section 102, which is so chosen that it vanishes at the tip of the profile. The differential equations are, see (102.09) and (102.03),

$$\psi_y = \rho u, \quad \psi_x = -\rho v, \tag{142.01}$$

$$u_y - v_x = 0. \tag{142.02}$$

Here ρ is to be expressed in terms of $\psi_x^2 + \psi_y^2$ through (142.01) and Bernoulli's equation (102.01). Since we assume the flow to be supersonic the Mach number $M_0 = q_0/c_0$ is greater than one.

The boundary conditions on the profile are

(142.03) $$\psi(x, \epsilon Y^{\pm}(x)) = 0.$$

In addition we should add as "initial" conditions the relations along the unknown shock line. To avoid the complications that result from the expansion of the shock conditions we proceed in a somewhat improper way. We assume first that the slope of the profile vanishes at the tip:

(142.04) $$Y_x^{\pm}(0) = 0.$$

Then we know that the shock begins only at some distance from the profile and this distance increases as the thickness parameter ϵ decreases. As "initial" conditions we then simply impose

(142.05) $$\psi = \rho_0 q_0 y, \quad \psi_x = 0 \quad \text{on} \quad x = 0,$$

without mentioning the shock. The results derived in this way will then be used for cases in which the slope Y_x does not vanish at the tip. This procedure gives the terms of first and second order correctly.

We expand the values of the quantities ψ, u, v, τ at each point (x, y) with respect to powers of ϵ, anticipating that $\psi = q_0 y$, $u = q_0$, $v = 0$, $\rho = \rho_0$, $p = p_0$ for $\epsilon = 0$; the subscript (0) refers to the undisturbed flow ahead of the airfoil. Then the expansions are

(142.06)
$$\psi = \rho_0 q_0 y + \epsilon \psi^{(1)} + \cdots,$$
$$u = q_0 + \epsilon u^{(1)} + \cdots,$$
$$v = \epsilon v^{(1)} + \cdots,$$
$$\tau = \tau_0 + \epsilon \tau^{(1)} + \cdots.$$

From relation (102.05) we have

(142.07) $$\tau^{(1)}/\tau_0 = M_0^2 u^{(1)}/q_0;$$

from relation (142.01) then

(142.08) $$(1 - M_0^2) u^{(1)} = \tau_0 \psi_y^{(1)}, \quad v^{(1)} = -\tau_0 \psi_x^{(1)}.$$

Inserting the expansions (142.06) into (142.01–.02) and using (142.07–.08), we obtain for $\psi^{(1)}$ the equation

$$\psi^{(1)}_{xx} + (1 - M_0^2)^{-1}\psi^{(1)}_{yy} = 0,$$

or by (141.05)

(142.09) $\qquad\qquad \psi^{(1)}_{xx} - t_0^2 \psi^{(1)}_{yy} = 0,$

which is the *classical wave equation*.

The boundary condition (142.03) gives by (142.06) the relation

$$\rho_0 q_0 \epsilon Y^{\pm}(x) + \epsilon \psi^{(1)}(x, \epsilon Y^{\pm}(x)) + \cdots = 0.$$

Expanding the function $\psi^{(1)}$ with respect to ϵ we obtain

$$\rho_0 q_0 \epsilon Y^{\pm}(x) + \epsilon \psi^{(1)}(x, 0) + \cdots = 0;$$

hence

(142.10) $\qquad\qquad \psi^{(1)}(x, 0) = -\rho_0 q_0 Y^{\pm}(x).$

This is the boundary condition for the expansion coefficient $\psi^{(1)}$. We note that this condition is to be satisfied at the axis $y = 0$, and not at the profile. This requirement is typical of all cases in which the boundary itself varies with the parameter. It should be emphasized that this requirement is not an additional condition or an artificial device but a natural result of the systematic perturbation procedure.

There still remains the initial condition (142.05) which gives

(142.11) $\qquad\qquad \psi^{(1)}(0, y) = 0, \ \psi^{(1)}_x(0, y) = 0.$

The general solution of the wave equation (142.09) is well known:

(142.12) $\qquad\qquad \psi^{(1)} = f(x - t_0^{-1} y) + g(x + t_0^{-1} y).$

The boundary and initial conditions (142.10–.11) are in particular satisfied by

(142.13) $\qquad \psi^{(1)}(x, y) = -\rho_0 q_0 Y^{\pm}(x \mp t_0^{-1} y) \text{ for } y \gtrless 0.$

Here $Y^{+}(x) = Y^{-}(x) = 0$ for $x < 0$. For $u^{(1)}$ and $v^{(1)}$ we find from (142.08) and (141.05)

$$u^{(1)} = \mp t_0 q_0 Y^{\pm}_x(x \mp t_0^{-1} y)$$
$$v^{(1)} = +q_0 Y^{\pm}_x(x \mp t_0^{-1} y) \qquad \text{for } y \gtrless 0.$$

These expressions agree with those derived by the first method, see (141.14).

F. Remarks about Boundary Value Problems for Steady Flow

143. Facts and conjectures concerning boundary conditions

The basic question in mathematical physics and mechanics is: how is a phenomenon determined by the general differential equations and the specific boundary conditions? For the field treated in the preceding chapter this question assumes a particularly intriguing and elusive aspect. We seek a characterization of appropriate boundary conditions that would ensure the existence and uniqueness of the solution of problems in steady flow. What can be said by way of an answer is still tentative and far from a clear-cut mathematical statement.

Any mathematical formulation of a problem of physics involves idealizing assumptions, part of whose justification may be found in the existence and uniqueness of the mathematical solution and, in addition, in its "stability," that is in its continuous dependence on the data. The difficulties that arise have their roots frequently not in physical reality but rather in imperfect mathematical idealizations. Strictly speaking, steady phenomena do not exist but correspond to such an idealization; for they must be considered as limiting states of transient phenomena arising under given boundary conditions from given initial states. The existence of such steady limiting states, however, is not at all obvious and probably cannot be deduced from the differential equations without taking some effect of viscosity into account; even then a rigorous proof seems beyond the present possibilities of analysis.

But let us start with the assumption that a steady limiting state exists. Then the question arises: can this state be determined intrinsically as a solution of the differential equations for the steady state by imposing proper boundary conditions, without reference to the transient phenomenon from which the steady flow asymptotically originates? Only when this question is answered may we consider the theory of steady state as satisfactory. In many instances of classical linear problems of mathematical physics the proper intrinsic boundary conditions for the steady state can be formulated immediately, being the same as in the underlying transient problem. Unfortunately, the situation is far more complicated in fluid dynamics. In many significant cases the boundary conditions which determine the transient flow are lost in the asymptotic process leading to a steady limiting state; hence the proper intrinsic boundary conditions must be found by more elaborate considerations. As a matter of fact, the search for these conditions has, in some cases, not been entirely successful. Neither intrinsic mathematical analysis nor physical intuition always provides a ready answer.

Even the simplest types of gas dynamical problems illustrate these difficulties. For example, we may consider the problem of finding the flow in an infinite duct which originally contains gases under different pressures at rest in different parts of the duct; these gases are separated by membranes which at the time $t = 0$ are suddenly removed. The effect of the resulting interaction process will spread through the whole duct and eventually reach every part of it. Therefore we cannot expect the asymptotic steady state to be determined by the same conditions that prevailed at the far ends originally. In particular, if there is only one membrane separating the tube into parts with quiet gas at different pressures, we have the simple case already studied in Section 80; the resulting steady state is a constant flow which at the far ends certainly does not satisfy any of the conditions inherent in the gas at the beginning; certainly the velocity of the steady flow is not zero at the ends and also pressures and densities differ from those prevailing initially there.

The situation is similar if originally, at the time $t = 0$, a steady flow exists in a duct whose shape is afterwards modified, e.g. by deforming the duct, or by introducing into the duct some obstacles (as in a wind tunnel). The effects of these disturbances will ultimately spread throughout the duct and may induce at the far ends boundary conditions different from the original ones. It should be noted, however, that sufficiently weak disturbances of the shape of the duct, at places where the original flow was supersonic, will never reach the upstream end of the duct. In this case it is proper to retain, for the steady state, the original boundary conditions at the upstream end.

To obtain the proper boundary conditions one should determine the limit flow at a definite place x as the time t approaches infinity and afterward let $|x|$ approach infinity. Mathematically speaking, the difficulty in formulating boundary conditions is thus connected with the fact that the passage to the limit $t \to \infty$ and the passage to the limit $|x| \to \infty$ are two processes generally not interchangeable with each other.

A peculiar feature of the intrinsic boundary conditions for problems of steady gas flow is that they depend on the nature of the solution. They are different if the steady flow which they are to characterize is subsonic throughout, or supersonic throughout, or partly subsonic and partly supersonic, and still different if the flow involves shocks.

We shall explain the various possibilities in the simple ideal case of flow through an infinite duct whose entrance section consists of two plane walls parallel to the x-axis between $y = 0$ and $y = b$. We assume that the exit section likewise consists of two parallel plane walls. The results of our discussion will also apply in a rather obvious manner to flow around an obstacle, such as an airfoil, placed in a duct or to flow around an obstacle in a free stream. We may characterize the flow by a stream function and assume that values of the stream function at the two walls are prescribed constants, $\psi = 0$ and $\psi = \psi_0$. This boundary condition need not be mentioned any more. We also assume that at $-\infty$, i.e. at the infinite end of the entrance section, the flow has a uniform velocity u_0 and a uniform density ρ_0 and pressure p_0. The value ψ_0 of the stream function at the wall is then given by $\psi_0 = \rho_0 u_0$. We shall then characterize the flow in terms of the "entrance Mach number" $M_0 = u_0/c_0$.

We shall first discuss the case in which the steady state flow is everywhere purely subsonic, a situation to be expected if the entrance Mach number is sufficiently small. The differential equation for the stream function is then elliptic. According to general experience, the same data may be prescribed for elliptic differential equations as for the potential equation. If ψ were a harmonic function it would be completely characterized by the condition that it assumes the values 0 and ψ_0 at the two walls and remains bounded at the entrance and exit. We therefore surmise that the same data determine a stream function of the compressible fluid. The velocity distribution would then be uniform across the infinite ends of the entrance and exit sections. In several other respects the flow would behave like a potential flow. Since sufficiently differentiable solutions of an elliptic differential equation are analytic, the flow is completely determined throughout the duct by its distribution over an arbitrarily small section; in particular, the flow velocity cannot be constant in any section unless it is constant in the whole field of flow. This fact is remarkable because a similar statement is not true for supersonic flow.

Next we consider the other extreme, that the flow is supersonic throughout the duct. In that case the differential equation is hyperbolic. The proper analogue is no longer the potential equation, but the wave equation with time corresponding to x, the distance along the axis of the duct. Accordingly, we expect that a uniquely determined flow exists if two initial conditions are imposed at the entrance and none at the exit. Since the behavior at the exit is not prescribed, we cannot, in general, expect the velocity to be uniformly distributed at the end of the exit section.

Continuous, purely supersonic flow will arise for properly shaped duct walls. We know that we can even achieve a situation in which the flow in the exit section has a uniform velocity by using the construction described in Section 113. On the other hand, we know from previous discussions, see, for example, Sections 117 and 123, that a continuous, purely supersonic flow does not exist if certain parts of the duct walls are strongly curved and force the occurrence of shocks.

Before discussing the boundary condition to be imposed in case shocks occur, we first consider the "mixed" continuous flows which are partly subsonic and partly supersonic. We shall first assume that the entrance flow is subsonic. There are then several possibilities of mixed flow. Suppose the duct walls are plane except for a small inward bulge at some section (see Fig. 82). If the entrance Mach number is not much below the value one, the flow becomes supersonic in a finite region adjacent to the bulge and is again purely subsonic throughout the exit section. In this case we expect that the same boundary conditions may be imposed as for subsonic flow. However, it is not known whether solutions exist for arbitrary shapes of the bulge, nor is it known whether or not a slight change in the shape of the bulge would induce a large change in the flow; there are strong indications of such instability, see [97] and [98]. When the entrance Mach number is increased, there is little doubt that a continuous flow will cease to exist under the given boundary conditions as soon as the entrance Mach number attains a certain critical value less than one.

Whether, upon increasing the entrance Mach number, discontinuities will develop, is still an open question, and is clarified neither by experiment nor by theory. It has been proved, see [99], that shocks do not originate as they do in one-dimensional flow on the tip of an incipient "limit line"; it has also been proved by H. Görtler, see [99], that in mixed flow two finite supersonic regions cannot merge when the

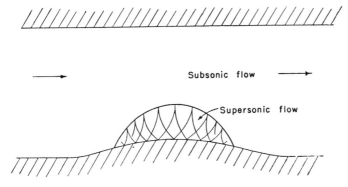

FIG. 82. A region of subsonic flow in which a region of supersonic flow is embedded. The Mach lines in the region of supersonic flow are indicated.

entrance Mach number is increased. Yet, how in detail the discontinuity starts, to what extent viscosity is necessary for an explanation, or whether instability alone in the non-viscous steady flow suffices to account for the formation of shocks, are important but unanswered questions.

Another possibility of mixed flow is that the flow changes across the whole duct from a subsonic into a supersonic state near the bulge and then remains supersonic. Such a type of flow occurs in exhaust nozzles, see Chapter V. For this mixed flow, only one boundary condition is appropriate at the entrance, as with purely subsonic flow, while, as with supersonic flow, none is allowable at the exit. It is not known, however, whether or not a flow is uniquely determined by such data. The problem of uniqueness of such mixed flow has been treated by Frankl [88–91] using certain methods of Tricomi [87]. Frankl has been able to prove the uniqueness of the flow out of an orifice in case the flow is subsonic up to the opening and supersonic outside of it.

Flows which are supersonic in the entrance section of a duct and then become subsonic owing to a constriction in the duct may be considered the reverse of flows that turn from a subsonic to a supersonic state. Continuous flows of this type can exist only for very specially designed duct walls, see Section 113. Flows which possess only finite subsonic regions because of the occurrence of an outward bulge do not seem to have been considered.

So far we have made the assumption that the flow is continuous. We know that in general this assumption is not tenable and that we must consider flows involving shocks. It is a question of great importance to know under what circumstances a steady flow involving shocks is uniquely determined by the boundary conditions and by the conditions at the entrance, and when further conditions at the exit are appropriate.

Shock fronts in the supersonic flow entering the duct begin at the cusps of envelopes of Mach lines of the incoming flow, as we assume in accordance with our previous analysis, see Section 117. The shock fronts beginning at these cusps can certainly be uniquely continued until they meet each other. Suppose they meet under such angles that "regular intersection" is possible, see Sections 125 and 126. Then again unique continuation is possible until the continued shock fronts meet the opposite walls. If they meet these walls at such angles that weak regular reflection is possible, then again continuation is determined provided that only weak reflections are admitted. Thus the flow can be uniquely continued if the situation is such that regular intersections and regular reflections are always possible when reflections or intersections are called for and if the weak character of the reflection is stipulated. Such a situation corresponds somewhat to one-dimensional unsteady flows in which the continuation of shock fronts in time is uniquely determined.

If regular reflection or intersection is not possible, Mach reflection or intersection will occur. Once reflection or interaction of the Mach type and of the strong regular type is admitted in the flow pattern, a completely different situation arises as to the unique determinacy of the flow. The important new point is: A new condition must be imposed at the exit end.

Consider, for example, a duct with an entrance section bounded by two straight parallel walls which at certain points turn sharply inward through a certain angle and then remain straight for some distance beyond these points, see Figure 83. Suppose these angles are

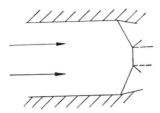

FIG. 83. Mach shock configuration in a duct.

so large that no regular intersection of the shocks issuing from the corners is possible. Then a stationary Mach intersection will take place in which the Mach shock front in general is curved unless the conditions for a stationary Mach configuration can be satisfied at the branch point. The position of the Mach shock is not determined by the data of the problem. This can best be seen in the case of a Mach configuration with straight shock lines. Then obviously the branch points of the configuration may have, within limits, an arbitrary distance from the wall in agreement with the shock conditions.

One may even imagine that the plane Mach shock front can be moved upstream so far that it begins at the walls themselves, and even beyond that position. In that case the configuration of Mach intersection has degenerated into a simple plane shock front across the duct. It is then known that in order to establish a definite position of the shock front a new condition at the exit must be imposed, such as, for example, the exhaust pressure.

Thus it is plausible that flow involving a single shock front is not uniquely determined unless the pressure at the exit is prescribed or some other appropriate condition is imposed. To illustrate this, we consider a flow involving a shock front completely spanning the duct at a place at which the duct walls are not parallel but inclined toward each other, such as in the exhaust section of a nozzle. In that case evidently the position of the shock is coupled with its strength and the strength in turn is coupled with the pressure of the flow in the exit section. It is then reasonable to expect the position of the shock front to be determined by prescribing the pressure at the infinite end of the exit.

It is not obvious, however, what one should prescribe at the infinite end of the exit for flows in which a great number of Mach intersections and Mach reflections occur and in which therefore the flow in the exit consists of several subsonic and supersonic sections. In an

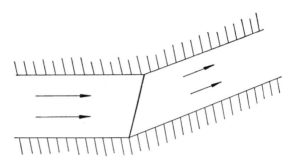

Fig. 84. Supersonic flow in a duct deflected by a strong shock.

actual flow viscosity will assure that the flow conditions are finally uniform over the cross section. Then again the pressure can be prescribed at the infinite end of the exit, but it seems that the location of Mach shock fronts cannot be determined without taking the effect of viscosity into account.

A significant illustration of the difficulties arising from the problem of uniqueness is given by the flow in a duct with a sharp corner, as in Figures 84, 85. An incoming parallel supersonic flow changes its direction through a shock S provided the angle of the corner is not too large. As we saw in Section 123 there are two shocks, a weak and a strong shock, compatible with the local conditions at the corner. According to our previous discussion the decision between a weak and a strong shock should depend on the boundary conditions imposed at the downstream end of the duct. One could design a duct so that for a given Mach number the incoming constant flow would be deflected into the new direction by a strong shock with a resulting constant parallel flow in the exit part of the duct. The same incoming flow, however, may also be deflected into the new direction by a weak shock in and near the inner curves. Then the angle between shock line and direction of incoming flow would be smaller than in the case of a strong shock. A rarefaction wave would issue from the opposite obtuse corner of the duct and a complicated flow pattern due to interactions and reflections

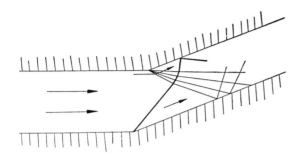

Fig. 85. Supersonic flow in a duct deflected by a weak shock and a rarefaction wave.

would result. There are even many more flow patterns, involving Mach configurations compatible with the assumption of a given incoming flow in the duct. Such mathematical indeterminacy indicates that the actual phenomenon singled out among the various conceivable ones will be determined by conditions imposed at the exit end. Certainly the various flows mentioned behave differently there.

Whether or not a flow compatible with the boundary conditions actually occurs depends moreover on its stability. Certainly, there exist unstable flow patterns, see e.g. [144, 147]; in particular, it has been frequently asserted that flows involving oblique strong shocks are unstable. Reasons given for this statement do not seem cogent. On the contrary, it seems plausible that the flow indicated in Figure 85 would become stable if the two sections of the duct were made slightly divergent in the direction of the flow; for strong shocks spanning a symmetrical divergent duct are known to be stable.

In the case of instability there may be another stable, actually occurring, flow which satisfies the given boundary conditions; on the other hand, there may be situations where no stable flow of the assumed type exists, for example if the unstable flow is uniquely determined by the boundary conditions. In that case the actual steady limiting flow, if it exists, does not satisfy the boundary conditions supposed to characterize it; other conditions should then be found.

It should be mentioned that there is still another possible occurrence in a flow which must be excluded in order to have the flow mathematically determined but which actually does take place in certain flows. This is the possibility of jet detachment from the walls. Such a detachment is liable to occur immediately behind the place at which a shock front meets the wall. Detachment is evidently compatible only with an oblique shock front, i.e. a shock front which does not meet the wall perpendicularly. Its occurrence and its strength in actual flows seem to depend decidedly on the viscous boundary layer which has been built up as far as the point of detachment.

In concluding we emphasize once more that the considerations of this section serve primarily to illuminate the altogether imperfect state of the theory and the need for much further investigation by experiment and mathematical theory.

CHAPTER V

Flow in Nozzles and Jets

144. Nozzle flow

In Sections 113 to 115 we discussed patterns of two dimensional flow in a duct or in a jet emerging from a duct. We shall now resume this discussion by studying, in greater detail, steady flow in two-dimensional as well as three-dimensional nozzles and jets in a more or less qualitative way.

Strictly speaking, a flow through a duct should be considered as a steady, isentropic, irrotational flow with cylindrical symmetry about the x-axis and should be determined by solving the differential equations (16.06) and (16.14) under appropriate boundary conditions. An approximate analysis of great practical value is, however, possible, avoiding a difficult study of such a boundary value problem.

The most important type of duct is the *de Laval nozzle*, which plays a fundamental role in the operation of turbines, wind tunnels, and rockets. The de Laval nozzle consists of a converging "entry" section and a diverging "exhaust" section. When a gas at rest in a container or chamber under high pressure escapes through such a nozzle, two possibilities arise. The first is that the flow, after being expanded in the entry section, is compressed in the exhaust section and remains subsonic throughout. This occurs when the ratio of chamber pressure to outside pressure remains below a certain "critical" value. When this pressure ratio exceeds the critical value, the other alternative occurs; the flow becomes supersonic on passing the throat and continues to expand from this point on.

145. Flow through cones

The important fact that subsonic flow is compressed, and supersonic flow is expanded, in a diverging section can best be understood by considering flow through a two-dimensional angular *sector* or through a *cone*. Since the flow in a sector has already been treated in Chapter IV, Sections 111 and 117, we may confine ourselves to the flow through a cone. We assume that the flow is steady and isentropic, that it is radially directed, and that its speed q, density ρ, and pressure p depend only on the distance r from the tip. Then the continuity equation can be written in the form

$$(r^2 \rho q)_r = 0,$$

or $r^2 \rho q$ = constant. Denoting the area intercepted by the cone on the sphere r = constant by $A = A_0 r^2$ and the rate of mass crossing this area per unit time by G, we have

(145.01) $$A \rho q = G.$$

Since the flow is irrotational, the only additional equations needed to determine the flow are the adiabatic relation

(145.02) $$p \rho^{-\gamma} = \text{constant},$$

and Bernoulli's law

(145.03) $$\mu^2 q^2 + (1 - \mu^2) c^2 = c_*^2$$

with $c^2 = \gamma p/\rho$, see equation (14.08).

The critical speed $c_* = q_*$ is always considered in what follows as a fixed parameter; then, by equations (145.01–.03) the values of pressure and of density at the critical speed, p_*, ρ_*, are also fixed, and likewise the critical value A_* of the cross section A, corresponding to the value $c = c_*$. (These critical values are well-defined quantities, whether or not they are attained in the actual flow.) The preceding equations may now be written in the following form, in which c/c_*, ρ/ρ_*, p/p_*, A/A_* are expressed in terms of $q/q_* = q/c_*$ or c/c_*:

(145.04)
$$\left(\frac{c}{c_*}\right)^2 = (1 - \mu^2)^{-1}\left(1 - \mu^2 \left(\frac{q}{q_*}\right)^2\right),$$
$$\frac{\rho}{\rho_*} = \left(\frac{c}{c_*}\right)^{2/\gamma-1}, \quad \frac{p}{p_*} = \left(\frac{c}{c_*}\right)^{2\gamma/\gamma-1},$$
$$\frac{A}{A_*} = \left(\frac{c}{c_*}\right)^{-(2/\gamma-1)} \left(\frac{q}{q_*}\right)^{-1}.$$

Accordingly, we may consider any one of the quantities A, q, r

as an independent variable (always for fixed $c_* = q_*$. p_* , ρ_* and A_*) and express all the other quantities in terms of it.

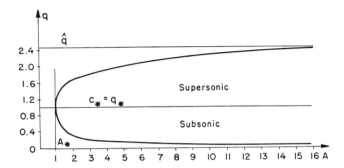

FIG. 1. The speed q of the flow through a nozzle depending on the intercepted area A (for air $\gamma = 1.4$).

For a further discussion we derive from (145.01–.03) the relations

(145.05) $$\frac{dA}{A} + \frac{d\rho}{\rho} + \frac{dq}{q} = 0,$$

(145.06) $$\frac{d\rho}{\rho} = \frac{2}{\gamma - 1} \frac{dc}{c} = \frac{1 - \mu^2}{\mu^2} \frac{dc}{c},$$

and

(147.07) $$\mu^2 q\, dq + (1 - \mu^2) c\, dc = 0;$$

hence we find

(145.08) $$\frac{dA}{A} = \left(\frac{q^2}{c^2} - 1\right) \frac{dq}{q}.$$

The last relation shows that for increasing area A the speed q increases when $q > c$ and decreases when $q < c$. Moreover, since increasing speed corresponds to decreasing density, it follows that *in the direction of increasing area A the flow is expanded when it is supersonic, compressed when it is subsonic.*

Another important consequence can be inferred from these formulas. The quantity A as a function of q has a minimum A_* for $q = q_* = c_*$. No flow with given c_* , p_* , A_* is possible in the

380 FLOW IN NOZZLES AND JETS CHAP. V

part of the cone with $A < A_*$, and a converging flow beginning with $A > A_*$ stops when the critical area $A = A_*$ is reached. *No transition from subsonic to supersonic flow is possible in a cone.*

146. De Laval nozzle*

Transition from subsonic to supersonic flow becomes possible, however, by the following modification. Two sections of cones or similarly shaped tubes with the same axis are placed opposite each other and connected as in Figure 2, thus forming a *de Laval nozzle*

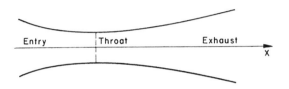

Fig. 2. DeLaval nozzle.

with *entry section, throat,* and *exhaust section*. Then a subsonic expanding flow in the entry section, on passing through the throat, can change into a supersonic expanding flow in the exhaust section.

The exact treatment of the flow through a cone can be modified in order to describe approximately the flow through a de Laval nozzle whose components are not necessarily conical. The approximate treatment described below is a slight modification of the "hydraulic" treatment due to O. Reynolds (1886). A set of surfaces of revolution which intersect the nozzle wall perpendicularly is introduced**; it is then assumed that the flow is orthogonal to these surfaces and that all relevant quantities are constant on them. If by A we denote the area on these surfaces intercepted by the nozzle wall, i.e. the cross section area, then (145.01) also holds for this flow. Hence we may apply the formulas (145.04–.08) derived above from (145.01–.03) for conical flow. In particular, we infer from relation (145.08) that if a change from a subsonic to a supersonic flow occurs at all, it occurs at

* With reference to what follows see [139–144] and [102] in the bibliography.
** In Reynolds' "hydraulic" treatment planes perpendicular to the axis are assumed instead of curved surfaces, which are naturally suggested by the consideration of flow through cones.

that surface on which the area A assumes its minimum. Then the flow becomes *sonic at the throat*, i.e. we have there $q = q_* = c_*$ and $A = A_t = A_*$. The assumptions on which this treatment is based are not exactly compatible with the irrotational character of the flow. Nevertheless, both experiment and a refined theoretical treatment, see [144], based on a more complete analysis of the differential equations show that the results of the hydraulic theory provide very good approximations.

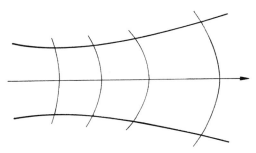

FIG. 3. Surfaces across a nozzle on which q, p, ρ are assumed constant.

It is evident that *two-dimensional* nozzle flow could be treated in much the same manner, and it might seem that an exact theory could be developed by investigating the *linear* differential equations characterizing the flow in the *hodograph plane*. Such a procedure, however, is not practicable because the image of the nozzle region in the (u, v)-plane is not simple; it forms a *fold*.

To see this, one need only visualize how the images of the streamlines would run in the hodograph plane. Consider a point of a streamline in the subsonic region with negative velocity component v. On the same streamline, v will eventually be positive in the supersonic region. Since $v = 0$ all along the axis of the nozzle, one arrives at the diagram shown in Figure 4, corresponding to the section of the flow with $y > 0$. One observes that the images of the streamlines intersect each other. Thus the image of the field of flow in the hodograph plane is not simple, and the functions representing x, y in their dependence on u, v become singular at the envelope of the images of the streamlines.

From the theory developed in Chapter II, Section 30 and Chapter

IV, Section 105, we know that this envelope, the edge of the fold, is a Γ-characteristic. It is evident that for the nozzle flow both Γ-characteristics starting at the point $u = c_*$, $v = 0$ are such edges.

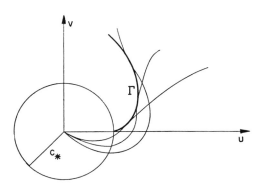

Fig. 4. Images of streamlines of a nozzle flow (for $y > 0$) in the hodograph plane with an envelope, which is a Γ characteristic.

The cusp region between them is covered three times by the mapping. The antecedents of these two branches of the edge in the (x, y)-plane, the *transition curves*, consist of two Mach lines starting at the point on the axis at which critical speed is reached. (Incidentally, a further discussion shows that *four*, and not two, Mach lines issue from this point.)

While it is not possible to use the velocity components, it is possible to use the stream function and the potential function as new independent variables. Thus one is led to a refined analysis of nozzle flow which can be carried out for both two- and three-dimensional cases [144]. Suppose that a distribution of the axial velocity along the axis is given. By expanding the velocity components and the distance from the axis in powers of the stream function, assumed to vary on the axis, and inserting these expansions in the appropriate differential equations, one is led to some simple formulas. The terms of first order agree with those obtained by the hydraulic method. Among the formulas of the next order we mention only

$$q_W = q_A(1 + \tfrac{1}{2}\kappa_W y_W)$$

for the flow speed q_W at a point W of the wall in terms of the speed q_A at the point A where the potential surface through W cuts the axis; κ_W is the curvature of the nozzle wall at the point W, and y_W is the distance of this point from the axis measured along the potential surface.

The results found by this method seem to agree well with those obtained by Taylor from expansions with respect to y up to the fifth power, or by the method of finite differences [143] and [146].

147. Various types of nozzle flow

In spite of its great simplicity, the hydraulic nozzle theory accounts for some peculiar types of nozzle flow that result under various conditions. For a proper understanding of these occurrences it is convenient to assume that the exhaust flow is discharged into a large *receiver* vessel in which an arbitrary pressure can be maintained. We then imagine that the *receiver pressure* p_r *is varied* while the *chamber pressure* p_c at the entrance of the flow into the nozzle *is kept fixed*. We assume that the flow speed q_c vanishes at the entrance; this corresponds to assuming that the cross section area of the chamber is infinite. From (145.03–.04) we can then determine the critical pressure p_*,

(147.01) $$p_* = (1 - \mu^2)^{\gamma/\gamma-1} p_c.$$

Let p be the pressure at the "cross section" of area A; then by formulas (145.04) the ratio A/A_* is a well-defined function of the ratio p/p_*, p_* and A_* being critical pressure and cross section area, respectively. While the critical pressure p_* is known, the critical area A_* is not determined by the given state in the chamber.

To visualize the variety of flows compatible with a fixed state in the chamber and various receiver pressures, it is advantageous to use graphs of the set of functions derived from (145.04)

$$A = A_* f\left(\frac{p}{p_*}\right)$$

for various values of the parameter A_*, see Figure 5.

All these curves have the lines $p = 0$ and $p = p_c$ as asymptotes and are loops reaching from $A = A_*$ to $A = \infty$. For any of these curves there are two values p attached to every $A > A_*$, the greater value of p referring to a subsonic, the smaller to a supersonic state, while for $A = A_*$ both states become identical, and for $A < A_*$ no flow is possible at all. In Figure 6 the pressure p is shown as function of the abscissa x along the axis for a given nozzle. It is obtained by inserting the value A as a function of x in $A = A_*f\left(\dfrac{p}{p_*}\right)$. The relation between pressures and areas along any flow in the nozzle from chamber to receiver is represented by arcs of these graphs.

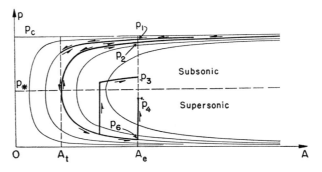

Fig. 5. Relation between pressure and cross section area for the various nozzle flows resulting from various receiver pressures p_1 to p_6.

When the receiver pressure p_r equals the container pressure p_c no flow results at all. When p_r is slightly less than p_c, a flow with low speed results (in Figure 5, $p_r = p_1$). To determine the flow we locate the point $p = p_r$, $A = A_e$ in the (A, p)-plane, A_e being the exit cross section area. For an appropriate value of A_*, $A_* = A_*(p_r, A_e)$, the curve $A = A_*f\left(\dfrac{p}{p_*}\right)$ passes through this point. We follow this curve until A assumes the value A_t of the throat cross section area. The section of the curve between A_t and A_e represents the flow from the throat to the exit, while the flow from the container to the throat is represented by the section from $A = \infty$ to $A = A_t$ of the same branch of this curve. The flow remains subsonic throughout. Evidently this description of the flow, characterized by the subscript $r = 1$, is valid only if the curve $A = A_*f\left(\dfrac{p}{p_*}\right)$ passing through the point (p_1, A_e) intersects the line $A = A_t$; i.e., if

$$A_t > A_*,$$

when $A_* = A_*(p_1, A_e)$ is the critical area associated with the specific curve under consideration, and when A_t is the given throat area of the nozzle.

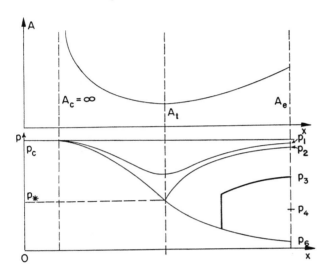

FIG. 6. The pressure as a function of the position along the axis of a nozzle for various flows resulting from various receiver pressures.

As the receiver pressure p_r is lowered, A_* decreases until finally the value $A_* = A_t$ is reached for a certain pressure $p_r = p_2$. For $p_r = p_2$, the flow just becomes sonic at the throat, but still remains subsonic elsewhere.

When now the receiver pressure p_r is lowered below p_2 a completely different type of flow arises, as indicated in Figures 5 and 6. From the chamber up to the throat the flow is subsonic and represented by the upper arc of the curve, $A = A_t f\left(\dfrac{p}{p_*}\right)$, coming from $A = \infty$ and belonging to $A_* = A_t$. This part of the flow is independent of the receiver pressure and is determined solely by A_t and p_* (or p_c by

(147.01)). After passing the throat the flow becomes supersonic and is represented by the lower branch of the same curve $A = A_t f\left(\dfrac{p}{p_*}\right)$; this curve intersects the line $A = A_e$ at a definite point with $p = p_6$; hence p_6 is determined from $A_e = A_t f\left(\dfrac{p_6}{p_*}\right)$. In other words, the flow is smooth with steadily decreasing pressure and density and steadily increasing speed which is sonic at the throat. If the receiver pressure happens to be exactly $p_r = p_6$, we thus have what is considered the ideal flow through the nozzle. The design of a nozzle for given pressures p_c and p_r usually means the selection of the dimensions A_t and A_e so that, in our notation, $p_r = p_6$, i.e., $\dfrac{A_e}{A_t} = f\left(\dfrac{p_r}{p_*}\right)$.

In our imagined experiment, however, when p_r is gradually lowered, we still have $p_r > p_6$, after first passing $p_r = p_2$. How does the flow, after having attained supersonic speed behind the throat, adjust itself to the prescribed receiver pressure p_r? The answer is that first the flow continues behind the throat as indicated by the lower branch of the specific curve $A = A_t f\left(\dfrac{p}{p_*}\right)$; but at a certain place in the diverging part of the nozzle a shock front intervenes, the gas is compressed and slowed down to subsonic speed. From there on the gas is further compressed and slowed down; the relation between pressure and area is then represented by the upper branch of the curve $A = A_* f\left(\dfrac{p}{p_*}\right)$ passing through A_e and p_r with an appropriate smaller value of p_*. The position and strength of the shock front are automatically adjusted so that the end pressure at the exit becomes p_r. In the diagram the place corresponding to the shock front indicates a jump from the supersonic branch of the curve with $A_* = A_t$ to the upper branch of the curve through p_r, A_e.

When the receiver pressure p_r is lowered from $p_r = p_2$, the shock front moves from the throat toward the exit. It reaches the exit for a value $p_r = p_4 > p_6$. In other words, for $p_r < p_4$ no adjustment of the flow to the receiver pressure is possible by a shock in the nozzle. Again a new type of flow pattern must be found to describe what happens under the condition $p_r < p_4$.

Certainly we must now expect that *in the nozzle* the flow is the same as that in the ideal case $p_r = p_6$. The whole curve $A = A_t f\left(\dfrac{p}{p_*}\right)$ (precisely, the subsonic branch $A_t < A < \infty$ and the supersonic branch $A_t < A < A_e$) indicates the flow in the nozzle. Now it is in the jet *outside the nozzle* that the adjustment to the outside pressure p_r takes place. There are two types of phenomena, according as $p_4 > p_r > p_6$ or $p_r < p_6$; the intermediate case $p_r = p_6$ is the ideally adjusted continuous flow considered before.

148. Shock patterns in nozzles and jets

To understand the adjustment of the flow in the jet it is better first to revert to the shock patterns occurring within the nozzle for

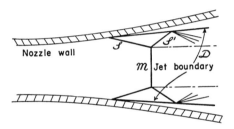

FIG. 7. Jet detachment and shock pattern inside a nozzle.

$p_4 < p_r < p_2$. In the simplest description these shock fronts would be curved discs across the nozzle perpendicular to the nozzle wall. The actual shape of such a shock front is, however, different. The shock front is actually oblique and consequently changes the direction of the flow abruptly, that is, the shock front leads to *jet detachment*. The situation is represented in Figure 7.

Thus at the wall the shock front begins obliquely as a conical surface and is followed by jet detachment. The shock front is cut off by a "Mach shock disc" perpendicular to the axis, presenting approximately the picture envisaged in the simplified description of the preceding section. Behind the incident and Mach shock front a reflected shock front S' and a discontinuity surface $\mathcal{D}$ develop.

When the receiver pressure p_r decreases to the value p_4, the place of detachment moves toward the rim of the nozzle and remains there as p_r becomes less than p_4, while the shock front leaving the rim becomes longer, see Figure 8. If the receiver pressure is de-

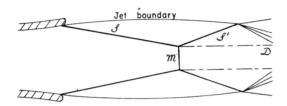

Fig. 8. Shock pattern in a jet emitted with a pressure less than the receiver pressure.

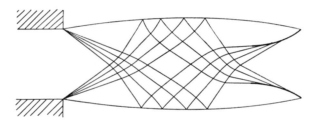

Fig. 9. Wave pattern in a jet resulting when a parallel flow enters a region of lower pressure.

creased to the value p_6, the strength of the shock leaving the rim becomes zero. If the receiver pressure p_r is further decreased below p_6 a new set of phenomena begins.

To explain what happens when the receiver pressure p_r is below the exhaust pressure p_6, first suppose that the jet is emitted from the nozzle with *constant axial velocity*. If the flow is two-dimensional, the flow pattern consists of two centered rarefaction waves leaving the rim, intersecting each other and being reflected as converging waves at the boundaries of the jet, see Section 115. If the pressure difference $p_6 - p_r$ is small, the reflected waves are approximately mirror images of the incident waves and hence converge approximately in a point at the opposite boundaries; an approximately periodic wave pattern (first described by Prandtl) therefore occurs. Actually, however, the reflected waves converge before reaching the opposite boundaries and give rise to shock fronts, see Figure 9 and [153]. If the receiver pressure further decreases, the tips of these shock fronts approach each other; eventually the two shocks intersect each other. A similar wave pattern* occurs in three-dimensional jets emitted from the nozzle with constant axial velocity, see Hartmann and Lazarus [152, Figure 3].

A new phenomenon occurs in a jet emitted from a nozzle *with a diverging opening* whenever the receiver pressure p_r is below the exhaust pressure p_6. Again we describe it in terms of two-dimensional flow. At the rim of the opening there develop rarefaction waves tending to lower the value of the pressure to that in the receiver, but somewhere on the outer border of these waves a shock front develops which cuts across them and intercepts them. The flow patterns that may result are indicated in Figures 10a and 10b. If the receiver pressure p_r is raised toward the exhaust pressure p_6, the tips of these intercepting shock fronts approach the rim of the opening. If the receiver pressure is raised further, the tip of the shock front remains at the rim. Thus there is a continuous transition from what we have called intercepting shock fronts to the shock fronts developing at the rim which were discussed earlier.

The following reasoning shows that such intercepting shocks are quite plausible: The flow continues uninfluenced by the state in the receiver until it meets the first Mach line issuing from the rim, i.e., the inner border of the rarefaction wave. Owing to the divergence of the jet, the pressure decreases along the axis. The decrease of pressure is considerable when the Mach number $\frac{q}{c}$ of the emerging jet is noticeably greater than unity, as follows from the relation $dp/p = \gamma(q/c)^2 \, dq/q$, see (145.06–.07). Across the rarefaction wave the pressure further decreases to receiver pressure at the rim, and hence to below receiver pressure farther out. In other words, the pressure is below p_r at the outer border of the rarefaction wave while it equals p_r at the jet boundary. Consequently there is a pressure gradient acting from the jet boundary toward the interior of the jet. Clearly,

---

* As to the difficulty in understanding conical shock intersection theoretically see Section 157.

this pressure gradient makes the jet curve inward. All the Mach lines issuing from the boundary make the same angle with the boundary, since at the boundary the pressure, and hence also sound speed and velocity remain constant. Thus these Mach lines tend to converge

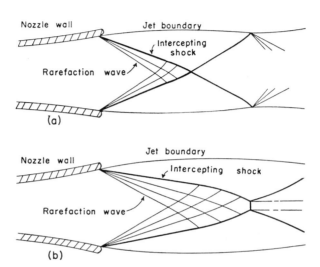

Fig. 10. Shock patterns in a diverging jet emitted with a pressure greater than the receiver pressure.

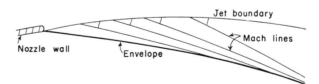

Fig. 11. Envelope of Mach lines issuing from the jet boundary at equal angles.

and very likely would have an envelope if they were not intercepted by a shock front. To prevent the envelope singularity, an "intercepting" shock is therefore necessary, see Figure 11.

If there is little divergence of the jet, the intercepting shock occurs only in the reflected wave and merges with the shock front occurring in a non-diverging jet, as explained above. In the jet various shock patterns may result, depending on the degree of divergence of the exhaust flow, its Mach number M_e, and the ratio of its pressure p_e to the receiver pressure p_r. A typical case is that shown in Figure 8 for $p_r > p_6$. The pattern is similar when $p_r < p_6$ except that the shock front does not begin at the rim but occurs as an intercepting shock. It involves conical "incident" and "reflected" shock fronts connected by a *Mach shock front* perpendicular to the axis. The characterizations "incident" and "reflected" are used only to identify parts of the shock pattern with the corresponding parts for shock waves striking obliquely. In a certain sense one may say that in the present problem the "reflected" shock is the primary phenomenon, the "incident" shock being determined by it. In jets with slight divergence and small pressure difference $p_6 - p_r$ only a section of a "reflected" shock wave occurs, as explained above; Mach shock front and "incident" shock are absent, see Figure 9. It should also be noted that the "reflected" shock is of the *strong* variety, an occurrence not normally observed in the reflection of impinging shock waves. Similar patterns occur in three-dimensional jets.

The configuration shown in Fig. 8 is observed, for instance, if M_e is approximately 3 and the half-angle of divergence is greater than 15°. The pressure in front of the Mach disc is found to be of the order of magnitude $0.1\ p_r$ to $0.03\ p_r$, and hence is very low. The strength of the Mach shock is very great, its excess pressure ratio being of the order of magnitude 20 to 50.

A few words about the continuation of the jet may be added. The jet boundary curves inward up to the place where the intercepting shock meets the boundary. There this shock front is reflected as a rarefaction wave and the jet boundary diverges again. The whole process repeats itself. By the action of viscosity at the jet boundary this periodic jet pattern is eventually blurred and dies out.

It is desirable to determine the *position of* the various *shock fronts* in nozzles and jets theoretically, in particular to determine the position of the first Mach shock disc because this disc is easily observed. From the discussion in Section 143, it is clear that this position depends essentially on the prescribed situation at the exhaust end of the jet.

The position of a shock front within the nozzle can be determined when one assumes that it consists of a disc across the duct and that the exhaust pressure equals the receiver pressure. The results of such a calculation agree roughly with experimental evidence. The details of the much more complicated shock pattern that actually occurs within a nozzle depend essentially on the action of the boundary layer which develops along the wall because of the action of viscosity.

For the determination of the position of the first shock disc in a jet no such simple assumptions are available. It is to be noted, however, that the air in the jet gradually acquires the receiver pressure owing to the action of the mixing layer developed at the interface between outside and jet air. It thus appears that the position of the first shock disc depends essentially on this mixing process.

149. Thrust

The exhaust flow out of a nozzle is the basis of the operation of a rocket motor. The burnt gases which are formed in the combustion chamber under high pressure acquire considerable momentum when they are ejected through the nozzle. Accordingly, as a reaction to this momentum flux, a thrust results which acts against the rocket in the direction opposite to that of the exhaust flow. The total thrust against the rocket can easily be expressed in terms of the quantities characterizing the exhaust flow, and conditions can be formulated for the shape of the nozzle to provide a maximum thrust.

It is customary to define the total thrust F as the difference

$$(149.01) \qquad F = F_i - F_a$$

between the internal thrust F_i, resulting from the pressure acting against the wall of the chamber and the nozzle, and the external counter-thrust F_a that would result if atmospheric pressure p_a were acting against the outer surface of the body in which the nozzle is imbedded.

To evaluate the thrust we consider the surface S through the exit rim on which the speed and hence the pressure are constant. The internal thrust F_i (counted positive when acting against the stream) is equal to the sum of the axial component of the momentum M

transported through S to the outside per unit time and the resultant pressure force P exerted against the surface S from the inside. In other words, the total pressure force, $F_i - P$, exerted against the volume of gas enclosed by chamber, nozzle, and surface S, equals the momentum M transported in unit time through S.

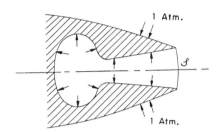

Fig. 12. Pressure forces producing the thrust.

We denote by A the projection of the surface S on a plane perpendicular to the axis; then $P = pA$ and the internal thrust is

(149.02) $$F_i = M + pA.$$

The external thrust is clearly

(149.03) $$F_a = p_a A,$$

counted positive when acting in the direction of the stream.

We consider the total thrust,

(149.04) $$F = F_i - F_a = M + (p - p_a)A,$$

as a function of position of the rim, by imagining that the nozzle may be continued or cut off at various places. In particular, we are interested in discovering for which position of the rim the total thrust is a maximum. The answer can be given completely.

The total thrust is a maximum when the nozzle is cut off at such a place that the pressure at the exit rim just agrees with the outside pressure. In that case the total thrust is just given by the momentum transport

(149.05) $$F_{\max} = M.$$

To prove this statement we consider two different surfaces S on which the pressure p is constant, $p = p_1$ and $p = p_2$. The change $[M + pA]_{p_1}^{p_2}$ of the sum of the momentum transport and pressure force is clearly equal to the axial component of the pressure force against the section of the nozzle wall cut out between the two surfaces S. Accordingly, letting the two surfaces coalesce, we find

$$d(M + pA) = pdA.$$

Consequently

$$dF = (p - p_a)dA.$$

It is thus shown that F is an extremum when $p = p_a$, or $dA = 0$. It can be shown that F is a minimum at the throat, where $dA = 0$, while F is a maximum when $p = p_a$.

The maximum thrust so determined still depends on the shape of the nozzle contour. We seek the condition that this shape in turn maximizes the maximum thrust for all nozzles delivering the same mass flux.

The position of the exit rim of a fixed nozzle contour for which the maximum thrust is obtained may be characterized by the subscript m. The surface S through this rim is S_m; the speed q_m on S_m is determined by formulas (145.04) through the condition that the pressure p_m on S_m equals p_a. The maximum thrust is given by

$$(149.06) \qquad F_m = M_m = q_m \int_{\mathsf{S}_m} \cos\theta\, dG,$$

where θ is the angle of the flow direction with the axis and dG the element of the mass flux per unit time G. Clearly $F_m \leq q_m G$ and $F_m = q_m G$ only if $\theta = 0$ on S_m. In other words, *the maximum thrust is a maximum for fixed mass flux G if the exhaust velocity is constant and has axial direction.*

150. Perfect nozzles

A nozzle which produces a constant axial exhaust flow may be called a *perfect nozzle*. A perfect nozzle through which the gas flows in the opposite direction is a *perfect diffuser*.

PERFECT NOZZLES

Two-dimensional perfect nozzles can be designed without difficulty. As a matter of fact, whenever a diverging exhaust flow is given, it is possible to make the flow "perfect" by re-routing only a section of it so that it eventually acquires constant axial velocity. Every streamline of such a perfect flow yields a perfect nozzle.

It is true that the actual designer of exhaust nozzles of rockets will not strive for "perfection" in this sense since the loss in thrust when the exhaust pressure differs from the outside pressure and the exhaust flow is not axial is, in general, balanced by a gain in simplicity of design. Nevertheless, there are cases, in particular those involving large Mach numbers, in which this perfection is desired. The possibility of constructing a perfect two-dimensional flow, indicated by Prandtl and Busemann [3], [102], [141], was explained in Chapter IV, Section 113.

An approximate method of constructing three-dimensional perfect nozzles has been given by Frankl*; it is not at all obvious but it is true that exact perfect diffusers and nozzles exist.** In a nozzle at the exhaust section the shock front would begin at the point on the axis at which the desired exhaust velocity is reached. A diffuser design in which no shock front develops at the rim of the entrance opening will be described in Section 157; in this design the shock front at the rim is avoided by giving the diffuser contour the direction of the axis at the rim; by shaping the contour properly, shocks can be entirely avoided.†

If the diffuser contour is inclined at the rim into the incoming flow, a conical shock front begins at the entrance rim. The form of this shock front was determined by A. Ferri [176], using a method of characteristics to determine the flow behind the front. It is interesting that even if the angle between the contour and the direction of the axis is only 5° at the rim, the resulting shock front tends to become perpendicular near the axis and forms a disc at which Mach reflection

---

* See the report by Kisenko [142]; (the authors had no access to the original paper by Frankl). See also Busemann [141] and [144].

** As proved by W. Y. Chen in an unpublished paper.

† The extreme sensitivity of the flow to changes in the shape of the contour is strongly indicated by the linearized treatment of diffuser flow by H. Ludloff in an unpublished paper.

occurs. Ferri also presents a calculation of the reflected shock front. If the angle at the rim becomes greater the disc near the axis at which the shock is strong becomes much larger and the reflected shock disappears as claimed by these calculations.

It should be mentioned that perfect diffusers of a different type can be constructed by inserting an appropriate "spindle" into the inlet, as proposed by Oswatitsch, see Section 113, Figure 24, and [176].

CHAPTER VI

Flow in Three Dimensions

Flow in one dimension and steady isentropic flow in two dimensions are amenable to a fairly systematic theory because of the special character of the underlying differential equations and because of the existence of simple waves. Flow in three dimensions, however, even under restricting assumptions of symmetry which make a reduction to two independent variables possible, presents a mathematical problem of a more complicated type. It may be that extensive numerical computations of special examples could produce valuable clues for a general theoretical attack. As matters stand now, however, one has to be content with a theoretical analysis of some specific types of problems or with the numerical treatment of individual cases unless approximate linearization of the problem is justified.

In this chapter we shall consider three distinct types of flow: in Part A, cylindrically symmetric steady flow against slender projectiles; in Part B, steady conical flow, in particular, flow against a conical obstacle; and in Part C, non-steady flow with spherical (or cylindrical) symmetry.

A. Steady Flow with Cylindrical Symmetry

151. Cylindrical symmetry. Stream function

While in Chapter V we gave a merely qualitative description of certain flows with cylindrical symmetry, we shall in this section present a more specific mathematical treatment of another kind of flow with cylindrical symmetry: a linearizing perturbation method for flow along slender bodies. It should be stated, however, that a rigorous justification and an appraisal of the validity for these procedures has not yet been accomplished.

Flow with cylindrical symmetry, see Section 16, is characterized by the condition that all pertinent quantities depend only on the abscissa x along the axis and the distance y from the axis, and that the flow vector at a point lies in the plane through this point and the axis. The axial and radial components of velocity are denoted by u and v. The differential equations for steady irrotational flow are then, see (16.06) and (16.13),

$$(151.01) \qquad u_y - v_x = 0,$$

$$(151.02) \qquad (y\rho u)_x + (y\rho v)_y = 0,$$

in which ρ is a given function of the speed $q = \sqrt{u^2 + v^2}$ by Bernoulli's law (14.02) and the adiabatic equation (7.04).

Instead of the *potential function* $\varphi(x, y)$ we prefer to use Stokes' *stream function* $\psi(x, y)$ defined so that

$$(151.03) \qquad \psi_x = -y\rho v, \qquad \psi_y = y\rho u,$$

see (16.15).

152. Supersonic flow along a slender body of revolution

When a slender body faces a supersonic stream the resulting flow can be approximately determined by an expansion with respect to the thickness of the body, just as the supersonic flow around a slender two-dimensional air-foil was approximately determined, Section 142. We describe the surface of the body by

$$(152.01) \qquad y = \epsilon Y(x), \ x \geq 0.$$

The function $Y(x)$ is supposed to be positive for $x > 0$, while $Y(0) = 0$. Ahead of the body, for $x < 0$, the flow velocity is constant and has axial direction; also pressure and density are constant there, and the Mach number is greater than one. We intend to determine the expansion of the solution of the flow problem with respect to powers of the thickness parameter ϵ by a procedure similar to that indicated in Section 142.

Assuming the flow to be cylindrically symmetric, we want to find a stream function $\psi(x,y)$ to satisfy the equations (151.03) and (151.01). The conditions for the state ahead of the body can then be formulated as

(152.02) $\quad \psi = \frac{1}{2}\rho_0 q_0 y^2, \quad \psi_x = 0, \quad x \leq 0.$

The speed q_0, the density ρ_0, and the pressure p_0 of the incoming stream are subjected to the condition

(152.03) $\quad M_0 = q_0/c_0 > 1,$

in which the sound speed c_0^2 is given by $c_0^2 = \gamma p_0/\rho_0$.

The condition that the surface of the body be a stream surface is

(152.04) $\quad \psi(x, \epsilon Y(x)) = 0, \quad x \geq 0.$

We expect a shock front to develop at the tip of the body if the angle of the conical tip lies below an appropriate limit, see Section

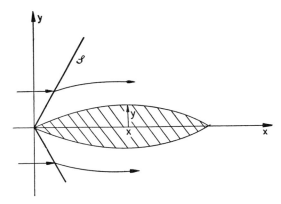

FIG. 1. Supersonic flow past a slender body of revolution.

154. In order to avoid the necessity of considering the shock transition relations, we first assume that the slope of the profile $y = \epsilon Y(x)$ vanishes at the tip, $Y'(0) = 0$. In that case the shock front begins at some distance from the surface of the body. The smaller the thickness parameter ϵ is, the further away the shock begins. We therefore disregard the initial conditions resulting from the shock and assume instead that the flow is isentropic and irrotational

throughout and that velocity, density, and pressure on the plane $x = 0$ through the tip have the same constant values as the incoming stream. The results obtained on the basis of these assumptions are then applied to cases in which $Y'(0) \neq 0$, a procedure which is correct in the lowest order in ϵ.

It is far from obvious whether or not an expansion with respect to integral powers of ϵ is possible. We assume that there is an expansion beginning with a term of order zero followed by a term proportional to ϵ^2. At least to a certain extent we shall motivate this assumption later on. We insert the expansions

$$(152.05) \quad \psi(x, y) = \psi^{(0)}(x, y) + \epsilon^2 \psi^{(1)}(x, y) + \cdots,$$

$$u = u^{(0)} + \epsilon^2 u^{(1)} + \cdots, \quad v = v^{(0)} + \epsilon^2 v^{(1)} + \cdots,$$

$$\rho = \rho^{(0)} + \epsilon^2 \rho^{(1)} + \cdots, \quad p = p^{(0)} + \epsilon^2 p^{(1)} + \cdots$$

in the differential equations (151.01–.03) and in the initial and the boundary conditions, (152.02) and (152.04).

Clearly, $\psi^{(0)}$ is

$$(152.06) \qquad \psi^{(0)}(x, y) = \tfrac{1}{2}\rho_0 q_0 y^2,$$

the stream function of the constant parallel flow ahead of the body and

$$(152.07) \quad u^{(0)} = q_0, \quad v^{(0)} = 0, \quad \rho^{(0)} = \rho_0, \quad p^{(0)} = p_0.$$

In order to obtain the terms of next order we first express $\rho^{(1)}$ by means of $u^{(1)}$ using the relation

$$(152.08) \qquad -c^2 \, d\rho/\rho = u \, du + v \, dv,$$

see (102.05), which follows from the assumed isentropic and irrotational character of the flow. Relation (152.05) then yields the relation

$$-c_0^2 \rho^{(1)}/\rho_0 = u^{(0)} u^{(1)} + v^{(0)} v^{(1)},$$

from which by (152.03) and (152.07)

$$(152.09) \qquad \rho^{(1)} = -\rho_0 M_0^2 u^{(1)}/q_0.$$

Inserting the expansions (152.05) into (151.03) we obtain
$$\psi_y^{(1)} = y(\rho_0 u^{(1)} + \rho^{(1)} u^{(0)}) = y\rho_0(1 - M_0^2)u^{(1)},$$
$$\psi_x^{(1)} = -y\rho_0 v^{(1)},$$

or with the abbreviation

(152.10) $$s_0^2 = M_0^2 - 1,$$

(152.11) $$u^{(1)} = -s_0^{-2} y^{-1} \rho_0^{-1} \psi_y^{(1)},$$
$$v^{(1)} = -y^{-1} \rho_0^{-1} \psi_x^{(1)}.$$

Equation (151.01) yields by (152.05)
$$u_y^{(1)} - v_x^{(1)} = 0,$$

which becomes by (152.11)

(152.12) $$y(y^{-1}\psi_y^{(1)})_y - S_0^2 \psi_{xx}^{(1)} = 0.$$

From (152.02) we derive the initial condition

(152.13) $$\psi^{(1)} = \psi_x^{(1)} = 0 \quad \text{for} \quad x = 0.$$

The process of expanding all quantities in powers of ϵ automatically determines the boundary condition for the function $\psi^{(1)}$. We assume that $\psi^{(1)}$ is defined for $y \geq 0$ as a continuous function with continuous derivatives up to $y = 0$. The success of this assumption is all that can be offered as its justification. Expanding the boundary condition (152.04) in powers of ϵ then yields
$$\psi^{(0)}(x, \epsilon Y(x)) + \epsilon^2 \psi^{(1)}(x, \epsilon Y(x)) + \cdots = 0.$$

By inserting (152.06) and expanding $\psi^{(1)}(x,\epsilon Y)$ in powers of ϵ, we obtain
$$\tfrac{1}{2}\epsilon^2 \rho_0 q_0 Y^2(x) + \epsilon^2 \psi^{(1)}(x, 0) + \cdots = 0,$$

from which

(152.14) $$\psi^{(1)}(x, 0) = -\tfrac{1}{2}\rho_0 q_0 Y^2(x).$$

This result may at the same time serve as a justification of the statement, anticipated until now, that the "first" order term in the expansion of $\psi(x, y)$ is proportional to ϵ^2.

Equation (152.12) is closely related to the *wave equation* for wave motion with cylindrical symmetry; here the abscissa x takes the place of the time. As a matter of fact, the potential function $\varphi^{(1)}$ corresponding to $\psi^{(1)}$ satisfies the wave equation, but the boundary condition for $\psi^{(1)}$ is simpler than that for $\varphi^{(1)}$. Equation (152.12) with the boundary condition (152.14) and the initial conditions (152.13) can be solved explicitly.

With the abbreviation

$$(152.15) \qquad k(x) = \tfrac{1}{2} Y^2(x) \quad x \geq 0,$$
$$ = 0 \quad x \leq 0,$$

the solution $\psi^{(1)}$ is

$$(152.16) \qquad \psi^{(1)}(x, y) = -\rho_0 q_0 \int_0^{x-s_0 y} \sqrt{(x - \xi)^2 - s_0^2 y^2} \; k''(\xi) \, d\xi.$$

This formula is used even if the slope of the profile does not vanish at the tip, $Y'(0) \neq 0$. We verify immediately that the differential equation (152.12) and the initial condition (152.13) for $x = 0$ are satisfied. To show that the boundary condition (152.14) is also satisfied we use the relations $k'(0) = k(0) = 0$ which follow from (152.15) and $Y(0) = 0$; then we derive

$$\psi^{(1)}(x, 0) = -\rho_0 q_0 \int_0^x (x - \xi) k''(\xi) \, d\xi$$
$$= -\rho_0 q_0 \int_0^x k'(\xi) \, d\xi = -\rho_0 q_0 k(x),$$

and thus (152.14) follows. The first order terms of the velocities are

$$(152.17) \qquad u^{(1)}(x, y) = -q_0 \int_0^{x-s_0 y} \frac{k''(\xi) \, d\xi}{\sqrt{(x - \xi)^2 - s_0^2 y^2}},$$

$$v^{(1)}(x, y) = y^{-1} q_0 \int_0^{x-s_0 y} (x - \xi) \frac{k''(\xi) \, d\xi}{\sqrt{(x - \xi)^2 - s_0^2 y^2}}.$$

They become infinite on the axis $y = 0$.

As an example, we consider the case of a conical body $Y(x) = \Theta x$, $x \geq 0$, Θ = constant. We find from (152.16)

$$\psi^{(1)}(x, y) = -\tfrac{1}{2}\rho_0 q_0 \Theta^2 \left[x \sqrt{x^2 - s_0^2 y^2} - s_0^2 y^2 \operatorname{arccosh} \frac{x}{s_0 y} \right]$$

(152.18)
$$\text{for } x \geq s_0 y,$$

$$\psi^{(1)}(x, y) = 0 \quad \text{for } x \leq s_0 y ;$$

thus $\psi^{(1)}$ is not regular at $y = 0$ since the expression in the bracket behaves like $x^2 + s_0^2 y^2 \log y$ at $y = 0$.

The solution $\psi^{(1)}(x, y)$ resulting from any profile function $Y(x)$ behaves somewhat like the function given by (152.18).

On the Mach cone through the tip, $x = s_0 y$, we evidently have

$$\psi^{(1)} = 0, \quad u^{(1)} = v^{(1)} = 0,$$

even if the slope $Y'(0)$ of the profile at the tip does not vanish. In other words the flow given by $\psi = \psi^{(0)} + \epsilon^2 \psi^{(1)}$ for $x \geq s_0 y$ is continuously connected with the flow $\psi = \psi^0$ for $x \leq s_0 y$. This fact indicates that *the strength of the shock* which begins at the tip *vanishes in first order*. This behavior is quite different from that of the solution of the corresponding two-dimensional problem, in which the shock strength in first order is proportional to the slope of the profile at the tip.

The theory presented here corresponds to a simplified version of the theory of v. Kármán and Moore, indicated by Ferrari, see [155–164] and [171–177].

In order to improve the approximation already obtained we could determine terms of higher order in the expansion. It is far from obvious which power of ϵ corresponds to the next term. Occasional references to such terms of next order are found in the literature, see Maccoll [157], Busemann [169], Chien [163], Lighthill [156].

Without employing terms of higher order we might try to obtain an improved approximate description of the flow by satisfying the boundary condition for $\psi^{(1)}$ not on the x-axis (which is consistent with the idea of expanding the problem in powers of the thickness parameter ϵ) but on the actual surface of the body. This procedure is the method of v. Kármán and Moore; it probably gives better results than the more systematic method we have outlined although it is not so easily worked out. v. Kármán and Moore proceed by replacing the body by one consisting of a head cone and a sequence of truncated cones.

For the limitations of the approximation method discussed here, see [164].

153. Resistance

In conclusion we calculate the resistance exerted by the pressure of the air against the body. We assume that as x tends to infinity the radius of the surface of the body approaches a finite value and the slope approaches zero:

(153.01) $\qquad Y(x) \to Y_\infty, \ Y'(x) \to 0 \quad \text{as} \quad x \to \infty.$

As is customary, we are interested only in the force F resulting from the excess pressure $p - p_0$; here p_0 is the atmospheric pressure. We have

$$(153.02) \qquad F = 2\pi \int_0^\infty (p - p_0) y \frac{dy}{dx} dx$$
$$= 2\pi \epsilon^2 \int_0^\infty (p - p_0) k'(x) \, dx;$$

the integration is to be performed over the surface of the body. From $p = p_0 \left(\dfrac{\rho}{\rho_0}\right)^\gamma$ and Bernouilli's law we obtain the approximation

(153.03) $\qquad p = p_0 - \epsilon^2 \rho_0 q_0 u^{(1)} - \tfrac{1}{2} \epsilon^4 \rho_0 (v^{(1)})^2 \cdots$

Now, the term of lowest order in the expansion $F = \epsilon^4 F^{(2)} + \cdots$ is

$$(153.04) \qquad F^{(2)} = -\rho_0 \lim_{\epsilon \to 0} 2\pi \int_0^\infty \left\{ q_0 \, u^{(1)}(x, \, \epsilon Y(x)) \right. $$
$$\left. + \frac{\epsilon^2}{2} (v^{(1)}(x, \, \epsilon Y(x)))^2 \right\} k'(x) \, dx.$$

From (152.17.1) and (153.03) we obtain

$$-q_0 u^{(1)}(x, y) = q_0^2 \left\{ k''(x - s_0 y) \operatorname{arccosh} \frac{x}{s_0 y} \right.$$
$$\left. + \int_0^{x-s_0 y} \frac{k''(\xi) - k''(x - s_0 y)}{\sqrt{(x - \xi)^2 - s_0^2 y^2}} d\xi \right\}$$
$$\sim q_0^2 \left\{ k''(x) \log \frac{2x}{s_0 y} - \int_0^x \frac{k''(x) - k''(\xi)}{x - \xi} d\xi \right\},$$

from which

(153.05) $$-q_0 u^{(1)}(x, \epsilon Y) \sim q_0^2 \left\{ k''(x) \log \frac{2}{\epsilon s_0} + k''(x) \log \frac{x}{Y(x)} \right.$$
$$\left. - \int_0^x \frac{k''(x) - k''(\xi)}{x - \xi} d\xi \right\}.$$

From (152.17.2) we find $\epsilon v^{(1)}(x, \epsilon Y) \sim Y^{-1}(x) q_0 k'(x)$.

The term of lowest order which results when we insert this formula and (153.05) in (153.04) is, except for a constant factor,

$$\log \frac{2}{\epsilon s_0} \int_0^\infty k''(x) k'(x) \, dx = \tfrac{1}{2} \log \frac{2}{\epsilon s_0} \, [k'(x)^2]_0^\infty = 0$$

by assumption (153.01). The remaining term is

(153.06) $$F^{(2)} = 2\pi \rho_0 \, q_0^2 \int_0^\infty k'(x) \left\{ k''(x) \, (\log x - \tfrac{1}{2} \log 2k(x)) \right.$$
$$\left. - \int_0^x \frac{k''(x) - k''(\xi)}{x - \xi} d\xi - \tfrac{1}{4} k^{-1}(x) \, (k'(x))^2 \right\} dx$$

The second term can be integrated by parts and cancels with the fourth term; the remaining terms can be combined and put in symmetrical form:

(153.07) $$F^{(2)} = -\pi \rho_0 q_0^2 - \int_0^\infty \int_0^\infty k''(x) k''(\xi) \log |x - \xi| \, dx \, d\xi,$$

as can be shown by calculations given below. Aside from the somewhat different interpretation of $k'(x)$, the present formula agrees with that given by v. Kármán [155]; see Lighthill [156] and compare also (157–164).

We supply the derivation of (153.07) from (153.06). Setting $k'(x) = f(x)$ we have

$$\int_0^\infty \int_0^\infty f'(x) f'(\xi) \log |x - \xi|\, dx\, d\xi$$

$$= 2 \int_0^\infty f'(x) \int_0^x f'(\xi) \log (x - \xi)\, d\xi\, dx$$

$$= 2 \int_0^\infty f'(x) f(x) \log x\, dx - 2 \int_0^\infty f'(x) \int_0^x \frac{f(\xi) - f(x)}{\xi - x}\, d\xi\, dx$$

$$= -2 \int_0^\infty f'(x) f(x) \log x\, dx + 2 \int_0^\infty f(x) \int_0^x \frac{f'(\xi) - f'(x)}{\xi - x}\, d\xi\, dx,$$

since

$$\frac{d}{dx} \int_0^x \frac{f(x) - f(\xi)}{x - \xi}\, d\xi = \left[\frac{f(x) - f(\xi)}{x - \xi}\right]_{\xi=x}$$

$$+ \int_0^x \frac{(x - \xi)f'(x) - f(x) + f(\xi)}{(x - \xi)^2}\, d\xi$$

$$= f'(x) + \left[\frac{(x - \xi)f'(x) - f(x) + f(\xi)}{x - \xi}\right]_0^x - \int_0^x \frac{-f'(x) + f'(\xi)}{x - \xi}\, d\xi$$

$$= \int_0^x \frac{f'(x) - f'(\xi)}{x - \xi}\, d\xi + \frac{f(x)}{x}.$$

B. Conical Flow

154. Qualitative description

Conical flow, the second type of flow treated in this chapter, permits a fairly explicit analysis on the basis of the differential equations. Conical flow is steady, isentropic, irrotational flow with cylindrical symmetry about the x-axis like the flow treated in Part A, but has in addition the property that the quantities u, ρ, p, q retain constant values on cones (considered infinite) with a common vertex, the origin. Flow satisfying this condition may occur, for instance, at the conical nose of a projectile facing a supersonic stream of air.*

* Flows in which ρ, p, and q are constant on rays through the center but which are not assumed to have cylindrical symmetry have been extensively treated by Busemann and others [178–185], by means of a perturbation scheme involving cones of small angle of opening.

Suppose the projectile is a body of revolution with a sharp tip, as considered in the preceding part, see Figure 1. The flow against such a projectile is analogous to the flow against a wedge and, as in the case of a wedge, see Section 123, two cases must be distinguished.

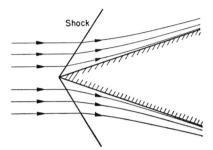

FIG. 2. Conical shock front and conical flow resulting from supersonic flow against a sufficiently small angle.

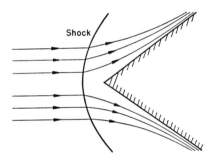

FIG. 3. Curved shock front in supersonic flow against a cone with a large angle.

If the angle of opening at the tip of the projectile is not too large, the deflection of the flow is achieved by an "attached" shock·front, which begins at the tip, see Figure 2. If the nose of the projectile consists of a section of a cone, the shock front and the flow behind it are conical until the expansion wave coming from the bend of the projectile interacts with the conical flow and the shock front and modifies their character.

On the other hand, if the angle of opening at the tip exceeds a certain extreme value, no attached shock front is possible. Instead, a "detached" shock front stands ahead of the projectile, see Figure 3. For projectiles with the same angle of opening the distance of this detached shock front from the tip of the projectile increases with the thickness of the body of the projectile and would move out to infinity as the thickness of the body increased indefinitely.

In particular, for the limiting case of an infinite cone with a large opening angle, no detached shock wave is possible. An attached shock front is, however, possible if the angle of opening is small enough. This shock front is then also an infinite cone. Since it makes everywhere the same angle with the incoming flow, its strength is constant and hence the state immediately behind the shock front is also constant. Therefore the flow is everywhere isentropic behind the shock front. The flow can be continued so that the state of the air is constant on each concentric cone between the shock cone and the obstacle cone. The angle between any such cone and the flow direction approaches zero when this cone approaches the obstacle cone. For the general case in which a shock is attached to the tip, the flow can be obtained mathematically*; we shall confine ourselves to this case.

155. The differential equations

For the mathematical construction of such flow patterns,** let x be the abscissa along the axis, y the distance from the axis, u and v the components of the flow velocity $\bar{q}$ in the direction of the axis and in the direction perpendicularly away from the axis respectively. The differential equations for isentropic flow, as derived in Section 16, see (16.13–.14), are then

(155.01) $$v_x = u_y ,$$

(155.02) $$\left(1 - \frac{u^2}{c^2}\right) u_x - 2 \frac{uv}{c^2} v_x + \left(1 - \frac{v^2}{c^2}\right) v_y + \frac{v}{y} = 0,$$

* For the other case, see [168].
** See the fundamental papers by Busemann, Taylor and Maccoll, and Bourquard [165–169].

THE DIFFERENTIAL EQUATIONS OF CONICAL FLOW

in which c can be expressed in terms of u and v by Bernoulli's law

(155.03) $$\mu^2(u^2 + v^2) + (1 - \mu^2)c^2 = c_*^2.$$

The basic assumption of conical flow now implies that u, v, and hence c depend only on the ratio

(155.04) $$\sigma = \frac{x}{y}.$$

Equation (155.01) then becomes

(155.05) $$v_\sigma + \sigma u_\sigma = 0,$$

while (155.02) is reduced to

(155.06) $$\left(1 - \frac{u^2}{c^2}\right)u_\sigma - 2\frac{uv}{c^2}v_\sigma - \left(1 - \frac{v^2}{c^2}\right)\sigma v_\sigma + v = 0.$$

Equations (155.05) and (155.06) are a pair of differential equations of first order for two functions u and v of σ. Clearly this pair is equivalen to one equation of second order for one function. This equation of second order assumes a form which is particularly amenable to treatment when v is introduced as a function of u. From (155.05) we have

(155.07) $$\sigma = -v_u.$$

Differentiating this relation with respect to σ yields

(155.08) $$u_\sigma = -\frac{1}{v_{uu}}.$$

This relation together with (155.07) and (155.05) leads to

(155.09) $$v_\sigma = -\frac{v_u}{v_{uu}}.$$

Inserting equations (155.07–.09) into (155.06) gives

$$\left(1 - \frac{u^2}{c^2}\right) + \left(1 - \frac{v^2}{c^2}\right)v_u^2 - vv_{uu} - 2\frac{uv}{c^2}v_u = 0$$

or

(155.10) $$vv_{uu} = 1 + v_u^2 - \frac{(u + vv_u)^2}{c^2}.$$

An elegant geometric interpretation of equation (155.10) by Busemann may be mentioned. This equation may be written in the form

$$(155.11) \qquad R = \frac{N}{1 - \dfrac{U^2}{c^2}},$$

in which R is the radius of curvature of an integral curve in the (u, v)-plane at the point $P = (u, v)$, while the meaning of N and U is obvious from Figure 4.

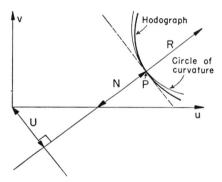

Fig. 4. Concerning the geometrical interpretation of equation (155.11).

Every section of a solution $v = v(u)$ of equation (155.10) represents a flow provided that the condition

$$(155.12) \qquad v_{uu} \neq 0$$

is satisfied, for x and y can then be introduced as independent variables by $v_u = -\sigma = -x/y$. Thus the ray through the origin in the (x, y)-plane to which the values of u and v are to be attached is determined. This ray in the (x, y)-plane is evidently parallel to the normal to the curve $v = v(u)$ at the point (u, v) in the hodograph plane.

The flows so obtained are in a certain way analogous to centered simple waves for two-dimensional flows. However, while in the case of two-dimensional steady flow the simple waves are represented in the hodograph plane by two families of fixed characteristics (epicycloids), the images in the hodograph plane of the conical flows considered here correspond to a greater variety of curves, namely, a whole one-parametric family through each point.

For the discussion in the following sections it is necessary to formulate the condition that the fluid streams along a given cone. This condition is evidently that the flow has the direction of the ray $x/y = \sigma$ traced by the cone in the (x, y)-plane,

(155.13) $$u/v = \sigma.$$

By (155.07) this condition becomes

(155.14) $$v_u = -u/v.$$

A point on an integral curve at which condition (155.14) is satisfied may be called an "end point." The normal to the integral curve at an end point evidently passes through the origin.

156. Conical shocks

The relations governing transitions through *conical shocks* are the same as those for plane oblique shocks; the curvature of the shock cone does not enter. If the shock cone is straight, as is assumed, the jumps of u, v, p, and of the entropy are constant along each ray under the assumption of conical flow on one side; as a consequence this assumption remains satisfied on the other side. The flow may continue as a conical flow with constant entropy after crossing the shock. In other words, proper conical shocks are compatible with the assumption of conical flow.

Suppose a flow characterized by p_0, ρ_0, u_0, v_0 crosses such a conical shock. Note that this is possible only if the speed $q_0 = \sqrt{u_0^2 + v_0^2}$ is supersonic, i.e., if $q_0 > c_0$. The velocity $q_1 = (u_1, v_1)$ immediately past the shock front is located on the loop of the shock polar in the (u, v)-plane, see Section 121. The inclination of the ray which generates the shock cone is perpendicular to the straight line joining (u_0, v_0) and (u_1, v_1). The positions of the cones corresponding to the cases (a): $v_1 > v_0$ and (b): $v_1 < v_0$ are indicated in Figure 5.

When the flow on either the front or the back side of the shock is to be continued according to the differential equation (155.10),

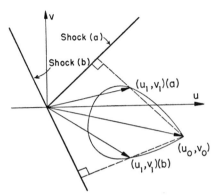

Fig. 5. Indicating a transition through a conical shock front for cases (a): $v_1 > v_0$ and (b): $v_1 < v_0$.

the slope of the integral curve is to be so determined that the ray given by (155.13) coincides with the trace of the flow behind the shock front in the (u, v)-plane. Since this ray is to be normal to the integral curve on the one hand and perpendicular to the straight segment connecting (u_0, v_0) with (u_1, v_1) on the other hand, the integral curve should begin at the point (u_1, v_1) in the direction of this segment. The initial slope of this curve is thus given by

$$(156.01) \qquad v_u = \frac{v_1 - v_0}{u_1 - u_0}.$$

The discussion of conical shock fronts by Taylor and Maccoll [165, 167] is restricted to the case in which $u_1 > v_0$ (case (a)), and $u_0 = q_0 > 0$ and $v_0 = 0$. This case, see Figure 6, occurs when a constant axial flow is deflected by a conical projectile. We shall indicate briefly how Busemann treats this problem. Through the shock relations the flow velocity (u_1, v_1) past the shock cone is given (the third transition relation guarantees that the Bernoulli constant, and hence c_*, are the same before and after the shock). A solution of equation (155.10) is to be found whose graph passes through the point (u_1, v_1) on the shock polar. The initial slope v_u of this curve at this point is given by (156.01). The solution is now to be so continued that $\sigma = x/y$ increases and hence, according to (155.07), that v_u decreases up to a point at which the flow and the ray have the same

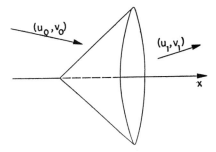

FIG. 6. Flow through a conical shock front with $v_1 > v_0$.

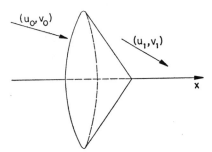

FIG. 7. Flow through a conical shock front with $v_1 < v_0$.

direction. Such an "end point" is characterized by condition (155.14). This end point depends on the choice of the point (u_1, v_1) on the shock polar associated with the double point $(q_0, 0)$. The manifold of end points that can be reached from $(q_0, 0)$ forms a curve which Busemann calls the "apple curve" because of its peculiar shape, see Figure 8. In this procedure the shock is prescribed and the end direction is found. If the end direction is prescribed one may find the corresponding point on the apple curve by intersecting is with the appropriate ray through the origin. In general, there are two intersections of which the one corresponding to the weaker shock is presumably more likely to occur in reality.

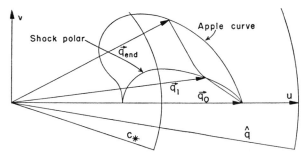

FIG. 8. Apple curve showing all possible end velocities that can be reached after crossing a conical shock front from a state with a given incoming velocity.

The values of pressures and angles calculated on the basis of the preceding theory agree exceedingly well with experimental values, see Taylor and Maccoll [167, 168]. An extensive tabulation of all conical flows was given by Kopal [170].

157. Other problems involving conical flow

Another problem that can be solved by the methods of conical flow is the following: A duct bounded by a surface of revolution is to be designed so that a supersonic flow entering it in the axial direction with a given Mach number is first continuously compressed and decelerated, and then instantaneously further compressed and decelerated through a shock front, beyond which it is again to have axial direction. To design such a *diffuser* one need only determine a flow with the described properties and then choose any "stream surface" as the wall of the duct.

OTHER PROBLEMS INVOLVING CONICAL FLOW 415

As Busemann [169] has observed, a flow with the desired properties can be constructed as a conical flow; then the shock front is also a cone, see Figure 9. To find such a conical flow it is convenient to begin with the state behind the shock cone represented by a point $(u_2, 0)$ on the u-axis. Then we draw the backward branch

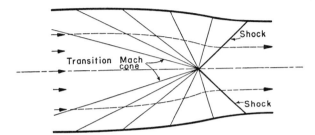

FIG. 9. Cylindrically symmetric diffuser flow.

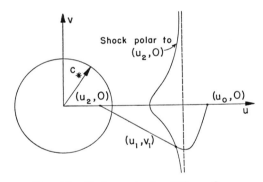

FIG. 10. Hodograph of a diffuser flow.

of the shock polar in the (u, v)-plane whose points represent all possible front states for which $(q_0, 0)$ represents the backward state. In case $u_2 < c_*$, the branch is as indicated in Figure 10. On this branch we choose a point (u_1, v_1) with $v_1 < 0$ and pass through this point the integral curve of equation (155.06) whose slope at (u_1, v_1) equals $v_1/(u_1 - u_2)$. The point $(u_0, 0)$ at which this integral curve intersects the u-axis then represents the entrance flow. In order that this construction should represent a flow it is necessary that $\sigma = x/y$ increase with decreasing u or, by (155.08), that v_{uu} be positive on our integral curve.

It is worth noticing that equations (155.10) and (155.03), after differentiation, imply $v_u v_{uu} > 0$ at a point $(u_0, 0)$ so that $v_{uu} > 0$ or < 0 according as $v_u > 0$ or < 0 there. Otherwise, the condition $v_{uu} > 0$ for the construction just described could never be satisfied. At the same time it follows from this fact that *no purely conical shockless diffuser flow is possible*. For, such a flow would be represented by an integral curve starting at a point $(u_0, 0)$ with $v_u > 0$ and entering a point $(u_1, 0)$ with $u_1 < u_0$ and $v_u < 0$; the condition $v_{uu} > 0$ would not be satisfied at the latter point.

Another problem which at first sight seems to involve conical flow is that of *conical shock reflection*. A constant parallel flow is deflected toward the axis through an "incident" conical shock front and made parallel again after having crossed a "reflected" conical shock front. Naturally, one would construct such a flow assuming that it is purely conical between the two shock fronts and past the reflected front. Flow of this type however, *does not exist* as can be shown by arguments involving the sign of v_{uu}. The conjecture has been advanced that observed conical shock reflections are in reality Mach reflections, see Section 148, with Mach discs too small to be observed. (To a certain extent this is indicated by Ferri's calculations, see Section 150 and [176].)

C. Spherical Waves

158. General remarks

The study of spherically symmetric motion* is important for the theory of explosion waves in water, air, and other media. In *spherical motion* the velocity is radial, and its magnitude as well as that of density, pressure, temperature, and entropy depends only on the

---

*With slight modifications the following considerations apply also to *cylindrical waves*.

distance r from the origin and on the time t. Such motion might be considered somewhat analogous to one-dimensional motion in a tube under the influence of a piston. In the three-dimensional space the piston is replaced by an expanding (or contracting) sphere which impresses a motion on the medium inside or outside.

The simplest model is that of a "spherical piston" pushing at constant velocity into an infinite surrounding medium, as we shall consider in Section 162. Such a flow pattern corresponds to the uniform "piston motion" in one dimension as studied in Chapter III, in particular in Section 53. One should bear in mind, however, that in three-dimensional space an energy supply at an increasing rate is required to maintain constant speed of the piston.

In better agreement with actual situations is the assumption that the total energy available for the motion is given. This is the case for *spherical blast waves* caused by the explosion of a given mass of explosive.

While in the first of these two models the shock wave racing ahead of the piston has constant speed so that the shock conditions are compatible with the assumption of isentropic flow on both sides of the discontinuity surface, this is no longer true of blast waves. In the latter the strength of the shock, and hence the change of entropy, rapidly decreases so that behind the shock front the flow is *no longer isentropic*. Moreover, after the air or water in a blast wave has crossed the shock front and thereby been compressed, it expands again rapidly to a pressure in general even below that in front of the shock wave. This *suction phase* is an important feature of motion caused by explosions.

A phenomenon of major importance is that of *reflection of spherical shock fronts;* a contracting spherical wave preceded by a shock front may be "reflected" at the center and result in an enormous increase in pressure behind the reflected shock front.

At the present state of knowledge all that can be done along the lines of mathematical analysis is to find and discuss some particular solutions of the differential equations of spherical waves which are approximately in agreement with the additional conditions of the problems. One may hope that these solutions display, at least qualitatively, important features of reality. It is remarkable that such an unambitious approach seems sufficient for a certain degree of understanding and control of actual phenomena.

159. Analytical formulations

The radial velocity u, the pressure p, and the density ρ, as functions of r and t, satisfy the differential equations, see (17.07–.09),

(159.01)
$$u_t + uu_r + \frac{1}{\rho} p_r = 0,$$
$$\rho_t + u\rho_r + \rho\left(u_r + \frac{2u}{r}\right) = 0,$$
$$(p\rho^{-\gamma})_t + u(p\rho^{-\gamma})_r = 0,$$

assuming that the medium is polytropic with the adiabatic exponent γ. The third equation expresses the fact that the entropy is constant along the path of a particle. It cannot be stipulated that the entropy is constant throughout since, as stated before, the entropy does not in general remain constant behind a shock front. If in an outward-going wave the position of the head of the wave is given as a function of the time,

(159.02)
$$r = R(t),$$

the *total energy* carried by the wave motion is expressed as

(159.03)
$$E = 4\pi \int_0^R \left\{\tfrac{1}{2}\rho u^2 + \frac{1}{\gamma - 1} p\right\} r^2 \, dr.$$

E is a function of the time t.

Another important quantity, the *impulse* I per unit area received by a section of the surface of the sphere at distance r, is given by

(159.04)
$$I = \int_T^\infty p \, dt$$

where $T = T(r)$ is the time at which the wave front arrives at the place r. $T(r)$ is connected with $R(t)$ through $r = R(T(r))$. I is a function of r.

160. Progressing waves

According to classical procedure, one may obtain particular solutions of the differential equations by assuming a specific form for the solution in order to reduce the problem to one involving ordinary differential equations. In this way solutions are obtained which have been called *progressing waves*.* These are solutions conveniently assumed to be of the form

$$(160.01) \quad u = \alpha r t^{-1} U(\eta), \quad c = \alpha r t^{-1} C(\eta),$$
$$\rho = r^{\kappa} \Omega(\eta), \quad p = \alpha^2 r^{\kappa+2} t^{-2} P(\eta),$$

in which η is the combination

$$(160.02) \quad \eta = r^{-\lambda} t,$$

while

$$(160.03) \quad \lambda = \alpha^{-1}$$

is an appropriate positive constant and

$$(160.04) \quad \gamma P / \Omega = C^2.$$

In other words, a progressing wave is a special gas flow for which the quantities $u r^{-1} t$, $c r^{-1} t$, $p r^{-\kappa}$ appear constant for an observer who moves so that $\eta = r^{-\lambda} t$ is constant. The factor $\alpha r t^{-1}$ of U and C in (160.01) is the velocity of such an observer counted positive for $t > 0$, negative for $t < 0$. Consequently, the condition that the gas moves with supersonic or subsonic speed relative to such an observer is simply expressed by $|1 - U| > |C|$ or $|1 - U| < |C|$ respectively.

Inserting equations (160.01) and (160.02) into equations (159.01) and eliminating Ω by (160.04), we obtain by straightforward computation the equations

$$(160.05) \quad \begin{aligned} D\eta U_\eta &= A(U, C), \\ D\eta C_\eta &= B(U, C)C, \\ D\eta P_\eta &= E(U, C)P, \end{aligned}$$

* Concerning a more general concept of "progressing wave" see, for instance, Courant-Hilbert [32, p. 448].

with

$$D = (1 - U)^2 - C^2,$$
$$A = (1 - U)(1 - \alpha U)U - [3\alpha U - \gamma^{-1}(2 - 2\alpha - \kappa\alpha)]C^2,$$
$$B = (1 - U)(1 - \alpha U) - (\gamma - 1)[(1 - \alpha U) - \tfrac{3}{2}(1 - \alpha)]U$$

(160.06)
$$- \left[\alpha + \gamma^{-1}\left(1 - \alpha + \frac{\gamma - 1}{2}\kappa\alpha\right)(1 - U)^{-1}\right]C^2,$$
$$E = (1 - U)[(2 + 2\gamma + \kappa)(1 - \alpha U) - (3\gamma + \kappa)]$$
$$+ \gamma(1 - \alpha U) - (2 + \kappa)\alpha C^2.$$

From the first two equations of (160.05) we can express the ratio dC/dU in terms of C and U and thus obtain a differential equation of the form

(160.07)
$$\frac{dC}{dU} = \frac{B(U, C)}{A(U, C)} C.$$

Once this equation is solved, the quantities η and P can be determined from the first and third equation of (160.05) by quadratures. Thus by introducing appropriate quantities we can reduce the problem to solving an ordinary differential equation and two quadratures. This remarkable fact was observed by Guderley [193] for $\kappa = 0$.

We shall see that solutions of the differential equations (160.05) may represent various types of flow. A solution may represent a flow of gas moving toward the center until it is stopped or modified by a reflected wave which issues from the center at the time $t = 0$, or it may represent just a wave coming from the center. If one is interested only in such an outward moving wave one may find it more convenient to employ instead of η the variable

(160.08)
$$\xi = rt^{-\alpha} = \eta^{-\alpha} \text{ for } t > 0,$$

setting $\alpha = \lambda^{-1}$ by (160.03).

Suppose the head of such an outward-going wave is given by

(160.09)
$$r = R(t) = \Xi t^\alpha,$$

see (159.02), or

(160.10)
$$\xi = \Xi = H^{-\alpha},$$

Ξ and H being constants. The *energy* contained in the wave at the time t is, by (159.03),

(160.11) $\quad E = 4\pi\alpha^2 t^{(5+\kappa)\alpha-2} \int_0^{\Xi} \left\{ \tfrac{1}{2} U^2(\eta)\Omega(\eta) + \dfrac{1}{\gamma-1} P(\eta) \right\} \xi^{\kappa+4}\, d\xi.$

The *impulse* per unit area transmitted by the wave motion to a sphere at distance r given by (159.04) is more conveniently expressed in terms of η:

(160.12) $\quad\quad\quad I = \alpha^2 r^{\kappa-\lambda+2} \int_H^{\infty} P(\eta)\eta^{-2}\, d\eta.$

161. Special types of progressing waves

We shall discuss the simplifications resulting from several special assumptions.

If the flow is isentropic, $p\rho^{-\gamma}$ is constant; relations (160.01–02) require

(161.01) $\quad\quad\quad\quad \kappa = 2\dfrac{1-\lambda}{\gamma-1}$

and

(161.02) $\quad\quad\quad\quad \eta^{-2} P(\eta)\Omega^{-\gamma}(\eta) = \text{constant}.$

If in particular $\alpha = \lambda = 1$, $\kappa = 0$, all the quantities u, c, p, ρ are constant along rays $r/t = $ constant. For an outward-moving wave, the head of the wave, $r = \Xi t$, moves with constant velocity Ξ, and ρ and p are constant behind it. Such wave motion is therefore compatible with a constant shock front ahead of it. We shall discuss in Section 162 various isentropic wave motions with $\lambda = 1$, $\kappa = 0$.

It is natural to ask under what conditions a shock front ahead of a progressing wave can move so that $\xi = rt^{-\alpha} = \Xi = $ constant, see (160.08). Relations (160.01) would then imply that the ratio u/c is constant on the back side of the shock front. This condition is compatible with the Rankine-Hugoniot conditions (54.08–.11) only if the strength of the shock remains constant. This is the case exactly only for $\alpha = 1$ and $\kappa = 0$.

Approximately, however, *strong shocks* are compatible with progressing waves provided that $\kappa = 0$. Denoting by ρ_0 the density ahead of the wave, we obtain relations approximately valid for strong shocks by setting the pressure ahead of the wave equal to zero; the shock transition conditions then reduce to

(161.03) $$\rho = \mu^{-2}\rho_0, \qquad p = (1 - \mu^2)\dot{R}^2\rho_0,$$
$$c = \mu\sqrt{1 + \mu^2}\,\dot{R}, \qquad u = (1 - \mu^2)\dot{R},$$

see (69.02–.04). We assume that the shock front is moving outward and is given by $r = R(t) = \Xi t^\alpha = (t/\mathrm{H})^\alpha$ with constant $\Xi = \mathrm{H}^{-\alpha}$. It is then evident that relations (161.03) are compatible with (160.01), (160.02), and (159.02) only if

(161.04) $$\kappa = 0,$$

i.e., if the density remains constant on the paths $t = \eta r^\lambda$, or $r = \xi t^\alpha$, $\eta = \xi^{-1} = $ constant, see (160.01–.03). For the state behind the shock front we then obtain from (160.01) the values

(161.05) $$U(\mathrm{H}) = (1 - \mu^2), \quad C(\mathrm{H}) = \mu\sqrt{1 + \mu^2},$$
$$\Omega(\mathrm{H}) = \mu^{-2}\rho_0, \qquad P(\mathrm{H}) = (1 - \mu^2)\rho_0.$$

For a contracting shock $r = (t/\mathrm{H})^\alpha$, $\mathrm{H} < 0$, $t < 0$, we find $C = -\mu\sqrt{1 + \mu^2}$, while U, Ω, P are the same as in (161.05).

The condition that the *energy* carried by the wave *remains constant* leads by (160.11) to the condition

(161.06) $$\nu = 2\lambda - 5.$$

If, in addition, the wave is to possess a strong shock at its head, so that $\kappa = 0$, we have

(161.07) $$\lambda = \tfrac{5}{2}, \quad \alpha = \tfrac{2}{5}.$$

The motion of the shock front is then given by

(161.08) $$r = R(t) = \Xi t^{2/5}.$$

The solution of the problem for $\kappa = 0$, $\lambda = 5/2$ was first formulated and determined numerically by G. I. Taylor [190] and [192] in an attempt to describe *blast waves* resulting from an explosion at the center. Obviously, the energy carried by the blast wave equals the energy supplied by the explosive and thus remains constant. Objections have, however, been raised to using Taylor's solution for the description of blast waves. First, one notes from (161.03), (161.05),

SPECIAL TYPES OF PROGRESSING WAVES 423

and (161.08) that the pressure behind the shock front,

(161.09) $\quad p = \frac{4}{25}(1 - \mu^2)\rho_0 \Xi^2 t^{-6/5} = \frac{4}{25}(1 - \mu^2)\rho_0 \Xi^5 R^{-3}$,

approaches zero as $t \to \infty$. Consequently the assumption that the shock is strong will eventually be violated. Furthermore, the very special property of Taylor's wave that, for example, $pt^{6/5}$ is constant on $rt^{-2/5}$ = constant does not agree with the actual pressure distribution near the center of the explosion. Nevertheless, it seems reasonable to assume that Taylor's solution gives a good description of a blast wave at distances from the center which are large enough to ensure that the details of the explosion process no longer influence the wave shape and which at the same time are small enough to ensure that the shock is still strong.

To construct Taylor's progressing wave one first finds the solution of the differential equation (160.07) with $\lambda = 5/2$, $\kappa = 0$, with the initial values $U = (1 - \mu^2)$, $C = \mu\sqrt{1 + \mu^2}$ for $\eta = $ H, as given by (161.05). This solution is followed in the direction of increasing η up to $\eta = \infty$. If $\gamma < 7$, it can be shown that U approaches a finite limit value while C becomes infinite like a power η^β of η with an exponent β greater than α. Consequently, when r approaches zero while t remains fixed, $c = t^{\alpha-1}\eta^{-\alpha}C$ also approaches infinity while $u = t^{\alpha-1}\eta^{-\alpha}U$ approaches zero. At the origin, therefore, the gas is at rest while the sound speed and, consequently, also temperature, density, and pressure are infinite.

A certain solution of equation (160.07) can in every case be given explicitly, namely the "trivial" solution $U = $ constant, $C = $ constant. These two constants are solutions of the equations

(161.10) $\quad\quad\quad A(U, C) = B(U, C) = 0$.

It is interesting to ask under what circumstances this solution is compatible with a strong shock. Inserting (161.05), $U = (1 - \mu^2)$, $1 - U = \mu^2$, $C^2 = \mu^2(1 + \mu^2) = \gamma\mu^2(1 - \mu^2)$, and $\kappa = 0$ into (161.10), we find after some computation the condition

$$\mu^2 = \tfrac{3}{4} \text{ or } \gamma = 7$$

for μ or γ and $\lambda = 5/2$. It is remarkable that this value of λ is just the one that corresponds to the condition of *constant energy* while the value 7 for γ can to a good approximation be used for *water*, see Section 3. Thus *progressing blast waves of Taylor's type in water can be given explicitly:*

The motion of the shock front ahead of such a blast wave is

$$r = \Xi t^{2/5}$$

with an appropriate constant Ξ; velocity and sound speed in the wave are

$$u = \tfrac{1}{10} t^{-1} r, \qquad c = \frac{\sqrt{21}}{10} t^{-1} r, \qquad 0 \leq r \leq \Xi t^{2/5};$$

the pressure distribution is found from the third equation of (160.05), (160.08), and (161.05),

$$p = \tfrac{1}{25} \rho_0 r^2 t^{-2} \xi / \Xi = \tfrac{1}{25} \rho_0 r^3 t^{-12/5} / \Xi, \qquad 0 \leq r \leq t^{2/5}.$$

This explicit solution was discovered by H. Primakoff.

162. Spherical quasi-simple waves

As was suggested earlier, we now discuss in detail the spherical flows in which velocity, sound speed, pressure, and density are constant on rays r/t = constant. Such flows would be triple waves in the terminology introduced in Section 32. We prefer to call them quasi-simple because they have many features in common with simple waves. These waves are special cases of the progressing waves considered in Section 160; they correspond to $\lambda = \alpha = 1$, $\kappa = 0$. The differential equations (160.05) reduce to

$$(162.01) \quad \begin{aligned} D\eta U_\eta &= \{(1 - U)^2 - 3C^2\} U, \\ D\eta C_\eta &= \{(1 - U)^2 - (\gamma - 1)U(1 - U) - C^2\} C, \\ D\eta P_\eta &= 2\{(1 - U)^2 - \gamma U(1 - U) - C^2\} P, \end{aligned}$$

with

$$(162.02) \qquad D = (1 - U)^2 - C^2.$$

We recall, see (160.08) and (160.01–.02), that $\eta = tr^{-1} = \xi^{-1}$ and

$$(162.03) \quad u = \eta^{-1} U(\eta), \quad c = \eta^{-1} C(\eta), \quad p = \eta^{-2} P(\eta), \quad \rho = \Omega(\eta).$$

Solutions of these differential equations can be obtained by first finding C as function of U from the differential equation

(162.04) $$\frac{dC}{dU} = \frac{(1-U)^2 - (\gamma-1)U(1-U) - C^2}{(1-U)^2 - 3C^2} \frac{C}{U}$$

and then determining η and P as functions of U from the two remaining equations (162.01). The behavior of the solutions of equation (162.04) can be read off from the vector field in the (U, C)-plane representing this equation. This vector field, shown in Figure 11, has singularities at various points, D_+, D_-, C_+, C_-, A, B, as illustrated. The region $C > 0$ corresponds to $\eta > 0$, i.e. to $t > 0$, while the region $C < 0$ corresponds to $\eta < 0$, $t < 0$. The arrow indicated in the figure gives the direction of increasing η, i.e., of increasing time t for a fixed value of r. We note that this direction is reversed on crossing the "critical" lines $D = 0$ or $|1 - U| = |C|$. No integral curve crossing these lines can therefore represent a real flow since the time must be increasing. Thus such a flow would lead to a "limiting" surface moving in space.

We first investigate the wave motion produced by a sphere which expands with constant speed ξ_0. At the surface of the expanding sphere, the velocity of the gas equals that of the surface, $\xi = \xi_0$; hence $u = \xi = \xi_0$ or, by (162.03) $U = 1$ there. We assume that the wave is preceded by a shock front moving with the constant speed Ξ, the value of which is to be determined. Gas and sound velocity behind the shock front are denoted by u_S and c_S respectively. As transition relations across the shock front we use the relations

(162.05) $$\mu^2(u_S - \Xi)^2 + (1 - \mu^2)c_S^2 = -\Xi(u_S - \Xi)$$
$$= \mu^2\Xi^2 + (1 - \mu^2)c_0^2,$$

which result from Prandtl's relation and the expression for c_*^2 in terms of the shock velocity Ξ and the sound speed c_0 ahead of the shock front, see (119.06) and (119.01).

Dividing equations (162.05) by Ξ^2 we obtain the formulas

(162.06) $$(1 - \mu^2)C_S^2 = (1 - U_S) - \mu^2(1 - U_S)^2$$
$$= (1 - \mu^2)(1 - U_S)^2 + U_S(1 - U_S),$$

(162.07) $$U_S = (1 - \mu^2)\left(1 - \frac{c_0^2}{\Xi^2}\right),$$

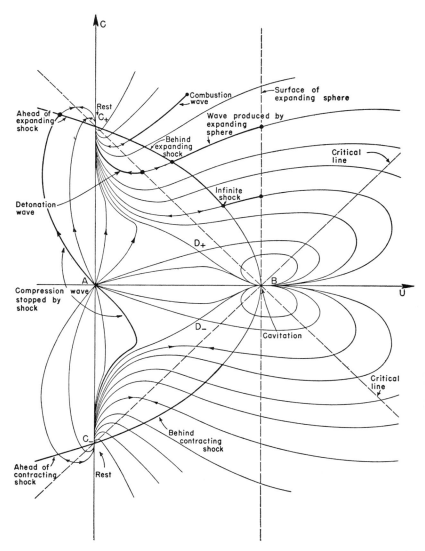

Fig. 11. Vector field representing the differential equation for $U = rt^{-1}u$ and $C = rt^{-1}c$ in spherical quasi-simple waves.

which establish relations between U_S, C_S and c_0/Ξ. The points (U_S, C_S) which satisfy (162.06) lie on an ellipse in the (U, C)-plane. Relation (162.07) imposes the condition $U_S < (1 - \mu^2)$ so that only a section of the "shock ellipse" represents states behind a shock front. For the state behind an expanding, outward-facing shock front the relations $u_S < \Xi$ and $c_S > \Xi - u_S$ hold, the former expressing that the gas crosses the front from the right to the left, the latter expressing that the flow behind the shock front is subsonic relative to it. These two conditions give

(162.08) $$0 < 1 - U_S < C_S$$

for an expanding forward-facing shock front. Both relations are satisfied on the section $0 < U_S < (1 - \mu^2)$ of the shock ellipse as seen from (162.06).

Suppose now that the speed Ξ of the leading shock front is given; then the state behind the front determines a point $U = U_S, C = C_S$ on the admitted section of the shock ellipse in the (U, C)-plane. Through this point an integral curve of equation (162.04) passes. We should follow this curve in the direction of decreasing ξ or increasing η. As seen from Figure 11, the integral curve continued in that direction ends somewhere at the line $U = 1$ without crossing the critical lines $|1 - U| = |C|$. The value ξ_0 of ξ at this end point can now be determined by integrating the first equation (162.01) from $U = U_S$ to $U = 1$. Thus ξ_0 is a given function of Ξ. It turns out that this function is monotone and that, accordingly, the shock speed Ξ can be determined as function of the prescribed velocity ξ_0 of the surface of the sphere. The solution of the problem, first worked out by G. I. Taylor, is thus established.

The dependence on the shock velocity of the velocity of the gas both at the surface of the sphere and also behind the shock front is shown in Figure 12, see also [195]. Note that these quantities would be equal for the corresponding one-dimensional problem of gas flow produced by the motion of a piston with constant speed.

An interesting fact is that the gas particles, which on passing across the shock acquire a positive velocity $u_s = \Xi U_s$, are further accelerated during their motion and attain asymptotically the velocity ξ_0 of the expanding sphere; at the same time the sound speed $c = \xi C$ *increases*, which indicates that the gas suffers a further *compression* in its motion behind the shock front.

It may be mentioned that Taylor has compared the flow calculated by the theory just discussed with the flow calculated on the basis of sonic approximation as treated in the theory of sound in case the speed of the sphere is $0.2c_0$. He found the agreement very good except for the distribution of the pressure near the surface of the sphere, for which the theory of sound gave too high a value.

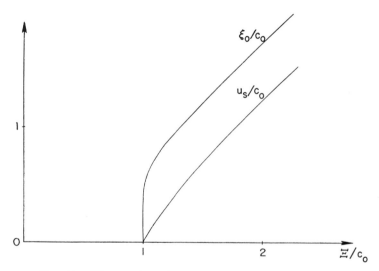

FIG. 12. The speed ξ_0 of the expanding sphere as a function of the speed Ξ of the initiating shock front. The flow speed u_s behind the shock front as a function of Ξ is added for comparison; it would represent the speed of a piston producing a shock of the same strength in one-dimensional motion.

It is not obvious beforehand that every flow of the quasi-simple wave type resulting from the expansion of a sphere is preceded by a shock front. Nevertheless this is the case as follows from the fact that every integral curve which begins at the line $U = 1$ with $C > 0$ enters the point $U = 0$, $C = 1$ with increasing values of C and $dU/dC = 0$ which can be shown by a discussion of the singularity of the vector field at this point. Consequently these curves intersect the shock ellipse before they intersect the critical line $C = 1 - U$ at which the direction of ξ would be reversed.

163. Spherical detonation and deflagration waves

Other interesting quasi-simple waves, as solutions of the differential equation (162.04), are *spherical detonation waves*. Let us recall the situation for one-dimensional detonation waves. It was seen in Chapter III, Part E, that the process of detonation, considered as a discontinuity like a shock, is uniquely determined by the nature of the burnt gas and by the energy available for release; the process is independent of the conditions behind it provided that we adopt the Chapman-Jouguet hypothesis, which states that the velocity of the gas immediately behind the detonation front is sonic. It was seen that the burnt gases behind the detonation front could adjust themselves through a centered simple wave so as to come to rest, see Section 90. The problem now arises whether or not the same is possible for the flow of the burnt gases behind a spherical detonation front, if we assume again that the Chapman-Jouguet hypothesis is valid. This condition in terms of front speed Ξ, gas and sound velocity u_D and c_D behind the front, is $\Xi - u_D = c_D$ or

$$(163.01) \qquad 1 - U_D = C_D.$$

In other words, the state behind the front is represented by a point $U = U_D$, $C = C_D$, determined by the nature of the detonation, on the critical line $1 - U = C$. The integral curve through this point is then to be continued in the direction of increasing η. From Figure 11 one sees that such an integral curve ends at the point $U = 0$, $C = 1$ provided that the point (U_D, C_D) lies on a certain section $0 < U < U_E$, $1 > C > C_E$ of the critical curve.

A simple discussion of the vector field in the neighborhood of the point $U = 0$, $C = 1$ shows that η approaches a finite value η_0 on approaching this point on any integral curve. Hence the detonation wave is terminated by an inner core which expands from the origin with sound speed $c_0 = \xi_0 = \eta_0^{-1}$. Inside this core the gas is at rest while the sound speed has the constant value $c_0 = \xi_0$ there. The pressure distribution in the wave can be found by solving the third equation (162.01). This problem was first formulated and solved by G. I. Taylor.

It appears from this solution that spherical detonation waves are possible in which the detonation front expands with constant speed as determined by the Chapman-Jouguet condition. At the same time, however, these waves show features which contradict certain basic assumptions of the detonation theory. Differential equations (162.01) show that $dU/d\eta$ and $dC/d\eta$ are infinite at a point at which $1 - U = C$, because then $D = 0$. In other words, the rates of change of the quantities u and c, and also p, are infinite immediately behind the detonation front. Now, one of the basic assumptions of the discontinuity theory of the detonation process is that the changes of the significant quantities in the flow process behind the front over a distance of the width of the detonation front are negligible. This assumption is evidently violated in the spherical detonation waves just discussed. We may draw the conclusion that, in spherical detonation waves, the detonation process is no longer independent of the rarefaction process behind it and that this rarefaction process interferes with the internal mechanism of the detonation process. As a matter of fact the result of this interference, in an actual spherical detonation process, is that the front is propagated with a speed less than that derived from the Chapman-Jouguet hypothesis.

We mention that spherical *combustion or deflagration waves* can also be described by our quasi-simple waves. We assume that the combustion process can be interpreted as a discontinuity progressing with a given burning speed relative to the unburnt gas ahead of it, see Section 91. Then the combustion process sends out a compression wave ahead of it which is preceded in its turn by a shock front. The compressed gas, on crossing the deflagration front, burns and comes to rest.

The pre-compression wave is of the same nature as the wave produced by an expanding sphere and hence is represented by the same branches of the integral curves in the (U, C)-plane. The only difference is that this compression wave terminates at a value U_c of U less than one. This value U_c is then to be so adjusted that $\xi_c - u_c = \xi_c(1 - U_c)$ is just the prescribed burning speed relative to the state of the unburnt gas ahead of it. The absolute speed of the burning front ξ_c can be determined from the condition that the burnt gas is at rest.

164. Other spherical quasi-simple waves

There are other interesting types of spherical quasi-simple waves. These are *contracting compression waves terminated by an expanding "reflected" shock front behind which the gas comes to rest*. In the (U, C)-plane these waves are represented by curves starting either at the point $U = 0$, $C = -1$, or at $U = 1$, $C = 0$, crossing the point $U = 0$, $C = 0$, and ending at the "shock ellipse" somewhere in the quadrant $U < 0$, $C > 0$. A simple discussion shows that η begins with a finite value η_A on an integral curve starting at $U = 0$, $C = -1$ or at $U = 1$, $C = 0$. Thus these waves of the first type possess a contracting inner core inside of which the gas is at rest, while there is cavitation in the case of waves of the second type. A further discussion shows that on crossing the point $U = 0$, $C = 0$, the value of η becomes zero in such a way that the quantities $u = \eta^{-1}U$ and $c = \eta^{-1}C$ pass through finite values. We note that u is negative and c positive throughout the wave. At the point at which the integral curve meets the shock ellipse we have $1 - U > C$ or $\xi - u > c$; in other words, the flow relative to the front is supersonic there, as it should be ahead of a shock front. The shock front is therefore crossed in the proper direction. By letting r tend to infinity while t is kept constant, we have $\eta \to 0$. Hence the quantities u and c approach for $r \to \infty$ those values that they assume for $t = 0$; in other words at an infinite distance the flow has a finite negative velocity and a finite sound speed.

It is natural to ask whether or not there are quasi-simple waves preceded by a contracting shock front and terminated by an expanding "reflected" shock front. The state behind a contracting shock front is represented in the (U, C)-plane by a point on the lower branch of the shock ellipse with $C < 0, 0 < U < 1 - \mu^2$. An integral curve through such a point continued in the direction of increasing η intersects the critical line $1 - U = -C$. Consequently, no such wave flow exists behind a contracting shock front which could be continued to times $t > 0$. Flows involving reflected shock fronts can therefore not be quasi-simple waves.

165. Reflected spherical shock fronts

We add a few remarks about a treatment of the problem of reflected spherical shock fronts, by Guderley [193]. Guderley assumes the flow behind the contracting front to be of the progressing type considered in Section 161, with the leading shock ahead of the wave so strong that the relations (161.03) and (161.05) for the state behind it may be assumed; also $\kappa = 0$ is assumed according to (161.04). The vector field in the (U, C)-plane corresponding to the differential equation (160.07) for values $\lambda \neq 1$ is more complicated than that for $\lambda = 1$, see Figure 11, but has several features in common with it. There are several singular points at which $A = B = 0$, see (160.06). From the relation

$$2(1 - U)B - (\gamma - 1)A = [2(1 - \alpha U) - 3(\gamma - 1)\alpha U]D$$

(to which one is naturally led in deriving the expressions (160.06)), one sees that $A = B = 0$ implies either $2(1 - \alpha U) - 3(\gamma - 1)\alpha U = 0$ or $D = 0$. The singular points on the critical lines $D = 0$ are obtained by setting $C^2 = (1 - U)^2$ in the equation $A = 0$. There are five, three, or one singular points depending on whether the quadratic equation $(1 - \alpha U)U - [3\alpha U - \gamma^{-1}(2 - 2\alpha - \kappa\alpha)](1 - U) = 0$ has two, one, or no real roots. These singular points are of the saddle point type and it turns out that through each of them just one integral curve passes so that the sign of $d\eta$ is not reversed on crossing the critical line $D = 0$. Thus these points offer the possibility of crossing from a subsonic to a supersonic flow. Guderley has found that the integral curve which starts at the point $U = (1 - \mu^2)$, $C = -\mu\sqrt{1 + \mu^2}$ corresponding to the state behind an *infinitely strong contracting shock*, see (161.05), just passes through one of these singular points if α has the value $\alpha = .717$ ($\alpha = .834$ in the case of cylindrical flow). The curve then crosses the point $U = C = 0$ thus leading from points corresponding to $t < 0$ to points corresponding to $t > 0$. It is then not difficult to fit a reflected expanding shock front into the resulting flow in such a way that the gas behind this shock front is at rest in the center and, outside of the center, comes gradually to rest as the time increases indefinitely. The fact that $\alpha = .717$ is less than one implies that the contracting strong shock front acquires infinite velocity when it reaches the center. Also, the reflected shock leaves the center with infinite velocity, but then becomes gradually weaker.

One of Guderley's important results is that the pressure behind the reflected shock front is about 26 times the pressure behind the incident shock front, as compared with the ratio 17 for cylindrical and the ratio 8 for one-dimensional strong shock reflections, see Chapter III, Section 70.

Is this pattern of shock reflection an accidental possibility or is it typical in the sense that every process of shock reflection would involve a flow which behaves similarly in the neighborhood of the origin? The answer does not seem certain at present.

166. Concluding remarks

The discussions of the preceding chapter exhibit again the preliminary character of the theory of nonlinear wave motion. All that has been achieved is the construction of particular solutions adapted to special situations, and we are far from a theory which determines the flow uniquely under given boundary and initial conditions. Nevertheless, it is an encouraging fact that some of the flow patterns, obtained mathematically, agree well with physical phenomena. But it is conceivable that in some cases other solutions exist which may well represent observable flows. The confidence of the engineer and physicist in the result of mathematical analysis should ultimately rest on a proof that the solution obtained is singled out by the data of the problem. A great effort will be necessary to develop the theories presented in this book to a stage where they satisfy both the needs of applications and the basic requirements of natural philosophy.

BIBLIOGRAPHY

For a comprehensive bibliography of literature concerning flow of compressible fluids, see

[1] Michel, L. R. Bibliography on flow of compressible fluids. Department of Mechanical Engineering, Massachusetts Institute of Technology, 2nd edition. Cambridge, Mass., 1946.
[2] Summaries of foreign and domestic reports on compressible flow. Headquarters Air Matériel Command, Wright Field, Dayton, Ohio, 1947.

Extensive bibliographies are also found in the three lectures by von Kármán referred to below. Hence, we may confine ourselves to giving a limited number of pertinent references. General presentations of gas dynamics are

[3] Busemann, A. Gasdynamik. Handbuch der Experimentalphysik, Vol. IV, Akademische Verlagsgesellschaft, Leipzig, 1931.
[4] Ackeret, J. Gasdynamik. Handbuch der Physik, Vol. VII, Springer, Berlin, 1927.
[5] Taylor, G. I., and Maccoll, J. W. The mechanics of compressible fluids. W. F. Durand's Aerodynamic Theory, Vol. III, Division H. Springer, Berlin, 1935. (Reprinted: California Institute of Technology, 1943.)
[6] Sauer, R. Theoretische Einführung in die Gasdynamik. Springer, Berlin, 1943.
[7] Liepmann, H. W., and Puckett, A. E. Introduction to Aerodynamics of a Compressible Fluid. Galcit Aeronautical Series, Wiley, New York, 1947.
[8] Lin, C. C. An introduction to the dynamics of compressible fluids. Monograph I. Headquarters Air Materiel Command, Wright Field, Dayton, Ohio, 1947.

General surveying reports are

[9] Taylor, G. I. Recent work on the flow of compressible fluids. Journal of the London Mathematical Society 5, 224–240 (1930).
[10] Prandtl, L. Allgemeine Überlegungen über die Strömung zusammendrückbarer Flüssigkeiten. Zeitschrift für angewandte Mathematik und Mechanik 16, 129–142 (1936), and literature given there.
[11] von Kármán, Th. The engineer grapples with non-linear problems. Bulletin of the American Mathematical Society 46, No. 8, 615–683 (1940), and literature given there.

[12] von Kármán, Th. Compressibility effects in aerodynamics. Journal of the Aeronautical Sciences *8*, No. 9, 337–356 (1941).
[13] Emmons, H. W. Shock waves in aerodynamics. Journal of the Aeronautical Sciences *12*, 188–194 (1945).
[14] von Kármán, Th. Supersonic aerodynamics—Principles and applications. Journal of the Aeronautical Sciences *14*, No. 7, 373–412 (1947)

As to work on gas dynamics in Germany see

[15] Döring, W., and Burkhardt, H. Beiträge zur Theorie der Detonation. Air Materiel Command, Wright Field, Ohio, Air Documents Division, FB1939, Film R 23,181.
[16] Bibliography on German supersonic research No. 1. Headquarters Air Materiel Command, Wright Field, Dayton, Ohio, 1946.

Chapter I. Compressible Fluids

Part A. General Dynamics of Fluids and Thermodynamic Notions

For a derivation and discussion of the general equations of fluid dynamics, see

[17] Lamb, H. Hydrodynamics. Cambridge University Press, 6th edition, Cambridge, 1932. (Reprinted: Dover, New York, 1945.)
[18] Milne-Thomson, L. M. Theoretical Hydrodynamics. Macmillan, London, 1938.
[19] Goldstein, S. Modern Developments in Fluid Dynamics; an Account of Theory and Experiment Relating to Boundary Layers, Turbulent Motion, and Wakes. (Fluid Motion Panel of the British Aeronautical Research Committee.) 2 volumes. Clarendon Press, Oxford, 1938

For an introduction into the notions of thermodynamics, see

[20] Epstein, P. S. Textbook of Thermodynamics. Wiley, New York, 1937.
[21] Zemansky, M. W. Heat and Thermodynamics. 2nd Edition, McGraw-Hill, New York, 1943.

Part B. Differential Equations for Specific Types of Flow

For a special treatment of non-isentropic flow, see

[22] Crocco, L. Una nuova funzione di corrente per lo studio del moto rotazionale dei gas, Atti Reale Accademia Nazionale dei Lincei *23*, 115–124 (1936). Eine neue Strömungsfunktion für die Erforschung der Gase mit Rotation, Zeitschrift für angewandte Mathematik und Mechanik *17*, 1–7 (1937).
[23] Vazsonyi, A. On rotational gas flows. Quarterly of Applied Mathematics *3*, No. 1, 29–37 (1945).

[24] Frazer, J. H., Hicks, B. L., Guenther, P. E., and Wasserman, R. H. Reports issued by the Ballistic Research Laboratories, Aberdeen Proving Ground, Maryland, 1946.
[25] Tsien, H. S. One dimensional flows of a gas characterized by van der Waals' equation of state. Journal of Mathematics and Physics 25, 301–324 (1947); 26, 76–77 (1947).

Appendix

As to experimental reports, see
[26] Einstein, H. A., and Baird, E. G. Progress report of the analogy between surface shock waves on liquids and shocks in compressible gases. Hydrodynamics Laboratory Report No. N-54, California Institute of Technology, Pasadena, 1946.

A detailed discussion of the shallow water theory with an extensive bibliography is given by
[27] Stoker, J. J. The formation of breakers and bores. Communications on Applied Mathematics 1, No. 1, 1–87 (1948).

Chapter II. Mathematical Theory of Flows Depending on Two Variables

The general theory of hyperbolic differential equations is developed in
[28] Hadamard, J. Leçons sur la propagation des ondes. Hermann, Paris, 1903.
[29] Lewy, H. Über das Anfangswertproblem bei einer hyperbolischen nichtlinearen partiellen Differentialgleichung zweiter Ordnung mit zwei unabhängigen Veränderlichen. Mathematische Annalen 98, No. 2, 179–191 (1927).
[30] Perron, O. Über Existenz und Nichtexistenz von Integralen partieller Differentialgleichungssysteme im reellen Gebiet. Mathematische Zeitschrift 27, No. 4, 549–564 (1928).
[31] Hadamard, J. Le problème de Cauchy et les équations aux dérivées partielles linéaires hyperboliques. Hermann, Paris, 1932. (In particular, the Appendix.)
[32] Courant, R., and Hilbert, D. Methoden der mathematischen Physik, Vol. II, Springer, Berlin, 1937. (Reprint: Interscience, New York, 1943.)
[33] Cinquini-Cibrario, M. Un teorema di esistenza e di unicità per un sistema di equazioni alle derivate parziali. Annali di matematica [4], 24, 157–175 (1945).

[34] Friedrichs, K. O. Nonlinear hyperbolic differential equations for functions of two independent variables. American Journal of Mathematics (in press, 1948).

Appendix

[35] Giese, J. H. Compressible flow with degenerate hodographs. Ballistic Research Laboratories, Aberdeen Proving Ground, Maryland.

Chapter III. One-dimensional Flow

Part A. Continuous Flow
Part B. Rarefaction and Compression Waves

For the first discussion of simple waves, see

[36] Poisson, S. D. Mémoire sur la théorie du son. Journal de l'école polytechnique, 14ᵐᵉ Cahier, 7, 319–392 (1808).
[37] Earnshaw, S. On the mathematical theory of sound. Transactions of the Royal Society of London *150*, 133–148 (1860).

For the theory of one-dimensional nonlinear waves in general, see

[38] Riemann, B. Über die Fortpflanzung ebener Luftwellen von endlicher Schwingungsweite. Abhandlungen der Gesellschaft der Wissenschaften zu Göttingen, Mathematisch-physikalische Klasse *8*, 43 (1860), or Gesammelte Werke, 1876, p. 144.
[39] Rayleigh, J. Aerial plane waves of finite amplitude. Proceedings of the Royal Society *84*, 247–284 (1910). Also Scientific Papers, Vol. V, Cambridge University Press, Cambridge, 1912, pp. 573–610.

Part C. Shocks

For a discussion of discontinuities in gas motion see, in addition to [38] and [39],

[40] Stokes, E. E. On a difficulty in the theory of sound. Philosophical Magazine [3], *33*, 349–356 (1848).
[41] Challis, J. On the velocity of sound. Philosophical Magazine [3], *32*, 494–499 (1848).
[42] Rankine, W. J. M. On the thermodynamic theory of waves of finite longitudinal disturbance. Transactions of the Royal Society of London *160*, 277–288 (1870).
[43] Hugoniot, H. Sur la propagation du mouvement dans les corps et spécialement dans les gaz parfaits. Journal de l'école polytechnique *58*, 1–125 (1889).

[44] Rüdenberg, R. Über die Fortpflanzungsgeschwindigkeit und Impulsstärke von Verdichtungsstössen. Artilleristische Monatshefte, No. 113 (1916).

[45] Becker, R. Stosswelle und Detonation. Zeitschrift für Physik *8*, 321–362 (1922).

This paper contains a theory of the thickness of a shock zone. For a refined treatment, see

[46] Thomas, L. H. Note on Becker's theory of the shock front. The Journal of Chemical Physics *12*, No. 11, 449–453 (1944).

Shocks in fluids with an arbitrary equation of state are discussed in

[47] Bethe, H. Office of Scientific Research and Development, Division B, Report No. 545, 1942.

[48] Weyl, H. Shock waves in arbitrary fluids. National Defense Research Committee, Applied Mathematics Panel Note No. 12 (Applied Mathematics Group—New York University No. 46), 1944.

Tables for shock transitions in air are given by

[49] Brinkley, S. R., Jr., Kirkwood, J. G., and Richardson, J. M. Office of Scientific Research and Development No. 3550, 1944.

Part D. Interactions

Lagrange's problem was treated by

[50] Love, A. E. H., and Pidduck, F. B. Lagrange's ballistic problem. Transactions of the Royal Society of London *222*, 167–226 (1922).

Interaction phenomena are mentioned in [64]. They were systematically investigated by

[51] von Neumann, J. Progress report on the theory of shock waves. National Defense Research Committee, Division 8, Office of Scientific Research and Development No. 1140, 1943.

[52] Courant, R., and Friedrichs, K. O. Interaction of shock and rarefaction waves in one-dimensional motion. National Defense Research Committee, Applied Mathematics Panel Report 38.1R (Applied Mathematics Group—New York University No. 1), 1943.

[53] Courant, R. Technical Conference on Supersonic Flow and Shock Waves. Naval Report 203–45, Bureau of Ordnance. 1945.

For the approximate treatment of interactions see

[54] Chandrasekhar, S. On the decay of plane shock waves. Ballistic Research Laboratories, Report No. 423, Aberdeen Proving Ground, Maryland, 1943.
[55] Friedrichs, K. O. Formation and decay of shock waves. Institute for Mathematics and Mechanics—New York University No. 158 (1947), or Communications on Applied Mathematics (in press, 1948).
[56] Leopold, A. Decaying shocks. Institute for Mathematics and Mechanics—New York University No. 167 (1947).

In particular, for reflection see

[57] Chandrasekhar, S. The normal reflection of a blast wave. Ballistic Research Laboratories, Report No. 439, Aberdeen Proving Ground, Maryland, 1943.
[58] von Neumann, J. Proposal and analysis of a new numerical method for the treatment of hydrodynamical shock problems. National Defense Research Committee, Applied Mathematics Panel Memo 38.7M (Applied Mathematics Group—New York University No. 18), 1943.
[59] Finkelstein, R. Normal reflection of shock waves. Explosives Research Report No. 6, Bureau of Ordnance, Navy Department, 1944.
[60] Martin, M. H. A problem in the propagation of shock. Quarterly of Applied Mathematics 4, No. 4, 330–348 (1947).
[61] National Defense Research Committee, Division 2, Office of Scientific Research and Development Nos. 4147, 4257, 4356, 4514, 4649, 4754, 4875, 5011, 5144, 5271, 5393, 6007.

For experimental results see

[62] Payman, W., and Shepherd, W. F. C. Explosion waves and shock waves. Part VI, The disturbance produced by bursting diaphragms with compressed air. Ministry of Supply, Advisory Council 735, Phys./Ex. 98 (W-12-157), 1941.
[63] Winckler, J. R., Van Voorhis, C. C., Panofsky, H., and Ladenburg, R. National Defense Research Committee, Division 2, Office of Scientific Research and Development No. 5204, 1945.

Part E. Detonation and Deflagration Waves

In addition to the references, Becker [45], von Neumann [51], Döring [15], and the literature referred to there, see

[64] Jouguet, E. Méchaniques des explosifs. O. Doin et Fils, Paris, 1917.
[65] Jouguet, E. La théorie thermodynamique de la propagation des explosions. Proceedings of the International Congress for Applied Mechanics, 1926, pp. 12–22.

[66] Semenov, N. N. The theory of the combustion process. Zeitschrift für Physik *48*, 571–582 (1928). Translated by I. Alcock. (Reprinted: Durand Reprinting Committee, California Institute of Technology.) Also: Thermal theory of combustion and explosion. National Advisory Committee on Aeronautics, Technical Memorandum No. 1024, 1026.

[67] Lewis, B., and von Elbe, G. Combustion, Flames and Explosion of Gases. Cambridge University Press, Cambridge, 1938.

[68] Jost, W. Explosions—und Verbrennungsvorgänge in Gasen. Springer, Berlin, 1939. (Reprinted: Edwards Brothers, Ann Arbor 1943.)

For experimental results see

[69] Payman, W., and others. Explosion waves and shock waves, Parts I–V. Proceedings of the Royal Society (A) *132*, 200–213 (1931); *148*, 604–622 (1935); *158*, 348–367 (1937); *163*, 575–592 (1937).

See further

[70] Taylor, G. I. Detonation waves. Ministry of Supply, Advisory Council on Scientific Research and Technical Development, Explosives Research Committee, Paper Advisory Council 639, Research Committee 178 (W-12-144), 1941.

[71] von Neumann, J. Office of Scientific Research and Development No. 549, 1942.

[72] Friedrichs, K. O. On the mathematical theory of deflagrations and detonations. Naval Report 79-46, Bureau of Ordnance, 1946.

[73] Gamow, G., and Finkelstein, R. Theory of the detonation process. Navy Department, Bureau of Ordnance, Navord 90-46 (103), 1947.

Part F. Wave Propagation in Elastic-plastic Material

See a great number of reports and memoranda by von Kármán and others issued through Division 2 of the National Defense Research Committee.

Chapter IV. Isentropic Irrotational Steady Plane Flow. Oblique Shock Fronts. Shock Reflections.

Part A. Hodograph Method

The hodograph method for two-dimensional steady flow is due to

[74] Molenbrock, P. Über einige Bewegungen eines Gases bei Annahme eines Geschwindigkeitspotentials. Archiv der Mathematik und Physik *9*, 157–195 (1890).

[75] Chaplygin, S. A. On gas jets. Scientific Annals of the Imperial University of Moscow, Physico-mathematical Division, No. 21, Moscow, 1904. (Translation by M. H. Slud available through the School of Mechanics, Brown University, Providence, Rhode Island.)

A great number of papers and reports employing this method for subsonic flow have been written. See for example [12] and

[76] Theodorsen, T., and Garrick, I. E. General potential theory of arbitrary wing sections. National Advisory Committee for Aeronautics, Report No. 452, 1933.
[77] Busemann, A. Hodographenmethode der Gasdynamik. Zeitschrift für angewandte Mathematik und Mechanik 17, 73–79 (1937), or Forschung 4, 186 (1933).
[78] Tsien, H. S. Two-dimensional subsonic flow of compressible fluids. Journal of the Aeronautical Sciences 6, 339–407 (1939).
[79] Ringleb, F. Exakte Lösungen der Differentialgleichungen einer adiabatischen Gasströmung. Zeitschrift für angewandte Mathematik und Mechanik 20, 185–198 (1940). Abstract in the Journal of the Royal Aeronautical Society 46, 403–404 (1942).
[80] Tollmien, W. Grenzlinien adiabatischer Potentialströmungen. Zeitschrift für angewandte Mathematik und Mechanik 21, 140–152 (1941).
[81] Bers, L., and Gelbart, A. On a class of differential equations in mechanics of continua. Quarterly of Applied Mathematics 1, 168–188 (1943).
[82] Garrick, I. E., and Kaplan, C. National Advisory Committee for Aeronautics, Report No. L4I29, 1944.
[83] Bers, L. National Advisory Committee for Aeronautics, Technical Note 969, 1945.
[84] Bers, L. On the circulatory subsonic flow of compressible fluid past a circular cylinder. National Advisory Committee for Aeronautics, Technical Note 970, 1945.
[85] Bergmann, S. On the two-dimensional flows of compressible fluids. National Advisory Committee for Aeronautics, Technical Note 972, 1945.
[86] Lin, C. C. On an extension of the von Kármán-Tsien method to two-dimensional subsonic flows with circulation around closed profiles. Quarterly of Applied Mathematics 4, No. 3, 291–297 (1946).

For transonic or mixed flow see

[87] Tricomi, F. Sulle equazioni lineari alle derivate parziali di 2° ordine, di tipo misto, Parts I–VII. Memorie della Reale Accademia Nazionale dei Lincei, classe di scienze fisiche, [5] 14, 133–247 (1922).

BIBLIOGRAPHY

[88] Frankl, F., and Aleksejeva, R. Zwei Randwertaufgaben aus der Theorie der hyperbolischen partiellen Differentialgleichungen zweiter Ordnung mit Anwendungen auf Gasströmungen mit Überschallgeschwindigkeit. Matematicheskiĭsbornik, (Moscow) *41*, 483–502 (1934/35). German résumé at end of article. See also review in Zentralblatt für Mechanik *3*, 131–132 (1935).

[89] Christianovitch, C. A. On supersonic gas flow. Reports of the Central Aerohydrodynamical Institute No. 543.

[90] Frankl, F. On Cauchy's problem for partial differential equations of mixed elliptico-hyperbolic type with initial data on the parabolic line. Akademiya Nauk S.S.S.R., Izvestiya, Seriya Matematicheskaya *8*, 195–224 (1944).

[91] Frankl, F. On the problems of Chaplygin for mixed sub- and supersonic flows. Akademiya Nauk S.S.S.R., Izvestiya, Seriya Matematicheskaya *9*, 121–148 (1945).

[92] Maccoll, J. W., and Codd, J. Ministry of Supply, Armament Research Department, Fort Halstead, Kent, 1945.

[93] Tsien, H. S. Galcit Publication No. 3, California Institute of Technology, 1946.

[94] von Kármán, Th. The similarity law of transonic flow. Journal of Mathematics and Physics *26*, No. 3, 182–190 (1947).

For the notion and the theory of the limiting line see, in addition to [9], [77], [78], and [79],

[95] Clauser, F. M. New methods of solving the equations for the flow of a compressible fluid. Doctorate Thesis, California Institute of Technology, 1937. Unpublished.

[96] Tsien, H. S. The limiting line in mixed subsonic and supersonic flow of compressible fluids. National Advisory Committee for Aeronautics, Technical Note No. 961, 1944.

[97] Liepmann, H. W. Investigations of the interaction of boundary layer and shock waves in transonic flow. Guggenheim Aeronautical Laboratory, California Institute of Technology.

[98] Guderley, G. The reason for the appearance of compressible shocks in transonic flow (Monograph On Transonic Flow). Translation by Kate Liepmann, Guggenheim Aeronautical Laboratory, California Institute of Technology.

[99] Friedrichs, K. O. On the non-occurrence of a limiting line in transonic flow. Institute for Mathematics and Mechanics—New York University No. 165, 1947.

Part B. Characteristics and Simple Waves
Part C. Oblique Shock Fronts

Simple waves and oblique shock fronts were studied by L. Prandtl and his pupils; see for example [3] and

[100] Meyer, Th. Über zweidimensionale Bewegungsvorgänge in einem Gas, das mit Überschallgeschwindigkeit strömt. Dissertation, Göttingen, 1908. Forschungsheft des Vereins deutscher Ingenieure, Vol. 62, Berlin, 1908, pp. 31–67.

[101] Steichen, A. Beiträge zur Theorie der zweidimensionalen Bewegungsvorgänge in einem Gas, das mit Überschallgeschwindigkeit strömt. Dissertation, Göttingen, 1909.

For the method of characteristics see

[102] Prandtl, L., and Busemann, A. Näherungsverfahren zur zeichnerischen Ermittlung von ebenen Strömungen mit Überschallgeschwindigkeit. Stodola Festschrift, Zürich and Leipzig, 1929, pp. 499–509.

Interactions of simple waves were studied by

[103] Taub, A. H. Refraction of plane waves. Physical Review [2], *72*, No. 1, 51–60 (1947).

Several useful tables have been issued, see

[104] Emmons, H. W. Gas Dynamic Tables for Air. Dover Publications, New York, 1947.

[105] Edmonson, N., Murnaghan, F. D., and Snow, R. M. The theory and practice of two-dimensional supersonic pressure calculations. Bumblebee Report 26, Applied Physics Laboratory, Johns Hopkins University, 1945.

[106] Shapiro, A. H., and Edelman, G. M. Method of characteristics for two-dimensional supersonic flow—graphical and numerical procedures. Journal of Applied Mechanics *14*, No. 2, A154–162 (1947).

Further see

[107] Schubert, F. Zur Theorie des stationären Verdichtungsstosses. Zeitschrift für angewandte Mathematik und Mechanik *23*, No. 3, 129–138 (1943).

[108] Thomas, R. N. Ballistic Research Laboratories, Report No. 483, Aberdeen Proving Ground, Maryland, 1944.

[109] Laitone, E. V. Exact and approximate solutions of two-dimensional oblique shock flow. Journal of the Aeronautical Sciences *14*, No 1, 25–41 (1947).

[110] Tsien, H. S. Flow conditions near the intersection of a shock wave with solid boundary. Journal of Mathematics and Physics *26*, No. 1, 69–75 (1947).

[111] Thomas, T. Y. On curved shock waves. Journal of Mathematics and Physics 26, No. 1, 62–68 (1947).

The stability of discontinuity surfaces was investigated by

[112] Landau, L. Stability of tangential discontinuities in compressible fluid. Akademiya Nauk S.S.S.R., Comptes rendus (Doklady), 44, No. 4, 139–141 (1944).

Part D. Interaction and Reflection

A few interaction configurations were mentioned by

[113] Preiswerk, E. Anwendung gasdynamischer Methoden auf Wasserströmungen mit freier Oberfläche. Mitteilungen aus dem Institut für Aerodynamik, Eidgenössische Technische Hochschule, Zürich, No. 7 (1938).

A systematic theory was developed by von Neumann; see [51] and

[114] von Neumann, J. Oblique reflection of shocks. Navy Department, Bureau of Ordnance, Explosives Research Report No. 12, 1943.

[115] Chandrasekhar, S. On the conditions for the existence of three shock waves. Ballistic Research Laboratories, Report No. 367, Aberdeen Proving Ground, Maryland, 1943.

[116] Polachek, H., and Seeger, R. J. Regular reflection of shocks in ideal gases. Navy Department, Bureau of Ordnance, Explosives Research Report No. 13, 1944.

For the graphical discussion presented in this book, see

[117] Friedrichs, K. O. Remarks on the Mach effect. National Defense Research Committee, Applied Mathematics Panel Memos Nos. 38.4M and 38.5M (Applied Mathematics Group—New York University Nos. 5 and 6), 1943. (Also Appendix.)

The theory was independently developed by

[118] Weise, A. Theorie des gegabelten Verdichtungsstosses. Institut für Gasdynamik der Deutschen Versuchsanstalt für Luftfahrt, E. V., Berlin-Adelershof, 1943.

[119] Weise, A. Über die Strömungsablösung durch Verdichtungsstösse. Institut für Gasdynamik der Deutschen Versuchsanstalt für Luftfahrt, E. V., Sonderdruck aus Technische Berichte 10, No. 2, 59–61 (1943).

[120] Weise, A. Die Herzkurvenmethode zur Behandlung von Verdichtungsstössen. Institut für Gasdynamik der Deutschen Versuchsanstalt für Luftfahrt, E. V., Berlin-Adlershof, 1943.

For translations of these papers see: British Intelligence Objectives Sub-committee Group II, Halstead Exploiting Centre Translations.

See further

[121] Bargmann, V. Applied Mathematics Panel Report No. 108.2R (Applied Mathematics Group—Institute for Advanced Study No. 2), 1945.

For experimental results see

[122] Cranz, C., and Schardin, H. Kinematographie auf ruhendem Film und mit extrem hoher Bildfrequenz. Zeitschrift für Physik, 56, 147–183 (1929).
[123] Smith, L. G. Division 2, Office of Scientific Research and Development No. 6271, 1945.
[124] Libessart, P. Ministry of Supply, Research Committee 417 (WA-2315-6), 1944.

For the approximate treatment of interactions see [55] and

[125] DuMond, J. W. M., Cohen, E. R., Panofsky, W. K. H., and Deeds, E. A determination of the wave forms and laws of propagation and dissipation of ballistic shock waves. Journal of the Acoustical Society of America 18, No. 1, 97–118 (1946).

Part E. Perturbation Method of Interaction. Airfoil Flow.

The perturbation theory of first order is due to

[126] Ackeret, J. Über Luftkräfte bei sehr grossen Geschwindigkeiten insbesonders bei ebenen Strömungen. Helvetica Physica Acta 1, 301–322 (1928).

that of second order to

[127] Busemann, A. Widerstand bei Geschwindigkeiten nahe der Schallgeschwindigkeit. Verhandlungen des 3. Internationalen Kongress für technische Mechanik, Stockholm, 1, 282–286 (1930).
[128] Busemann, A., and Walchner, O. Profileigenschaften bei Überschallgeschwindigkeit. Forschung auf dem Gebiet des Ingenieurwesens 4A, 87–92 (1933).
[129] Busemann, A. Aerodynamischer Auftrieb bei Überschallgeschwindigkeit. Reale accademia d'Italia, classe delle scienze fisiche, matematiche e naturali, Quinto Convegno Volta 13, 3–35 (1935).

that of third and fourth order to

[130] Donov, A. A plane wing with sharp edges in a supersonic stream. Izvestia Akademia Nauk U.S.S.R. Série Mathématique (1939).

See further

[131] Epstein, P. S. On the air resistance of projectiles. Proceedings of the National Academy of Sciences 17, 532–547 (1931).

[132] Taylor, G. I. Applications to aeronautics of Ackeret's theory of aerofoils moving at speeds greater than that of sound. British Aeronautical Research Committee, Reports and Memoranda No. 1467 (WA-4218-5a), 1932.

[133] Hooker, S. G. British Aeronautical Research Committee, Reports and Memoranda No. 1721, 1936.

[134] Crocco, L. Singolarità della corrente gassosa iperacustica nell' interno di una prora a diedro. (Singularity of supersonic gas flow in the neighborhood of leading edge of a wedge.) L'Aerotecnica 17, 519–534 (1937).

[135] Goldstein, S., and Young, A. D. The linear perturbation theory of compressible flow, with applications to wind-tunnel interference. British Aeronautical Research Committee Technical Report 6865, Reports and Memoranda No. 1909 (WA-4029-5), 1943.

[136] Lighthill, M. J. British Aeronautical Research Committee, Reports and Memoranda No. 1929, 1944, addendum, 1944. The conditions behind the trailing edge of the supersonic aerofoil. British Aeronautical Research Committee, Reports and Memoranda No. 1930, 1944.

See also

[137] Tukey, J. W. Linearization of solutions in supersonic flow. Quarterly of Applied Mathematics 5, No. 3, 361–364 (1947).

Experimental results are described by

[138] Ferri, A. Experimental results with airfoils tested in the high-speed tunnel at Guidonia. National Advisory Committee for Aeronautics, Technical Memorandum No. 946. From Atti di Guidonia, No. 17 (1939).

Part F. Remarks about Boundary Value Problems for Steady Flow

See in particular the references [88], [90] and [92] to [94] given in connection with transonic flow.

Chapter V. Flow in Nozzles and Jets

Among the extensive literature on this subject we mention

[139] Stodola, A. Dampf- und Gasturbinen, 6th edition, 1924. (Steam and Gas Turbines, translation by L. C. Lowenstein. McGraw-Hill, New York, 1927).

[140] Stanton, T. E. On the flow of gases at high speeds, Proceedings of the Royal Society of London (A) *111*, 306–339 (1926).
[141] Busemann, A. Lavaldüsen für gleichmässige Überschallströmungen. Zeitschrift des Vereins Deutscher Ingenieure *84*, 857–862 (1940).
[142] Kisenko, M. S. Comparative results of tests on several different types of nozzles. Central Aero-hydrodynamical Institute, Moscow, Report No. 478, 1940. (Translation, National Advisory Committee for Aeronautics, Technical Memorandum No. 1066, 1944).
[143] Emmons, H. W. The numerical solution of compressible fluid flow problems. National Advisory Committee on Aeronautics, Technical Note No. 932, 1944.
[144] Friedrichs, K. O. Theoretical studies on the flow through nozzles and related problems. National Defense Research Committee, Applied Mathematics Panel Report 82.1R (Applied Mathematics Group—New York University No. 43), 1944.

About unsteady nozzle flow see

[145] Schultz-Grunow, F. Nichtstationäre eindimensionale Gasströmung. Forschung *13*, 125–134 (1942).
[146] Green, J. R., and Southwell, R.V. High-speed flow of compressible fluid through a two-dimensional nozzle. Relaxation methods applied to engineering problems, Part IX. Philosophical Transactions of the Royal Society of London (A) *239*, No. 808, 367–386 (1944).
[147] Kantrowitz, A., and Donaldson, C. duP. National Advisory Committee on Aeronautics, No. L5D20, 1945.

About condensation shocks see

[148] Oswatitsch, K. Kondensationserscheinungen in Überschalldüsen. Zeitschrift für angewandte Mathematik und Mechanik *22*, No. 1, 1–14 (1942). (Translation reprinted by the Durand Reprinting Committee, California Institute of Technology.)
[149] Herrmann, R. Der Kondensationsstoss in Überschall-Windkanaldüsen. Luftfahrtforschung *19*, No. 6, 201–209 (1942).

About jets see

[150] Prandtl, L. Neue Untersuchungen über die strömende Bewegung der Gase und Dämffe. Physikalische Zeitschrift *8*, 23–32 (1907).
[151] Fraser, R. P Flow through nozzles at supersonic speeds. Four interim reports on jet research, 1940 to 1941. Ministry of Supply, D.S.R., Extra Mural Research F72/115 (WA-1513-1a).
[152] Hartmann, J., and Lazarus, F. The air-jet with a velocity exceeding that of sound. Philosophical Magazine [7], *31*, 35–50 (1941).

See also [63] and

[153] Pack, D. C. Branch for Theoretical Research, Fort Halstead, Kent, (WA-3903-3), 1944.

About the theory of rockets see

[154] Rocket fundamentals. Office of Scientific Research and Development No. 3992, Division 3, Section H, ABL-SR4, George Washington University, 1944.

Chapter VI. Flow in Three Dimensions

Part A. Steady Flow with Cylindrical Symmetry

[155] von Kármán, Th., and Moore, N. B. Resistance of slender bodies moving with supersonic velocities, with special reference to projectiles. American Society of Mechanical Engineers, Transactions *54*, 303 (1932).
[156] Lighthill, M. J. Fluid Motion Panel, Aeronautical Research Committee, 8040, F.M. 727, 1944.
[157] Maccoll, J. W. Theoretical Research Committee Report 15/45, British Aeronautical Research Committee 45/13, 1945.
[158] Sears, W. R. On compressible flow about bodies of revolution. Quarterly of Applied Mathematics *4*, No. 2, 191–192 (1946).
[159] Hayes, W. D. Linearized supersonic flows with axial symmetry. Quarterly of Applied Mathematics *4*, No. 3, 255–261 (1946).
[160] Sears, W. R. On projectiles of minimum wave drag. Quarterly of Applied Mathematics *4*, No. 4, 361–366 (1947).
[161] Sears, W. R. A second note on compressible flow about bodies of revolution. Quarterly of Applied Mathematics *5*, No. 1, 89–91 (1947).
[162] Laitone, E. V. The linearized subsonic and supersonic flow about inclined slender bodies of revolution. Journal of the Aeronautical Sciences *14*, No. 11, 631–642 (1947).

See also [137].

Refinements are indicated in [156] and by

[163] Chien, W. Z. Galcit Publication No. 3, Abstracts, California Institute of Technology, 1946.

The accuracy of the approximation is discussed by

[164] Kopal, Z. A few remarks on the limitations of linearized theory in supersonic flow. Massachusetts Institute of Technology, Technical Report No. 2, 1947.

Part B. Conical Flow

[165] Busemann, A. Drücke auf kegelförmige Spitzen bei Bewegung mit Überschallgeschwindigkeit. Zeitschrift für angewandte Mathematik und Mechanik *9*, No. 6, 496–498 (1929).
[166] Bourquard, F. Ondes ballistiques, planes obliques et ondes coniques. Comptes rendus *194*, 846 (1932).

[167] Taylor, G. I., and Maccoll, J. W. The air pressure on a cone moving at high speeds. Proceedings of the Royal Society (A) *139*, 278–311 (1933).
[168] Maccoll, J. W. The conical shock wave formed by a cone moving at high speed. Proceedings of the Royal Society (A) *159*, 459–472 (1937).
[169] Busemann, A. Die achsensymmetrische kegelige Überschallströmung. Luftfahrtforschung *19*, No. 4, 137–144 (1942).

Extensive tables about the flow against a conical projectile were issued by

[170] Kopal, Z. Tables of supersonic flow around cones. Massachusetts Institute of Technology, Technical Report No. 1, 1947.

As to flow with cylindrical symmetry in general see

[171] Ferrari, C. Determinazione della pressione sopra solidi di rivoluzione a prora acuminata disposti in deriva in corrente di fluido compressibile a velocità ipersonora. Atti della reale accademia delle scienze di Torino *72*, 140–163 (1936).
[172] Ferrari, C. Campi di corrente ipersonora attorno a solidi di rivoluzione. L'Aerotecnica *16*, No. 2, 121–130 (1936). Campo aerodinamico a velocità iperacustica attorno a un solido di rivoluzione a prora acuminata. L'Aerotecnica *17*, No. 6, 508–518 (1937).

See also:

[173] Ferrari, C. Campi di corrente ipersonora attorno a solidi di rivoluzione. Atti 1, Congr. Un. Mat. Ital., 1938, pp. 616–626.
[174] Ferrari, C. Sulla determinazione del proietto di minima resistenza d'onda, I and II. Atti della reale accademia delle scienze di Torino *74*, 675–693; *75*, 61–96 (1939).
[175] Sauer, R. Charakteriskenverfahren für raumliche achsen symmetrische Aberschallströnungen. Forschungsbericht, No. 1269, 1940.
[176] Ferri, A. Application of the method of characteristics to supersonic rotational flow. National Advisory Committee on Aeronautics, Technical Note No. 1135, 1946.
[177] Sauer, R. Überschallströmung um beliebig geformte Geschossspitzen unter kleinem Anstellwinkel. Luftfahrtforschung *19*, No. 4, 148–152 (1942).

Linearized flow without assuming cylindrical symmetry is treated in many papers:

[178] Schlichting, H. Airfoil theory at supersonic speed. Jahrbuch der deutschen Luftfahrtforschung *1*, 181–197 (1937). Translated by National Advisory Committee for Aeronautics, Technical Memorandum No. 897, 1939.

[179] Busemann, A. Deutsche Akademie der Luftfahrtforschung, 1942, pp. 455–470 and *7B*, No. 3, 105–122, 1943. Reprinted in translation by Douglas Aircraft Company, Inc.
[180] Stewart, H. J. The lift on a delta wing at supersonic speeds. Quarterly of Applied Mathematics *4*, No. 3, 246–254 (1946).
[181] Frankl, F. I., and Karpovich, E. A. Resistance of a delta wing in a supersonic flow. Prikladnaya Matematika i Mekhanika *11*, No. 4, 495–496 (1947).
[182] Galin, L. A. A wing rectangular in plane in a supersonic flow. Prikladnaya Matematika i Mekhanika *11*, No. 4, 465–474 (1947).
[183] Bonney, E. A. Aerodynamic characteristics of rectangular wings at supersonic speeds. Journal of the Aeronautical Sciences *14*, No. 2, 110–116 (1947).
[184] Puckett, A. E., and Stewart, H. J. Aerodynamic performance of delta wings at supersonic speeds. Journal of the Aeronautical Sciences *14*, No. 10, 567–578 (1947).
[185] Hayes, W. D. Linearized supersonic flow. Doctorate Thesis, California Institute of Technology, 1947.
[186] Evvard, J. C. Distribution of wave drag and lift in the vicinity of wing tips at supersonic speeds. National Advisory Committee for Aeronautics, Technical Note 1382, 1947. See also: The effects of yawing thin pointed wings at supersonic speeds. National Advisory Committee for Aeronautics, Technical Note 1429, 1947.
[187] Evvard, J. C., and Turner, R. L. Theoretical lift distribution and upwash velocities for thin wings at supersonic speeds. National Advisory Committee for Aeronautics, Technical Note 1484, 1947.

Part C. Spherical Waves

Waves of the type discussed in this part were treated by

[188] Bechert, K. Zur Theorie ebener Störungen in reibungsfreien Gasen. Annalen der Physik [5], *37*, 89–123 (1940); *38*, 1–25 (1940). See further: Über die Ausbreitung von Zylinder- und Kugelwellen in reibungsfreien Gasen und Flüssigkeiten. Annalen der Physik [5], *39*, 169–202 (1941). And also: Über die Differentialgleichungen der Wellenausbreitung in Gasen. Annalen der Physik [5], *39*, 357–372 (1941).
[189] Marx, H. Zur Theorie der Zylinder- und Kugelwellen in reibungsfreien Gasen und Flüssigkeiten. Annalen der Physik [5], *41*, 61–88 (1942).

See further [69] and

[190] Taylor, G. I. The formation of a blast wave by a very intense explosion. Ministry of Home Security, R.C. 210 (II-5-153), 1941.
[191] Landau, L. D. On shock waves. Akademiya Nauk S.S.S.R., Fizichaskiĭ Zhurnal, *6*, 229–230 (1942).

[192] Taylor, G. I. The propagation and decay of blast waves. British Civilian Defense Research Committee, 1944.
[193] Guderley, G. Starke kugelige und zylindrische Verdichtungsstösse in der Nähe des Kugelmittelpunktes bzw. der Zylinderachse. Luftfahrtforschung *19*, No. 9.
[194] Kuo, Y.H. The propagation of a spherical or a cylindrical wave of finite amplitude and the production of shock waves. Quarterly of Applied Mathematics *4*, No. 4, 349–360 (1947).
[195] Taylor, G. I. The air wave surrounding an expanding sphere. Proceedings of the Royal Society of London [A], *186*, 273–292 (1946).

For approximate treatments of types of spherical waves see

[196] Kirkwood, J. G., and Bethe, H. A. Progress report on "the pressure wave produced by an underwater explosion I." Division B, Serial No. 252, Office of Scientific Research and Development No. 588, 1942.
[197] Sauer, R. Charakteristikenverfahren für Kugel- und Zylinderwellen reibungsloser Gase. Zeitschrift für angewandte Mathematik und Mechanik *23*, No. 1, 29–32 (1943).
[198] Kirkwood, J. G., and Brinkley, S. R., Jr. National Defense Research Committee No. A-318, Office of Scientific Research and Development No. 4814, Cornell University—Division 2, 1945.

Additional reference may be made to a general lecture, "High speed gas flow" by M. J. Lighthill (which will appear in the Proceedings of the VII International Congress of Applied Mechanics, held in London, England, 1948) and to a third volume of the book by S. Goldstein [19], in preparation, which will be concerned with compressible fluid flow.

Index of Symbols

*Numbers refer to pages on which the symbols are defined.
The same symbol often denotes different concepts.*

A	6, 164, 378	p	3	z	12
c	5, 12, 41	P	124, 196, 419	α	42, 323, 419
c_v	6	$\mathcal{P}$	79	Λ	260, 378
c_*	23	q	16	Λ'	260
C	20, 42, 419	$\bar{q}$	19	β	42, 306
$\mathfrak{D}$	197	$\hat{q}$	22	γ	6
e	4	r	87	Γ	44
E	11, 205, 418	R	6, 196, 418	δ	304
f	4, 383	s	87	Δ	13
F	195, 363, 392, 404	S	4, 43	η	63, 199, 419
		$\mathfrak{S}$	60, 120, 176, 294	H	8, 422
g	4, 33, 205, 265	t	12, 74,	θ	250, 261, 306
G	363, 378	t_0	292, 329	λ	134
h	8, 30	T	3, 43, 418	μ	23, 134
H	43, 138, 209	$\mathfrak{T}$	176	ξ	63, 76, 110, 123, 329
i	4, 383	u	13	Ξ	422
I	207, 418	$\tilde{u}$	307	ρ	4
j	39	U	54, 78, 123, 307, 419	τ	4, 65, 74
J	39			ϕ	21, 183
l	87	v	13, 123	Φ	78, 249
l_0	93	w	13	ψ	27, 28, 185
L	38, 299	$\mathfrak{W}$	219	Ψ	249
m	123	x	12	ω	110, 264
M	18, 393	X	54, 98	Ω	21, 419
N	299	y	12	ζ	42, 330

Subject Index

A

Acoustic approximation, 18
Acoustic impedance, 5, 31, 81
Adiabatic equation, 5
Adiabatic exponent, 6, 9, 10
Adiabatic process, 5
Airfoil, 296, 350, 356, 369
Angle, extreme, 313, 327
Angle of incidence, 320
Angle of reflection, 320
Apple curve, 413
Approximation, acoustic, 18
  by finite differences, 57, 197
  by perturbation, 357, 360, 364, 398
  $\gamma = -1$, 10
  weak shock, 131, 132, 147, 155, 301, 312, 320, 350

B

Backward-facing simple wave, 93
Bend, concave, flow in, 273, 278, 317
  convex, flow around, 279
Bernoulli's constant, 22
Bernoulli's law, 22, 125, 146, 247, 262, 300
Berthelot, 204
Bethe, 141
Binding energy, 205
Blast waves, 417, 422
Body of revolution, flow along, 398
Boltzmann equation, 135
Bores, in gas flow, 118. See also *Shock front.*
  in shallow water, 132
Boundary conditions, for hyperbolic differential equations, 56, 57
  for steady plane flow, 367
Boundary layer, 345, 376
  Prandtl's theory, 137
Bourquard, 408
Bump, flow around, 278, 356
Burning velocity, 228, 232, 235
Busemann, 142, 364, 395, 403, 406, 408, 410, 412, 413, 415

C

Caloric equation of state, 4, 6, 121
Canonical hyperbolic differential equations, 44
Cavitation, 102, 262, 271, 274, 277, 291
Centered simple wave, 62, 103, 107, 194, 223, 241, 268, 277
Challis, 118
Chandrasekhar, 161, 339
Chapman, 204, 213
Chapman-Jouguet detonations and deflagrations, 211, 223, 231
Chapman-Jouguet hypothesis, 223, 231
Characteristic direction, 40, 42, 72
Characteristic equations, 43
  for cylindrical flow, 48
  for one-dimensional flow, 45, 70, 80
  for plane flow, 46, 259
  for spherical flow, 46
Characteristic initial value problem, 56
Characteristic parameters, 42
Characteristic shift rate, 239
Characteristic surfaces, 75
Characteristics, 40, 72. See also *Mach lines.*
  as separation line, 55
  C-, 42
  cross-, 97, 267, 271
  in centered rarefaction wave, 98, 104, 107, 273
  Γ-, 44
  straight, 60, 97
Chen, 395
Chien, 403
Circulation, conservation of, 19
Circulatory flow, 252
Collision. See *Interaction.*
Combustion, 204, 206, 429. See also *Deflagration.*
Complete energy, 205, 207
Complete enthalpy, 207

456 SUBJECT INDEX

Complete rarefaction wave, 101, 105, 270, 278, 291
Complete simple wave, 101, 105, 270, 278, 291
Compression wave, 94, 107, 269, 278, 352
 envelope in, 108, 110, 168
 velocity profile in, 96
Concave bend, flow in, 273, 278, 317
Condensation, 154
Condensation wave. See *Compression wave.*
Conduction, heat, 118, 134, 233, 345
Conical duct, flow in, 377, 414
Conical flow, 378, 406
 differential equations of, 408
Conical shock reflection, 416
Conical shocks, 411
Conservation of circulation, 19
Conservation of energy, 15, 122, 206, 297, 299
Conservation of mass, 14, 122, 134, 206, 297, 299
Conservation of momentum, 15, 122, 134, 206, 297, 299
Constant pressure deflagration, 209, 234
Constant state, 88
Constant volume detonation, 209
Contact discontinuities, 126, 177, 180, 197, 298, 333
Contact layer. See *Contact surface.*
Contact lines, 119, 333
Contact surface, 119, 126, 178, 298
Continuity, equation of, 14, 26. See also *Conservation of mass.*
Convex bend, flow around, 279
Critical curve, 63, 257
Critical speed, 23, 146, 308
Cross-characteristics in simple wave, 98, 104, 107, 267, 271
Cross Mach lines, 267, 271
Curve, apple, 413
 characteristic. See *Characteristic.*
 critical, 63, 257
 shock. See *Shock curve.*
 transition, 69, 258, 382
Cusp, at limiting line, 258
 of envelope in compression wave, 110, 168
Cylindrical body, flow past, 398
Cylindrical steady flow, 27, 38, 48, 397

Cylindrical symmetric flow, 29, 416
Cylindrical wave, 416

D

Decaying shock wave, 164, 354
Decay time, 203
Deflagration, 204, 206
 Chapman-Jouguet, 211
 constant pressure, 209, 234
 internal mechanism, 227, 232
 solution of flow problems involving, 224
 spherical, 429
 strong, 211, 218, 229
 weak, 211, 234
Deflagration wave, spherical, 429
Deflagration zones, width, 227, 232
Degree of indeterminacy, 221, 234
De Laval nozzle, 377, 380
Density, expansion in simple wave, 95
Dependence, domain of, 51, 58, 82, 83, 100
De Prima, 196
Descartes, Folium of, 308. See also *Shock polar.*
Determinacy of solution, 83, 138, 218, 234, 375
Detonation, 204, 206, 222, 226
 Chapman-Jouguet, 211, 223, 231
 constant volume, 209
 internal mechanism, 231
 strong, 211
 weak, 211, 218, 231
Detonation wave, spherical, 429
Detonation zone, width, 231
Differential equations, canonical hyperbolic, 44
 elliptic, 41
 homogeneous, 39
 hyperbolic, 41, 72
 linear, 44, 73
 non-linear, 74
 of conical flow, 408
 of elastic-plastic motion, 238
 of motion, 3, 12
 of Lagrange, 12, 30, 38, 106, 238
 of one-dimensional isentropic flow, 28, 30, 37, 45, 88
 of plane flow, 25, 38, 46, 247
 of spherical flow, 29, 30, 37, 46, 418
 of steady flow with cylindrical symmetry, 27, 38, 48, 398

SUBJECT INDEX

Differential equations, quasi-linear, 37
  reducible, 39, 44, 59
  several functions of two variables, 70
  totally hyperbolic, 70, 72
  two functions of several variables, 75
  two functions of two variables, 37
Diffuser, 394, 414
Direction, characteristic, 40, 42, 72
  exceptional, 63, 258
Direct Mach configuration, 335
Discontinuity. See also *Shock*.
  contact, 126, 177, 180, 197, 297, 298, 332
  development of, 108
  propagation along characteristics of, 53. See also *Sonic* and *Sound*.
  sonic, 116. See also *Sonic* and *Sound*.
Discontinuity conditions, 121. See also *Shock conditions*.
Discontinuity surfaces, 119. See also *Shock front* and *Contact discontinuities*.
Discontinuous motion, model of, 120
Disturbance, sonic. See *Sonic disturbance*.
Disturbance wave, 84
Domain of dependence, 51, 58, 82, 83, 100
Donov, 364
Double wave, 77
Drag, 363, 404. See also *Thrust*.
Duct, conical, 414
  cylindrical, 377
  two-dimensional, flow in, 282, 317, 368

E

Earnshaw, 1, 87, 118
Edge direction, 64
Elastic, 237
Elastic-plastic material, 10, 235
Elementary interactions, 174
  analysis, 182
Elliptic differential equations, 41
End point in hodograph plane, 411
Energy, binding, 205
  binding molecular, 205
Energy, carried by spherical wave, 418
  complete, 205, 207
  conservation of, 15, 122, 206, 297, 299
  internal, 4, 205
  of formation, 205, 208
  of progressing wave, 421
  separable, 6, 132
Engineering stress, 236
Enthalpy, 4, 17, 125, 207
  complete, 207
  specific, 4, 17
Entrance Mach number, 369
Entropy, 4, 7
  across a shock, increase of, 122, 141, 301
  in a Chapman-Jouguet process, 212
Envelope in compression wave, 108, 110, 168, 281
Epicycloid, 262, 275, 353
Equation of continuity, 14, 26
Equation of state, 6
  caloric, 4, 6, 121
Equations of motion. See *Differential equations of motion*.
Escape speed, 101. See also *Cavitation*.
Euler's equations of motion, 12
Exceptional direction, 63, 258
Excess pressure ratio, 154
Existence of solution of flow equations, 58, 59, 109, 219
Existence of steady state, 367
Exothermic character, 208
Expansion wave. See *Rarefaction wave*.
Extreme angle, 313

F

Ferrari, 403
Ferri, 395, 416
Finite differences, 57, 197
Flame front, 204
Flow, adjacent to region of constant state, 61. See also *Simple waves*.
  against projectile, 407
  along body of revolution, 398
  around bump, 278, 356
  around concave bend, 273, 278, 317
  around convex bend, 279
  circulatory, 252

458 SUBJECT INDEX

Flow, conical, 406
  cylindrical, 27, 29, 416
  in a cone, 377
  in a conical duct, 414
  in a duct, 282, 317, 368
  in jet, 289, 377, 387
  in nozzle, 377, 387
  irrotational, 19, 26
  isentropic, 18, 20, 88, 421
  isentropic irrotational steady plane, 26, 38, 46, 247
  mixed, 370
  nonsteady, 28
  one-dimensional, 28, 30, 37
    isentropic, 37, 38, 45, 79, 88
    non-isentropic, 38, 70
  parallel, 284
  past airfoil, 296, 350, 356, 369
  past body of revolution, 398
  past obstacle, 369
  past wedge, 277, 296, 317
  plane. See *Plane flow*.
  radial, 253
  Ringleb's, 255
  spherical, 29, 30, 37, 46, 396, 416
  spherical isentropic, 37, 46, 421
  steady, 18, 22, 25
    non-isentropic, rotational in two dimensions, 70
    with cylindrical symmetry, 27, 38, 48, 397
  subsonic, 10, 18, 250, 314, 369
  supersonic, 18, 47
  through a two-dimensional sector, 377
Flow equations. See *Differential equations of motion*.
Flow speed, 16
  relative to reaction front, 215
  relative to shock front, 141, 147, 304
Fluid motion, equations of. See *Differential equations of motion*.
Folium of Descartes, 308. See also *Shock polar*.
Formation, energy of, 205
Forward-facing simple wave, 93
Frankl, 371, 395
Front, flame, 204
  shock. See *Shock front*.

G

γ, 6, 9
$\gamma = -1$, 10
Γ-characteristics, 44, 80, 88, 262
Gases, ideal, 6, 8
  polytropic. See *Polytropic gases*.
Giese, 78
Görtler, 371
Guderley, 420, 432, 433

H

Hadamard, 2, 110
Harmonic function, 369
Hartmann, 389
Head of wave, 93
Heat conduction, 118, 134, 233, 345
Heat content. See *Enthalpy*.
Hodograph, 39, 62, 248, 381
Homogeneous differential equations, 39
Hooke's law, 11
Hugoniot, 1, 119, 121, 126
Hugoniot conditions. See *Shock conditions*.
Hugoniot function, 138, 142, 209
Hugoniot relation, 126, 138, 148, 208
Hydraulic method for nozzle flow, 382
Hyperbolic differential equations, 41
  boundary conditions for, 57
  canonical, 44
  Riemann theory of, 90, 192
Hypergeometric function, 90, 195
Hysteresis, 243

I

Ideal gases, 6, 8
Impact-loading, 240
Impact velocity, 241
Impedance, acoustic, 5, 31, 81
Impulse, of progressing wave, 421
  of spherical wave, 418
Incidence, angle of, 320
Incident shock front, 152, 319
Incomplete rarefaction wave, 101, 105, 270, 277
Indeterminacy, degree of, 221, 234
Influence, range of, 51, 82, 83
Initial value problem, 48, 56, 73
Instability, 367
Interaction, in elastic-plastic material, 245

SUBJECT INDEX

Interaction, one-dimensional flow, 173
plane flow, 283, 286, 318, 335, 350
Interior ballistics. See also *Detonations* and *Deflagrations*.
Lagrange's problem in, 173
Internal energy, 4, 205
Internal mechanism, of deflagration, 227, 232
of detonation, 232
of shock, 135
Intersection. See *Interaction*.
Invariants, Riemann, 87, 156, 192
Inverted Mach configuration, 335
Irreversible processes, 4, 116
Irrotational flow, 19
Irrotational steady plane flow, isentropic, 25, 38, 46, 247
Isentropic flow, 18, 20, 88, 421
Isentropic process, 5
Iteration, 49, 73

J

Jacobian, 39
Jet detachment, 376
Jets, 289, 387
Jouguet, 204, 212, 276
Jouguet-Chapman. See *Chapman*.
Jouguet's rule, 215
Jump conditions across shock front, 121, 299

K

v. Kármán, 403, 404, 405
Kinematic shock condition, 133
Kisenko, 395
Kopal, 414

L

Lagrange's equations of motion, 12, 30, 38, 106, 238
Lagrange's problem, 173
Lagrangian representation, 12, 30, 91, 106, 132, 236
Lazarus, 389
Le Chatelier, 204
Legendre function, 196
Legendre transforms, 249, 251
Lewy, 48
Liepmann, 345
Lift of airfoil, 363

Lighthill, 403, 405
Limiting cases of oblique shock, 313
Limiting line, 68, 250, 253, 254, 256
Linear differential equations, 44, 73
Linear waves, 2
Love, 173
Ludloff, 395

M

Maccoll, 403, 408, 412, 414
Mach, 332
Mach angle, 259, 260, 262, 264, 275
Mach configuration, 334
direct, 335
inverted, 335
stationary, 335, 373
Mach lines, 47, 54, 70, 259, 262, 304. See also *Characteristics*.
cross, 267, 271
Mach number, 18, 149, 154, 262, 338
Mach reflection, 331, 342, 372
Mach shock disc, 387
Mach shock front, 334, 373
Mach shock line, 336
Mallard, 204
Mass, conservation of, 14, 122, 134, 206, 297, 299
Mechanical models, 130
Mechanical shock conditions, 130, 206, 301
Mechanism, internal. See *Internal mechanism*.
Mixed flow, 370
Moderate shock, 353
Molecular binding energy, 205
Momentum, conservation of, 15, 122, 134, 206, 297, 299
Moore, 403, 404

N

n-tuple wave, 77
N-wave, 164
v. Neumann, 177, 178, 226, 227, 232, 321, 327, 332, 342
Non-exceptional direction, 64, 258
Non-isentropic flow, 38, 70, 199
Non-linear waves, 2
Non-steady flows, 28
Non-uniform shocks, 160
Normal shock, 313. See also *Shock, one-dimensional*.

Nozzle, 371, 377, 387
 de Laval, 377, 380
 hydraulic method for, 382
 perfect, 394
 with diverging opening, 389
 with non-diverging opening, 388

O

Oblique reflection, 318
Oblique shock fronts, 294
One-dimensional flow, 79
 boundary conditions for, 82
 characteristics for, 45, 80
 differential equations of, 28, 30, 37, 80
 shocks in, 116
Oswatitsch, 396
Overtaking of waves. See *Interaction*.

P

Parallel flow, duct to deflect, 284
 duct to induce, 285
Particle path, 98, 104. See also *Streamline*.
Path, 79
 of shock front, 162
Payman, 182
Penetration, of rarefaction waves, 179, 191
Permanent strain, 244
Perturbation method, 357, 360, 364, 398
Pidduck, 173
Piston, moving in gas, 79, 99, 111, 120, 218
Piston path, 79
Plane flow, 247
 boundary conditions for, 367
 characteristics for, 46, 259
 differential equations of, 25, 38, 46, 247
 shocks in, 294
 subsonic, 250, 314, 369
Plastic, 237
Poisson, 118
Polachek, 339
Polars, shock, 306, 353
Polytropic gases, 6
 Bernoulli's law, 22
 Hugoniot relation, 148
 shock polars, 306

Poyltropic gases, shock relations, 148, 302
 shock strength, 154
Potential, 19, 248, 250, 382
Potential equation, 369
Potential lines, 258, 259
Prandtl, 137, 142, 147, 395
Prandtl-Meyer waves, 267. See also *Simple waves*.
Prandtl's boundary layer theory, 137
Prandtl's relation, 146, 302
Pre-compression wave, 224
Pressure, expansion in shock strength, 158
 expansion in simple wave strength, 95, 292
Pressure–density relation, 5, 6, 10
Pressure ratio, excess, 154
 reflected, 154, 327, 342, 433
Primakoff, 424
Progressing waves, 419
Projectile, flow against, 407

Q

Quasi-linear differential equations, 37
Quasi-simple waves, spherical, 424

R

Radial flow, 253
Range of influence, 51, 82, 83
Rankine, 1, 119, 121
Rankine-Hugoniot conditions. See *Shock conditions*.
Rarefaction wave, 79, 94, 99, 275. See also *Simple wave*.
 and contact discontinuity, 180, 197
 cavitation, 102
 centered, 103, 104, 107, 194, 223, 241, 268
 complete, 101, 105, 270, 278, 291
 incomplete, 101, 105, 270, 277
 Lagrangian, 107
 overtaking shock, 161, 176, 180, 202, 350, 353
 shock overtaking a, 175, 180
 velocity profile, 96
Rayleigh, 2, 119
Reaction, internal mechanism, 232
Reaction front, determinacy of gas flow with, 218, 234
 flow speed relative to, 215

SUBJECT INDEX

Reaction processes, 204. See also *Detonations* and *Deflagrations*.
Reaction rate, 228, 232
Reaction zone, width, 232
Receding shock waves, 128, 129
Reducible differential equations, 39, 44, 59
Reflected shock front, 152, 319
Reflection, 152, 318
   angle of, 320
   conical shock, 416
   elastic-plastic material, 245
   head-on shock, 152, 178, 326
   Mach, 331, 342, 372
   oblique shock, 318
   one-dimensional shock, 152, 178
   regular, 319, 372
   of shock waves on contact surfaces, 178, 318
   simple wave, 288
   sonic, 154
   spherical shock fronts, 417, 432
   strong oblique, 323, 391
   weak oblique, 323
Refraction of shocks, 178, 197, 318
Regular reflection, 319, 372
   in polytropic gas, 329
Resistance, of airfoil, 363
   of cylindrical body, 404
Reversible process, 4
Revolution, flow along a body of, 398
Riemann, 1, 87, 90, 118, 181
Riemann invariants, 87, 156, 192
Riemann problem, 181
Riemann's function R, 196
Riemann theory of linear hyperbolic differential equations, 90, 192
Ringleb's flow, 255
Rocket, 392

S

Sector, flow through a two-dimensional, 277, 296, 377
Seeger, 327, 342
Separable energy, 6, 132
Shallow water theory, 32, 132
Shaw, 196
Shock, conical, 411
   determinacy of, 138, 317
   entropy change across, 122, 141, 301
   in elastic-plastic material 243
   internal mechanism, 135

Shock, in water, 132
   moderate, 161, 353
   non-uniform, 160
   normal, 313. See also *Shock, one-dimensional.*
   oblique, 294
      limiting cases, 313
   one-dimensional, 116
   sonic, 314, 323, 343. See also *Sonic* and *Sound.*
   spherical, 421
   strong, 154, 155, 314, 325, 421
   two-dimensional, 294
   unstable, 317
   weak, 131, 132, 147, 155, 301, 312, 320, 353
Shock condition(s), invariance of, 125
   kinematic, 133
   Lagrangian, 132
   mechanical, 130, 206, 301
   one-dimensional, 121, 134, 148
   plane, 299
   spherical, 421
Shock curve, 294. See also *Shock path.*
Shock discontinuities, 109. See also *Shock, Shock wave,* and *Shock front.*
Shock front, 1, 3, 79, 119. See also *Shock wave* and *Shock.*
   back side, 119, 127, 299
   configurations of three, 332
   conical, 411
   front side, 119, 127, 299
   incident, 152, 320
   Mach, 334, 373, 387
   meeting contact surface, 178, 197, 318
   meeting simple waves, 175, 191, 199, 350
   oblique, 294
   overtaken by rarefaction wave, 161, 175, 180, 202, 350, 353
   overtaking rarefaction wave, 175, 180, 350
   position of, 162, 353
   reflected, 152, 320
   reflection. See *Reflection.*
   refraction, 178, 197, 318
   speed, 119, 141, 159, 304
   spherical, 417, 421

SUBJECT INDEX

Shock front, stationary, 129, 298, 300, 320, 335, 373
   velocity, 119, 122, 127, 141, 159, 304, 422
Shock lines, 119, 316
Shock motion, models, 129
Shock path, 162. See also *Shock curve*.
Shock polar, 306, 353
Shock reflection. See *Reflection*.
Shock strength, 154
   expansions in, 158
Shock tubes, 181
Shock velocity, 119, 122, 127, 141, 159, 304, 422
Shock wave, 99, 119. See also *Shock front* and *Shock*.
   decaying, 164, 354
   in elastic-plastic material, 236
   in tube, 120
   receding, 128
   width, 164
Shock zone, width, 137
Simple wave, 59, 77, 88, 92, 99, 265, 273
   centered, 62, 103, 107, 194, 223, 241, 268, 277
   compared with weak shocks, 156, 350
   complete, 101, 105, 270, 278, 291
   compressive. See *Compression wave*.
   cross-characteristics in, 98, 104, 107, 267, 271
   elastic-plastic material, 236, 240, 243
   expansions for pressure, etc. 95, 292
   expansive. See *Rarefaction wave*.
   formula for, 95, 290
   incomplete, 101, 105, 270, 278, 291
   in Mach configuration, 332
   interaction, 176, 180, 189, 283, 286
   Lagrangian, 106
   meeting shocks, 161, 175, 199, 350
   particle path in, 98, 271
   reflection, 288
   straight characteristics in, 60, 97
   tables of quantities, 293
   velocity profile, 96
Smith, 342
Solution, determinacy of, 86, 138, 218, 234, 375
   existence and uniqueness, 51, 58, 85, 109, 219, 367

Solution, stability, 367
Sonic. See also *Sound*.
Sonic discontinuities, 116
Sonic disturbance, 84, 121, 301, 313
Sonic propagation, 116
Sonic reflection, 154
Sonic shock, 314, 323, 343
Sonic wave, 320
Sound. See also *Sonic*.
Sound speed, 2, 5, 7, 84
Sound waves, 82, 94, 131
Space-like curves, 57, 84, 219
Specific energy, 4
Specific enthalpy, 4, 17
Specific heat, 17
Specific volume, 4
Speed. See also *Velocity*.
   critical, 23, 146, 308
   escape, 101
   flow. See *Flow speed*.
   limit, 23
   of shock front. See *Shock velocity*.
   of sound, 2, 5, 7, 84
Spherical blast waves, 417, 422
Spherical detonation and deflagration waves, 429
Spherical flow, 29, 30, 37, 46, 416
Spherical quasi-simple waves, 424
Spherical shock front, 421
   reflection, 417, 432
Spherical waves, 416
Stability of solution, 367
Stagnation point, 318
Stationary Mach configuration, 335, 373
Stationary shock front, 129, 298, 300
Steady flow, 18, 22, 25
   with cylindrical symmetry, 27, 38, 48, 397
Steady plane flow, boundary conditions for, 367
   isentropic irrotational, 25, 38, 46, 247
Steady state, 88
   existence of, 367
Stoker, 34
Stokes, 1, 108, 118
Stokes' stream function, 398
Stopping shocks, 243
Straight characteristics of simple wave, 60, 97
Strain, 11, 235
   permanent, 244

SUBJECT INDEX 463

Stream function, 27, 28, 70, 248, 250, 369, 382
  Stokes', 398
Streamlines, 27, 70, 271
Stream surfaces, 28
Strength, shock, 154, 323
Stress, 11, 235
  engineering, 236
Stress-strain relation, 11, 236, 246
Strong deflagration, 211, 218, 229
Strong detonation, 211
Strong reflection, 323
Strong shock, 154, 155, 314, 325, 421
Subsonic flow, 10, 18, 250, 314, 367, 369
Supersonic flow, 18, 47
Surface, contact, 119, 126, 178, 298

T

Tail of wave, 93, 101
Taylor, 383, 408, 412, 414, 422, 423, 427, 428
Terminal state, 179
Thermodynamic notions, 3
Thomas, 137
Three-dimensional flow, 377
Throat, 380
Thrust, 392. See also *Drag*.
Time-like, 57, 84, 219
Total energy, 15
Tin spherical wave, 418
Totally hyperbolic differential equations, 70, 72
Transformation, hodograph, 39, 62, 248, 381
Transforms, Legendre, 249, 251
Transition curve, 69, 258, 382
Transition lines, 256
Tricomi, 371
Tubes, shock, 181
Two-dimensional flow. See *Plane flow*.

U

Uniform compressive motion, 150
Uniqueness of solution, 51, 58, 85, 109, 219, 367
Unstable shock, 317

V

Velocity. See also *Speed*.
  burning, 228, 232, 235

Velocity deflagration, 204, 209, 212, 215
  detonation, 204, 209, 212, 215
  impact, 241
  shock, 119, 122, 127, 141, 159, 304, 422
  sound, 2, 5, 7, 84
Velocity potential, 21
Velocity profile, 96
Vielle, 204
Viscosity, 118, 134, 345, 371, 374
Viscous boundary layer, 376
Volume, specific, 4
Vortices, 289, 298, 347

W

Water, 8, 32, 132, 424
Wave, backward facing. See *Backward-facing simple wave*.
  blast, 417, 422
  combustion. See *Deflagration wave*.
  compression. See *Compression wave*.
  condensation. See *Compression wave*.
  cylindrical. See *Cylindrical wave*.
  deflagration. See *Deflagration wave*.
  detonation. See *Detonation wave*.
  disturbance. See *Disturbance wave*.
  double, 77
  expansion. See *Rarefaction wave*.
  forward-facing. See *Forward-facing simple wave*.
  head of, 93
  isentropic progressing, 421
  linear, 2
  n-tuple, 77
  N-, 164
  nonlinear, 2
  pre-compression. See *Pre-compression wave*.
  progressing, 419
    in water, 424
  rarefaction. See *Rarefaction wave*.
  reflected. See *Reflected wave*.
  shock. See *Shock wave*.
  simple. See *Simple wave*.
  sound. See *Sound* and *Sonic*
  spherical, 416
  spherical blast, 417 422,
  spherical detonation or deflagration 429

Wave, spherical quasi-simple, 424
  tail of, 93
Wave equation, 19, 370, 402
Wave form, 96
Wave propagation in elastic-plastic material, 235
Weak deflagration, 211, 234
Weak detonation, 211, 218, 231
Weak reflection, 323
Weak shock, 131, 132, 147, 155, 301, 312, 320, 353
Wedge, flow past, 277, 296, 377

Weyl, 141, 142, 144, 215
Width of deflagration zone, 227, 232
  of detonation zone, 231
  of shock zone, 137
  of shock wave, 164

Z

Zone of deflagration, 227, 232
Zone of detonation, 231
Zone of penetration, 179
Zone of shock, 137